Waste Treatment Processes in Environmental Engineering

The Editors

Dr. S.N. Kaul has obtained his bachelor and master degrees from IIT, Kanpur and was awarded Ph.D. degree from the University of New Castle Upon Tyne, UK. Dr. Kaul joined CSIR-National Environmental Engineering Research Institute [NEERI] Nagpur in 1970 and worked for over three decades. Dr. Kaul superannuated as Acting Director and Director Grade Scientist from NEERI and was also associated with MIT College of Engineering, Pune as its Principal.

Dr.Kaul has published over 350 Papers in National and International journals of repute beside authorship of over 30 Books on Environmental Engineering and Science, 10 Patents and 150 conference papers. Dr. Kaul has designed and operated several waste treatment plants in India. Dr. Kaul has supervised over a dozen of Ph.D. students and 40 M Tech students. In addition, Dr. Kaul was responsible for organizing several National and International Conferences and has traveled extensively in India and aboard concerning various R and D programs. Dr. Kaul has received several National and International awards and is expert member in importants National committees.

Dr. D.R. Saini obtained his B Tech and M Tech from HBTI,Kanpur and PhD [Chem.Engg.] from IIT-Delhi. Presently working as Principal Scientist at CSIR-NCL, Pune. Dr. Saini is co-author of 8 books, six chapters and published seventy papers in international journals. He has guided three PhD students and few M.Tech. students.

Dr. B.N. Rai is presently working as Professor in the Deapartment of Chemical Engineering and Technology, IIT (BHU), Varanasi. Dr. Rai has done B.E. (Chem. Eng.) from Bangalore University and M.Tech. (Chem) and Ph.D. from Banaras Hindu University. He has more than 20 years of teaching, research and administrative experiences in various capacities. He has more than 40 publications in international journals, book chapters, conferences, etc. and delivered invited key note lectures in various conferences. He has guided more than 20 M.Tech. and two Ph.D. students. Dr. Rai was awarded by the Ministry of Non-Conventional Energy, Govt of India for writing a book in Hindi on fundamental topics. His research interests include Environmental Biotechnology, Industrial Pollution and Control. He has also organized international conferences, refresher courses and workshops on waste management. Besides above, he has served the Universitiy in various capacities such as President, IT Gymkhana, Coordinator, National Youth Festival, etc.

Er. Prateek Kaul has obtained M.Tech. from V.NIT, Nagpur. Er. Kaul travelled respectively in India and abroad. Er. Kaul has several research publications, books and patents to his credit. Er. Kaul has worked in several multinational industries and is presently associated with Symantec (USA), Pune.

Waste Treatment Processes in Environmental Engineering

Volume 3

Editors
Prof. (Dr.) S.N. Kaul
Principal
MIT College of Engineering, Pune
Former Acting Director and Director-Grade
Scientist, NEERI [CSIR], Nagpur, M.S.

Dr. D.R. Saini
Scientist
National Chemical Laboratory [CSIR],
Pune, M.S.

Dr. B.N. Rai
Professor, Department of Chemical Engineering & Technology,
Indian Institute of Technology (BHU),
Banaras Hindu University, Varanasi, U.P.

Er. Prateek Kaul
Sr. Manager,
Symantec Corporation (USA), Pune

2016
Daya Publishing House®
A Division of
Astral International Pvt. Ltd.
New Delhi – 110 002

Waste Treatment Processes in Environmental Engineering
(4 Volumes Set)
Volume 1: Page 0001-0306
Volume 2: Page 0307-0628
Volume 3: Page 0629-0908
Volume 4: Page 0909-1184

Cataloging in Publication Data—DK
Courtesy: D.K. Agencies (P) Ltd. <docinfo@dkagencies.com>

Waste treatment processes in environmental engineering / editors, Prof. (Dr.) S.N. Kaul, Dr. D.R. Saini, Dr. B.N. Rai, Er. Prateek Kaul.
4 volumes cm
Includes indexes.

ISBN 9789351309130 (Vol. 3)
ISBN 978-93-5124-372-4 (International Edition)

1. Sewage—Purification—India. 2. Sewage—Purification. 3. Environmental engineering—India. 4. Environmental engineering. I. Kaul, S. N., editor.

DDC 628.30954 23

Published by : **Daya Publishing House®**
A Division of
Astral International Pvt. Ltd.
– ISO 9001:2008 Certified Company –
4760-61/23, Ansari Road, Darya Ganj
New Delhi-110 002
Ph. 011-43549197, 23278134
E-mail: info@astralint.com
Website: www.astralint.com

Laser Typesetting : **Classic Computer Services**, Delhi - 110 035

Printed at : **Thomson Press India Limited**

Dedicated to the Sacred Memory of

Sri Sri Sarda Maa

&

Swami Vivekananda

Preface

The Editors of the present edition on Waste Treatment Processes in Environmental Engineering, have been associated with various educational institutes, research organizations, universities etc. It was felt that a comprehensive selective topics in Environmental Engineering concerning India with Indian Experiences from various part of the India with Indian data from various Indian experts must be collected to disseminate the authentic information to our readers.

Text books and reference books are already available in the market but most often they relate to information and experiences of the western world, which is not relevant to Indian conditions and economics. The editors have been contacting various experts throughout the country in order to present the Indian experiences on which the contributing authors have been working. Often the good and invaluable work gets published in foreign journals which are expansive and, therefore,beyond the reach of most of the readers to acquire. To reduce this gaps the editors have made some efforts in this direction, and the editors feel that readers will appreciate this and try to benefits from this compilation. It is also primarily with this view to make aware Indian readers about the work of Indian experts and their experiences.

Waste Treatment Processes in Environmental Engineering is a compendium of knowledge concerning Environmental Issues from several experts who are associated with prestigious institutes in India, *viz.*: IITs, CSIR, NITS, Central Universities, Pollution Control Board. This book endeavors to explore various important topics of immediate concern to all our stake holders, especially students, researches, practitioners, teachers, NGOs etc. This book also introduces the reader about areas of research of the experts in India.

Economic progress is directly connected with environment and ecology. Material progress goes linear while environmental degradation follows geometric progression. Therefore, environment and ecology is the rate determining step in the progress of the country. This aspect is extensively exhibited in this book.

Emphasis has been laid on environmental pollution and control and the editors feel that these this specific book will be immensely useful to all concerned with Environmental and Ecology encompassing all the three component of Environment, *viz.*: Air, Water and Solid Waste besides EIAs, cleaner technology, problem associated with implementation including environmental education etc.Finally, the editors would like to thanks all the contributing authors for accepting our invitation to contribute their valuable research and development work for our readers and editors would also like to thanks them for the time and efforts devoted while compiling their work.

The trust reposed by the Astral International (P) Ltd., New Delhi is thankfully acknowledged.

Prof. (Dr.) S.N. Kaul

Dr. D.R. Saini

Prof. (Dr.) B.N. Rai

Er. Prateek Kaul

Contents

Volume 2

Volume 3

Volume 4

Chapter 34

Steady State Dispersion of Air Pollutants in Atmosphere

Manju Agarwal, M.M. Agarwal and Samina Hasan

Global Environmental and Agriculture Center, Lucknow

ABSTRACT

A steady state two dimensional mathematical model of air pollutant dispersion in the atmosphere, taking vertical height dependent power law profiles for wind velocity and eddy diffusivity is solved. This dispersion is being studied for both gaseous (buoyant) and particulate (non buoyant) type of pollutants, and in order to discriminate their nature, inclusion of gravitational settling velocity is incorporated in the model. Keeping in mind, their mitigation from the atmosphere as the topmost priority, the concept of dry deposition at the ground and removal by other mechanisms have also been included. The concentration profiles of these pollutants is being analysed for velocity profiles, nature of pollutants, removal parameters etc. involved in the problem.

Keywords: *Gaseous pollutant, Particulate pollutant, Finite element, Basis functions, Weighing functions, Settling velocity, Deposition velocity.*

Introduction

Environmental problems are becoming more and more serious for our world and their importance may even increase in the future, among this high pollution of air (high concentrations and / or high deposition of certain harmful chemical species) is a problem of great concern for us because of their damage to plants, animals and humans. Moreover, some ecosystems may be damaged when the pollution levels are very high. This is why the pollution levels must be carefully studied in the efforts to make it possible the following:

- ☆ To predict the appearance of high pollution levels, which may cause different damages in our environment.
- ☆ To decide what can be done in order to prevent the exceedance of prescribed critical levels (or in other words, to attempt to keep the harmful concentrations and/or depositions under the prescribed acceptable limits).

Keeping this in mind, a major portion of the subject of air pollution is devoted to the description of contaminant within the atmosphere. Air pollutants are emitted in the atmosphere and while in the air, these compounds can react with other compounds and be transformed into secondary pollutants. Simultaneously, these compounds are transported by atmospheric motion, both horizontal and vertical, and are removed either by physical encounter with the surfaces (referred to as dry deposition) or through precipitation (Carmichael *et al*. (1980)). But the first step before studying their motion, chemical conversion, removal etc. is to study the nature of these air pollutants, hence the mathematical modeling should be done, keeping in mind, the structure of the individual air pollutants, as well as the domain of interest and its boundaries.

Air pollutants can be classified physically as gaseous and particulate air pollutants. Particulate air pollutants consist of solid particles or liquid droplets (aerosols) small enough to remain suspended in the air. Such particles have no general composition hence this term refer to all atmospheric substances that are not gases, examples include lead, mercury, RSPM, SPM etc. These particulate pollutants characterize by size range of approximately 0.002 μm to 500 μm in diameter. These particulate pollutants have a feature to settle down to the lower strata of the atmosphere, so for proper treatment of such types, their non buoyant nature must be taken into consideration, by prescribing a negative sink velocity, called gravitational settling, in the downward negative vertical direction. These pollutants emitted into the atmosphere through some source and can be removed by a number of natural or artificial removal processes (*e.g.* rain, fog droplets, chemical conversion, green belt, artificial sinks etc.) and also by dry deposition onto the surface of the earth as a result of gravitational settling and ground absorption by the soil, vegetation, buildings etc.

In this paper, steady state dispersion of air pollutants, both gaseous and particulate, in the atmosphere is being studied. The gaseous and particulate nature of the air pollutants is being discriminated through the use of settling velocity, in the model, taking its value to be zero for gaseous pollutants as done by Peterson and Seinfeld (1977) and Lee (1985), these authors had studied their dispersion under the effect of constant wind speed and eddy diffusivity, but here height dependent velocity profile and eddy diffusivity is taken into account as suggested by Lin and Hildemann (1996). As stated earlier, their removal from the atmosphere is a matter of great concern for us, hence to predict their effect on the concentration of pollutants, inclusion of first order removal term is also taken into consideration. Apart from this, at the boundaries especially ground, it has been assumed that partial uptake and removal of the pollutants is taking place. In this perspective, dry deposition at the ground is taking into consideration by utilizing the concept of deposition velocity.

The most striking feature that is being depicted here is the solution of the obtained mathematical model by use of numerical methods. Due to limitations of analytical solutions, these numerical methods are indispensable tools for researchers today, in order to predict the results for more realistic mathematical models of air pollution dispersion in the atmosphere. Instead of conventional Galerkin finite element method of weighted residuals, Petrov Galerkin finite element method (Evans and Abdullah (1986))is being used, taking local basis and quadratic asymmetric functions to solve the problem.

The main objective here is to explore the application of this above said finite element technique to these atmospheric pollutant transport problems and to analyze the effect of various parameters like velocity profiles, nature of pollutants, removal mechanisms etc. on the concentration profiles and dispersion of these air pollutants in the atmosphere.

Mathematical Formulation

The steady state concentration of air pollutant in the atmosphere for two dimension domain (x, z) is governed by the following advection diffusion equation:

$$u\frac{\partial C}{\partial x}+w\frac{\partial C}{\partial z}=\frac{\partial}{\partial x}\left(k_x\frac{\partial C}{\partial x}\right)+\frac{\partial}{\partial z}\left(k_z\frac{\partial C}{\partial z}\right)+R+S \tag{1}$$

where, C is the concentration of the pollutant, R represents the removal term and S is the source term. u,w are the components of wind velocity in x,z directions respectively and K_x,K_z represent the eddy diffusion coefficients in these directions. Here, x axis is taken in the mean wind direction and it is assumed that wind speed is sufficiently large so that diffusive transport in the x direction can be neglected in comparison to advection by the wind.

The effect of buoyancy on the trajectory of the motion of the heavy (or particulate) pollutants is modelled by prescribing a negative sink velocity '$-w_s$' in the z direction, *where,* $w_s=|w_s|$, taken to be constant for the given type of pollutant, in particular its value is zero for gaseous pollutants.

To account for the removal of these pollutants through wet deposition *i.e.* removal by natural, artificial removal mechanisms, chemical conversion etc., here

$$R=-\lambda C \tag{2}$$

where, λ is first order rate coefficient responsible for these removal processes.

Horizontal wind velocity u as well as eddy diffusivity K_z both are expressed as power law functions of vertical distance z as follows:

$$u=u(z)=u_H\left(\frac{z}{H}\right)^{\alpha} \tag{3}$$

$$K_z = K_z(z) = K_H \left(\frac{z}{H} \right)^{\beta} \tag{4}$$

where, u_H and K_H are the wind velocity and diffusivity respectively at a reference height H.

Source term is incorporated in the model through one of the boundary condition, the pollutant is released from an elevated source of strength Q located at $(0,h)$, hence the source boundary condition at $x = 0$ has the form.

$$C(0,z) = \frac{Q}{u(h)} \delta(z-h) \quad ,.0 \le h \le H \tag{4}$$

where, h is the height of the source and δ is the Dirac Delta function.

The boundary conditions at the ground $z = 0$ and inversion height $z = H$ have the following forms:

$$K_z \frac{\partial C}{\partial z} + w_s C = v_d C \quad \text{at } z = 0 \tag{5}$$

$$K_z \frac{\partial C}{\partial z} = 0 \quad \text{at } z = H \tag{6}$$

where, v_d is the deposition velocity at the ground level, responsible for dry deposition at the ground level.

Taking all this in account, we get the following form of steady state advection diffusion equation,

$$u(z) \frac{\partial C}{\partial x} - w_s \frac{\partial C}{\partial z} = \frac{\partial}{\partial z} \left(K_z(z) \frac{\partial C}{\partial z} \right) - \lambda C \tag{7}$$

along with boundary conditions (4) – (6).

Using the following dimensionless quantities

$$\bar{x} = \frac{x K_H}{u_H H^2}, \; \bar{z} = \frac{z}{H}, \; \bar{C} = \frac{C u_H H}{Q}, \; \bar{u} = \frac{u}{u(H)}, \; \overline{K_z} = \frac{K_z}{K_H},$$

$$\bar{\lambda} = \frac{\lambda H^2}{K_H}, \; \overline{w_s} = \frac{w_s H}{K_H}, \; \overline{v_d} = \frac{v_d H}{K_H}$$

the equations (7) along with its boundary conditions (4), (5) and (6), on omitting the bars, transformed into the following form:

$$u(z)\frac{\partial C}{\partial x} - w_s \frac{\partial C}{\partial z} = \frac{\partial}{\partial z}\left(K_z(z)\frac{\partial C}{\partial z} \right) - \lambda C \tag{8}$$

$$C(0,z) = \frac{\delta(z-h)}{u(h)} \quad ,0 \le h \le 1 \tag{9}$$

$$K_z \frac{\partial C}{\partial z} + w_s C = v_d C \text{ at } z = 0 \tag{10}$$

$$K_z \frac{\partial C}{\partial z} = 0 \quad \text{at } z = 1 \tag{11}$$

Method of Solution

Here, the numerical finite element approach is being used to solve the obtained mathematical model.

The horizontal distance x is being discretised with constant step size, whereas for the vertical distance z ($0 \le z \le 1$) Gaussian distribution of vertical grid point density, as suggested by Runca and Sardei (1975), is being used. As a result of this, the size of the vertical step size Δz_k ($z_{k+1} - z_k$) is variable and is determined by the given values of Δz_s and n , where Δz_s is the size of the first interval above and below the source and n is the number of vertical grid points.

Let $\Omega = \{(x,z) \in R^2 | 0 < x, 0 < z < 1\}$ be a domain and let

$$L(C) \equiv u(z)\frac{\partial C}{\partial x} - w_s \frac{\partial C}{\partial z} - \frac{\partial}{\partial z}\left(K_z(z)\frac{\partial C}{\partial z} \right) + \lambda C = 0 \tag{12}$$

For $(x,z) \in \Omega$, we have $C(x,z) \approx \hat{C}(x,z)$ (13)

In the formulation of the finite element equation, this approximate solution $\hat{C}$ in every element is assumed to be represented by a linear combination of the solutions at every node point of the element known to satisfy

$$C(x,z) \approx \hat{C}(x,z) = \sum_j C_j(x)\psi_j(z) \tag{14}$$

In this presentation of C by $\hat{C}$, the parameters involved in it are C_j's depend solely on x and ψ_j's are basis functions of vertical distance z .

On substituting this trial function $\hat{C}(x,z)$ into equation (12) we have

$$L\left(\hat{C}\right) \equiv u(z)\frac{\partial \hat{C}}{\partial x} - w_s \frac{\partial \hat{C}}{\partial z} - \frac{\partial}{\partial z}\left(K_z(z)\frac{\partial \hat{C}}{\partial z}\right) + \lambda \hat{C} = R \tag{14}$$

where, R is the residual arising from the fact that $\hat{C}(x,z)$ does not identically satisfy equation (12). So as per Galerkin approach of weighted residuals, we require the residual $R \equiv L(\hat{C})$ to vanish, hence in this formulation, weighing function $w(z)$ are chosen such that the residual is orthogonal to this weighing function.

$$\int_{\Omega} L(\hat{C}) w(z) dz = 0, \tag{15}$$

For the mathematical model considered here, the linear local basis functions, ψ_i's, and asymmetric quadratic weighing functions, w_i's, are used as done by Carmichael *et al.* (1980). The basis functions are of the form

$$\psi_i = 1 - \frac{z}{\Delta\rho} \tag{16}$$

$$\psi_{i+1} = \frac{z}{\Delta\rho} \tag{17}$$

where, $\Delta\rho$ is the distance between the two node points in vertical direction and $0 \le z \le \Delta\rho$

In order to find the form of asymmetric quadratic weighing functions, let the weighing functions w_i and w_{i+1} be expressed in the form

$$w_i(z) = \psi_i(z) - f(z) \tag{18}$$

$$w_{i+1}(z) = \psi_{i+1}(z) - f(z) \tag{19}$$

where, $f(z)$ is a piecewise quadratic function with undetermined coefficients a, b and c defined as follows:

$$f(z) = a\frac{z^2}{\Delta\rho^2} + b\frac{z}{\Delta\rho} + c \tag{20}$$

To determine the values of a, b and c we impose the following condition:

$$f(0) = 0 \tag{21}$$

$$f(\Delta\rho) = 0 \tag{22}$$

Substitution of these into equation (20) we get

$$f(z) = b\left(-\frac{z^2}{\Delta\rho^2} + \frac{z}{\Delta\rho} \right) \tag{23}$$

where, b is the remaining undetermined parameter. For convenience, we define $\theta = \frac{a}{3}$, hence the complete weighing functions can now be written as:

$$w_i(z) = \psi_j(z) + 3\theta \frac{z^2}{\Delta\rho^2} - 3\theta \frac{z}{\Delta\rho} = 1 - \frac{z}{\rho} + \frac{3\theta}{\Delta\rho^2}\left(z^2 - (\Delta\rho)z\right) \tag{24}$$

$$w_{i+1}(z) = \psi_{i+1}(z) - 3\theta \frac{z^2}{\Delta\rho^2} + 3\theta \frac{z}{\Delta\rho} = \frac{z}{\Delta\rho} - \frac{3\theta}{\Delta\rho^2}\left(z^2 - (\Delta\rho)z\right) \tag{25}$$

Here parameter $|\theta| \leq 1$ and it determines the degree of upwind differencing. For $\theta = 0$, the weighing function obviously reduces to the basis functions and the method becomes identical to the conventional Galerkin approximation.

In finite element formulation, the most commonly encountered η system is related to the above local z coordinate system through the relationship:

$$\frac{z}{\Delta\rho} = \frac{\eta + 1}{2}, -1 \leq \eta \leq 1 \tag{26}$$

Thus with this we can write equation (16), (17), (24) and (25) as:

$$\psi_i(\eta) = \frac{1}{2}(1 - \eta) \tag{27}$$

$$\psi_{i+1}(\eta) = \frac{1}{2}(1 + \eta) \tag{28}$$

$$w_i(\eta) = \frac{1}{4}\{(1 + \eta)(3\theta\eta - 3\theta - 2) + 4\} \tag{29}$$

$$w_{i+1}(\eta) = \frac{1}{4}\{(1 + \eta)(-3\theta\eta + 3\theta + 2)\} \tag{30}$$

On substituting $L(\hat{C})$ from equation (14) in equation (15) and taken the range of integration as $0 \le z \le 1$, we have

$$\int_0^1 \left(u(z)\frac{\partial \hat{C}}{\partial x} - w_s \frac{\partial \hat{C}}{\partial z} - \frac{\partial}{\partial z}\left(K_z(z)\frac{\partial \hat{C}}{\partial z} \right) + \lambda \hat{C} \right) w_i(z)dz = 0 \tag{31}$$

or

$$\int_0^1 u(z)\frac{\partial \hat{C}}{\partial x} w_i(z)dz - w_s \int_0^1 \frac{\partial \hat{C}}{\partial z} w_i(z)dz - \int_0^1 \frac{\partial}{\partial z}\left(K_z(z)\frac{\partial \hat{C}}{\partial z} \right) w_i(z)dz + \lambda \int_0^1 \hat{C} w_i(z)dz = 0$$

Now on integrating the third term by integration by parts we get

$$\int_0^1 u(z)\frac{\partial \hat{C}}{\partial x} w_i(z)dz - w_s \int_0^1 \frac{\partial \hat{C}}{\partial z} w_i(z)dz - \left\{ \left(K_z(z)\frac{\partial \hat{C}}{\partial z} \right) w_i(z) \right\}_0^1$$

$$+ \int_0^1 \left(K_z(z)\frac{\partial \hat{C}}{\partial z} \right) w_i(z)dz + \lambda \int_0^1 \hat{C} w_i(z)dz = 0$$

since by the boundary conditions (10) and (11) we have

$$\int_0^1 u(z)\frac{\partial \hat{C}}{\partial x} w_i(z)dz - w_s \int_0^1 \frac{\partial \hat{C}}{\partial z} w_i(z)dz - \left\{ 0 - (v_d - w_s)\hat{C}(0) w_i(0) \right\} +$$

$$\int_0^1 \left(K_z(z)\frac{\partial \hat{C}}{\partial z} \right) w_i(z)dz + \lambda \int_0^1 \hat{C} w_i(z)dz = 0$$

or we have

$$\int_0^1 u(z)\frac{\partial \hat{C}}{\partial x} w_i(z)dz - w_s \int_0^1 \frac{\partial \hat{C}}{\partial z} w_i(z)dz + \int_0^1 \left(K_z(z)\frac{\partial \hat{C}}{\partial z} \right) w_i(z)dz +$$

$$\lambda \int_0^1 \hat{C} w_i(z)dz = (v_d - w_s)\hat{C}(0) w_i(0) \tag{32}$$

Now transforming z to η through the relation (26) we get

$$\int_{-1}^{1}\left[u(\eta)\frac{\partial \hat{C}}{\partial x}w_i(\eta)-w_s\frac{\partial \hat{C}}{\partial \eta}\frac{d\eta}{dz}w_i(\eta)+K(\eta)\frac{\partial \hat{C}}{\partial \eta}\frac{d\eta}{dz}w_i(\eta)+\lambda\hat{C}w_i(\eta)\right]\frac{\Delta\rho}{2}d\eta$$

$$=\left[(v_d-w_s)\hat{C}w_i\right]_{\eta=-1} \tag{33}$$

Taking $\hat{C}(x,z)=\sum_j C_j(x)\psi_j(z)$ we have

$$\frac{\partial \hat{C}}{\partial x}=\sum_j \frac{dC_j}{dx}\psi_j$$

$$\frac{\partial \hat{C}}{\partial \eta}=\sum_j \hat{C}_j\frac{d\psi_j}{d\eta}\quad \frac{d\eta}{dz}=\frac{2}{\Delta\rho}$$

Now as basis functions and weighing functions are known, the unknown that we have to find are the $C_j(x)$ which can be obtained by solving a system of differential equations given by equation (33)

Now writing equation (33) in matrix form as

$$[M]\left[\frac{dC}{dx}\right]+[N][C]=0 \tag{34}$$

where

$$M_{ij}=\int_{-1}^{1}u(\eta)\psi_j(\eta)w_i(\eta)\frac{\Delta\rho}{2}d\eta \tag{35}$$

$$N_{ij}=-w_s\int_{-1}^{1}\frac{d\psi_j}{d\eta}w_i(\eta)d\eta+\frac{2}{\Delta\rho}\int_{-1}^{1}\left(K(\eta)\frac{d\psi_j}{d\eta}\frac{dw_i(\eta)}{d\eta}\right)d\eta+\lambda\int_{-1}^{1}\psi_j(\eta)w_i(\eta)d\eta-$$

$$\left[(v_d-w_s)\psi_j(\eta)w_i(\eta)\right]_{\eta=-1} \tag{36}$$

Now for considering the matrix equation in (34) at the mth level of x, we have

$$[M]_m \left[\frac{dC}{dx} \right]_m + [N]_m [C]_m = 0 \tag{37}$$

Using the logical extension of the backward difference approximation for the derivative involved in above equation we have

$$[M]_{m+1} \left[\frac{C_{m+1} - C_m}{\Delta x} \right]_{m+1} + [N]_{m+1} [C]_{m+1} = 0$$

or $\{[M]_{m+1} + \Delta x [N]_{m+1}\}[C_{m+1}] = [M]_{m+1}[C_m]$ (38)

Results and Discussion

The concentration of air pollutants (gaseous and particulate) in the atmosphere is being explored through the use of finite element approach; hence the obtained concentration profiles are plotted against different value of parameters involved in the model. The parameters values for the mathematical model are taken as follows:

$\alpha = 0.5$, $\beta = 1.0$ (Runca & Sardei (1975))

$K_H = 5\text{m}^2/\text{sec}$, $(w_s)_{gaseosus} = (w_s)_{particulate} = 0.03\text{cm/sec}$, $(v_d)_{gaseosus} = 0.3\text{cm/sec}$,

$(v_d)_{particulate} = 0.6\text{cm/sec}$. (Lee (1985)

$H = 1980$ m, $h = 396$ m (Sharan & Gupta (2002))

The parameter values for the numerical method involved are:

$\theta = 0.125$, $n = 81$, $\Delta z_s = \frac{1}{2}\left\{\frac{1}{2(n-1)} + \frac{1}{(n-1)}\right\} = 0.009375$, $\Delta x = 10^{-4}$,

$x = m\Delta x$ where m is the number of steps in downwind distance x.

For computation purposes, the dimensionless values of all the parameters have been used. These velocity profiles have been changed by changing the value of α in equation (3), where in particular for $\alpha = 0$ corresponds to the situation of constant wind speed i.e. wind speed independent of vertical height z. It can be inferred from the graph that at a constant wind speed the concentration curve of the gaseous pollutants as well as for particulate pollutants is more concentrated near the source (at $z = 0.2$), and it is incapable to spread downwind, whereas the concentration curves for variable velocity profiles for both type of pollutants shows that the pollutant concentration increases to spread downwind as vertical dependence of velocity profiles came into existence, confirming the significance of the vertical dependence of velocity profile in air pollution mathematical models as suggested by Lin and Hildermann (1996).

The vertical plume profiles for the gaseous as well as particulate pollutants at three downwind distances have been drawn in Figure 34.1. The concentration curves

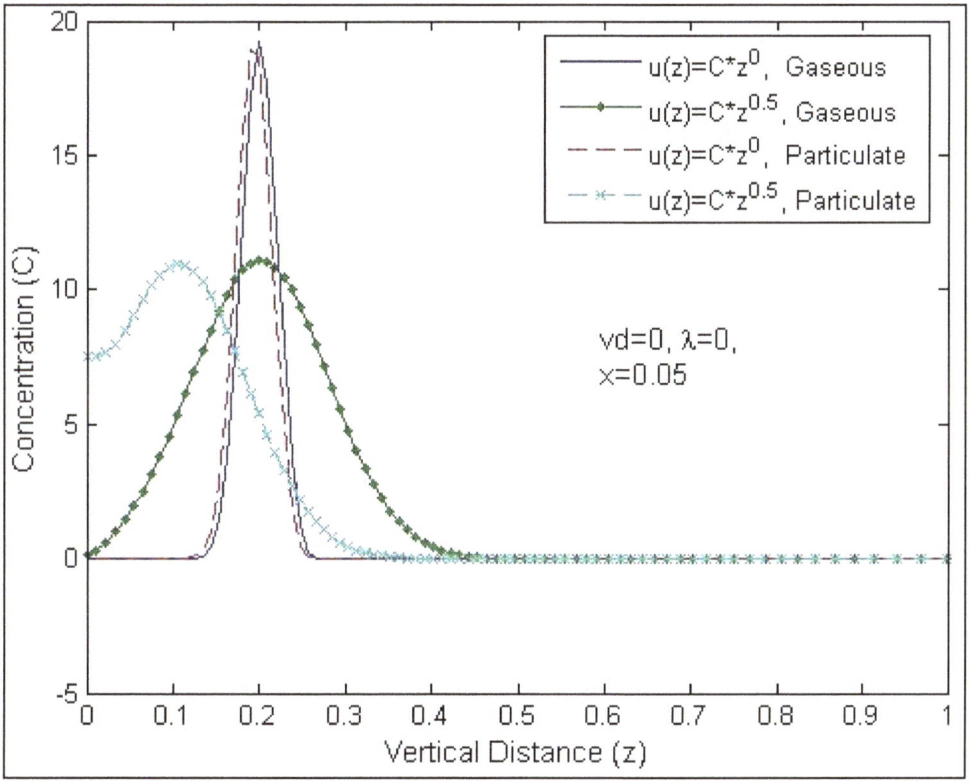

Figure 34.1: Concentration Profiles of Pollutants with Change in Velocity Profiles (u(z)=C(z$^{\alpha}$)).

corresponding to gaseous pollutants reveal that at $x = 0.03$ the plume is only a short distance downwind and the plume have not spread enough to touch the ground $(z = 0)$, but as the downwind distance increases the plume is beginning to touch the ground and deposition is beginning to occur. Due to ground deposition the peak of the concentration curve, which is maintained for all the downwind distance near the vicinity of the source, begins to descend as downwind distance increases. It is also observed that for gaseous pollutants the concentration level around lower vertical heights ($z < 0.1$) and for vertical heights beyond ($z > 0.3$), the concentration level increases as downwind distance increases, becomes negligible at greater vertical levels. Now in Figure 34.1, if we consider the plume profiles for particulate pollutants, it can be easily revealed that there is a higher concentration of the particulate pollutant in comparison to gaseous pollutant at lower vertical heights (breathing level) for all downwind distances. This implies that the particulate pollutant has a tendency to settle down at the ground level, which suggests the higher concentration of the particulate pollutant in the lower strata of the atmosphere, imposing a great threat to the human exposure. The dominant feature that can easily be drawn by comparing the concentration curves of gaseous and particulate pollutants is the shifting of the

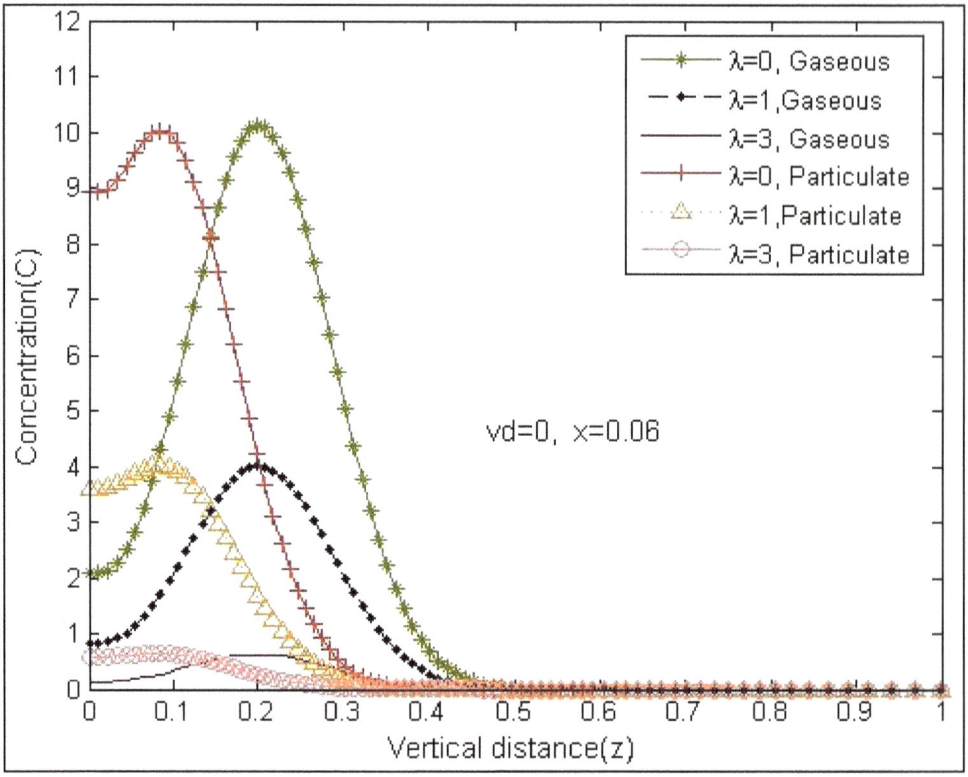

Figure 34.2: Concentration Profiles of Pollutants for Different Removal Rates.

concentration peak near the ground level for the particulate pollutants whereas for gaseous this peak at each downwind distance is maintained near the vicinity of the source. This shifting towards the ground go on increasing as downwind distance from the source increases.

In order to illustrate the effect of wet deposition through natural removal, artificial removal, chemical transformation etc. on the gaseous and particulate pollutants, concentration profiles have been plotted in Figures 34.2 for different values of first order removal rate coefficients, λ , and as anticipated it can easily visualized that these removals are effective and hence essential to remove the pollutants from the atmosphere especially to mitigate their effect in the lower strata of the atmosphere where there is more threat to human exposure.

The curves drawn in Figure 34.3 depict the effect of deposition velocities (dry deposition) on the vertical plume profiles of the gaseous and particulate pollutants. For both type of pollutants, the ground level concentration decreases when the dry deposition velocity increases as a direct result of the surface deposition. It has also been revealed from the graph of particulate pollutant that deposition velocity not only reduces the concentration at the ground level but it also moves the peak of the

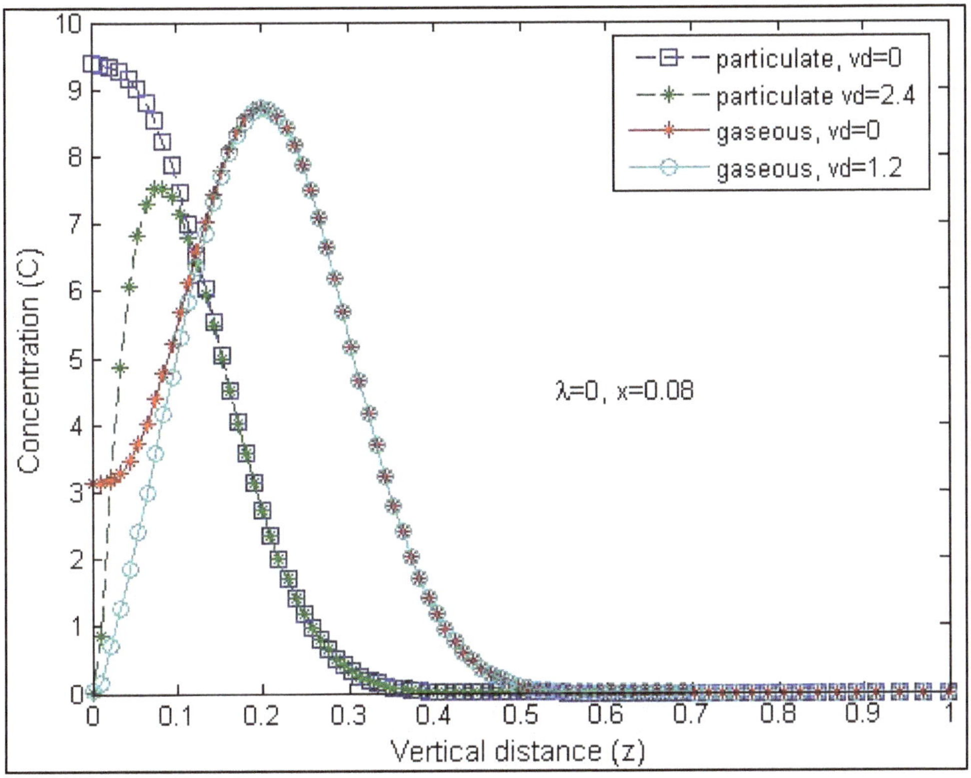

Figure 34.3: Effect of Deposition Velocity on Concentration Profiles of the Pollutants.

concentration curve towards the vicinity of the source, confirming the results given by Lee (1985) and Ermak (1977).

Conclusion

The analysis done in this paper helps us to reveal that gaseous (buoyant) as well as non buoyant pollutant (particulate) pollutant both are threats to the environment and hence to the human beings, in particular the latter one because of their accumulation in the lower strata of the atmosphere causing more possibility to human exposure. Taking their harmful nature in mind, the concentration profiles of these pollutants have further been studied under the effect of different removal strategies like natural, artificial removal mechanisms, removal by ground deposition etc. and it has been inferred that these are helpful in alleviating the effect of these pollutants.

The mathematical model considered here is tried to solve through numerical approach of finite element method and up to very much extent it has been successful in achieving it by confirming the standard results. So, this approach can be utilized in future to solve more complex and realistic mathematical models of air pollution, whose analytical solutions are not possible or difficult to obtain.

References

1. Carmichael, G. R., Kitada, T. and Peters, L. K., 1980, Application of a Galerkin Finite Element Method to Atmospheric Transport Problems, *Computers and Fluids,* 8, 155–176.
2. Runca E. and Sardei F., *1975,* Numerical treatment of time dependent advection and diffusion of air pollutants, *Atmospheric Environment,* 9, 69 – 80.
3. Peterson, T. W and Seinfeld, J. H., 1977, Mathematical Model for transport interconversion and removal of gaseous and particulate air pollutants – application to the urban plume, *Atmospheric Environment,* 11, 1171 – 1184.
4. Lee, H. N., 1985, Three dimensional analytical models suitable for gaseous and particulate pollutant transport, diffusion, transformation and removal, *Atmospheric Environment,* 19, 1951 – 1957.
5. Evans, D. J. and Abdullah A. R., 1986, Boundary Value Techniques for Finite Element Solutions of the Convection Diffusion Equation, *International Journal For Numerical Methods in Engineering,* 23, 1145 – 1159.
6. Ermak, D.L., 1977, An analytical model for air pollutant transport and deposition from a point source, *Atmospheric Environment,* 11, 231-237.
7. Ames, William F., 1992, *Numerical Methods for Partial Differential Equations,* Academic Press Inc., San Diego, California, USA.
8. Lin, J.S. and Hildemann, L.M., 1996, Analytic Solutions of the Atmospheric Diffusion Equation with Multiple Sources and Height dependent wind speed and eddy diffusivities, *Atmospheric Environment,* 30, 239-254.
9. Sharan, M. and Gupta, S. (2002), Two dimensional analytical model for estimated crosswind integrated concentration in a capping inversion: eddy diffusivity as a function of downwind distance from the source, *Atmospheric Environment,* 36, 97-105.

Chapter 35

Wastewater Management in Man-Made Fibre Industries in India

M. Karthik, S. Deotale, N.N. Rao and T. Nandy

NEERI-CSIR, Nagpur

Introduction

Manmade fiber (MMF) industry is a major contributor in the economic development of nation and is closely connected to agriculture and chemical sector where it produces various aspects of human activity such as textile, transportation, packaging and fashion industry[1]. The demand for MMF products have grown substantially over the last few decades as economy grew and emerged as a major substitute to natural fibres especially cotton[2]. MMF is the most dominant fibre with a share of 63 per cent of the total fibre consumption across the globe. It is classified into both synthetic and semi-synthetic fibres. Synthetic fibres are Polyester, Acrylic, Olefin, Nylon, Spandex and other propriety products derived from synthesised organic polymers. Semi-synthetic fibres include Rayon (Viscose), Acetate and Triacetate fibres are made from cellulose obtained from cotton or wood pulp[3,4].

There are more than 100 MMF industries in India from large to small scale operators. About 99 per cent of total manmade fibre production is dominated by the polyester, viscose, nylon and acrylic fibres producing filament yarn, tyre cord along with staple fibers[4]. Currently, India produces about 3.4 billion kg, which is around 6.6 per cent of global MMF capacity. India's share in the global exports is miniscule at around 2.25 per cent. The demand is anticipated to be about 4.2 billion kg in the year 2015 on the basis of 8 per cent growth rate[1, 3 and 4]. The per capita consumption of MM fibres in India is 2 kgs vis-à-vis China (12 kgs) and Europe (11 kgs) respectively[5]. Per cent contribution of man-made fibre from India is presented in Figure 35.1.

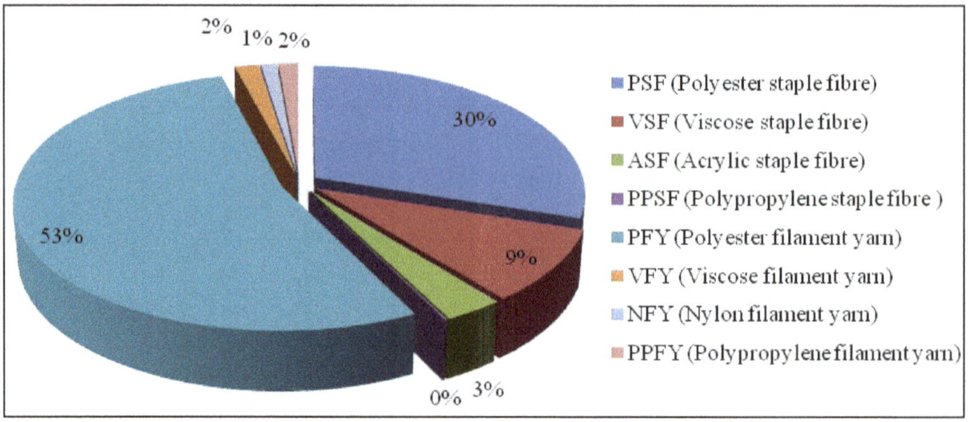

Figure 35.1: Per cent Contribution of MMF Production in India.

Raw Material Required for Production

Chemicals

Production of different MMF products needs extremely pure raw materials and chemicals including monomers, catalysts, and solvents as presented hereunder.

- Acrylic fibre–Acrylonitrile from hydrogen cyanide with acetylene
- Nylon fibre–Caprolactam from caproic acid
- Polyester fibre–Pure Terepthalic Acid (PTA), Mono-Ethylene Glycol (MEG)
- Rayon (Viscose) fibre–wood and cotton linters from sulphite or sulphate process

The requirement of raw materials for various MMF and their quantity are mentioned in Table 35.1.

Table 35.1: Raw Material Requirement for Production in MMF[1,5]

MMF Product	*Raw Materials*	*Quantity (kg/t)*	*Water (m^3/t)*
Acrylic	Acrylonitrile	500-900	120-200
Nylon	Caprolactam	700-1100	
Polyester	Pure Terepthalic Acid (PTA)	700-1000	
	Mono-Ethylene Glycol (MEG)	300-500	
Rayon (Viscose)	Wood pulp	900-1200	150-1000

Water

MMF industry is a large consumer of water specially semi synthetic fibres where the entire quantity reappears as wastewater particularly from viscose, polymerisation, spinning, washing and finishing sections[6,7]. The water from different units is recycled to varying extent the total intake of water per ton of fibre is very large. It is estimated that water required for synthetic fibre industry is about 120-200 m^3/t and for semi-synthetic fibre industry 150-1000 m^3/t (Table 35.1). Besides, water consumption for steam generation domestic purposes is also to be accounted and is negligible in comparison to process water requirement[7].

Energy

After raw material, energy is the single largest input contributing to about 25 per cent of the production cost[1,4]. Energy costs in India are US $ 0.112 per kWh which is higher than other countries. Energy requirement for manufacture is much higher. Large mills entirely depend on the captive power plants to complete their needs. The power consumption for production of MMF in India is presented in Table 35.2. It was observed from the data among MMF units power requirement was highest among synthetic manmade fibre units and subsequent its carbon emissions.

Table 35.2: Power Requirements for Production and Subsequent Carbon Emissions

Product	*Power, Kwh/kg*	*Carbon Emissions, kg/t*
Acrylic	48.65	12.37
Nylon	69.50	15.50
Polyester	34.75	9.52
Rayon	27.80	8.61

Impact on Water Environment

Manufacturing processes require inputs of different raw materials, chemicals, machineries to produce both semi-synthetic and synthetic fibres[13]. The impacts as a result of use of these chemicals are elaborated hereunder on the water environment.

Synthetic Fibre–Nylon Fibre

Caprolactam is the major ingredient in manufacture of nylon fibre. Additives like acetic acid and titanium dioxide (TiO_2) are added as delusterant and heated at temperature between 250-270°C to produce molten polymer which is extruded into ribbons. The polymer melt is converted into chips which are washed to remove the monomeric and oligomeric impurities. The wastewater is generated during the monomer and polymer recovery processes during centrifuging and monomer stripping. Polymer chips are melted at 250-260°C. The molten polymer flows into a conical section to form a pool, which feeds a spinning pump to spinnerets. The pool is kept under nitrogen atmosphere to prevent decomposition by air. Therefore polymerization section is identified as a critical area from the perspective of wastewater management. Spinning and finishing section of the nylon plant generates wastewater with high BOD and COD apart from solid waste generation, condensation products and negligible quantity of Caprolactam. The molten polymer is filtered before entering the spinnerets where it is drawn, twisted and subjected to secondary processes like crimping and twisting. In textile section of nylon dipping process resorcinol, formaldehyde and latex solution are used which contributes to the organic pollution.

The nylon fibre manufacturing process with ensuing effluent generation is presented in Figure 35.2.

Synthetic Fibre–Polyester Fibre

In polyester manufacturing process, continuous polycondensation involves mixing of two raw materials *viz.*, pure Terepthalic acid (PTA) and mono-ethylene glycol (MEG). The slurry which is formed due to continuous mixing is supplied to esterification reactor at 240-270°C under set environment using heat exchanger and

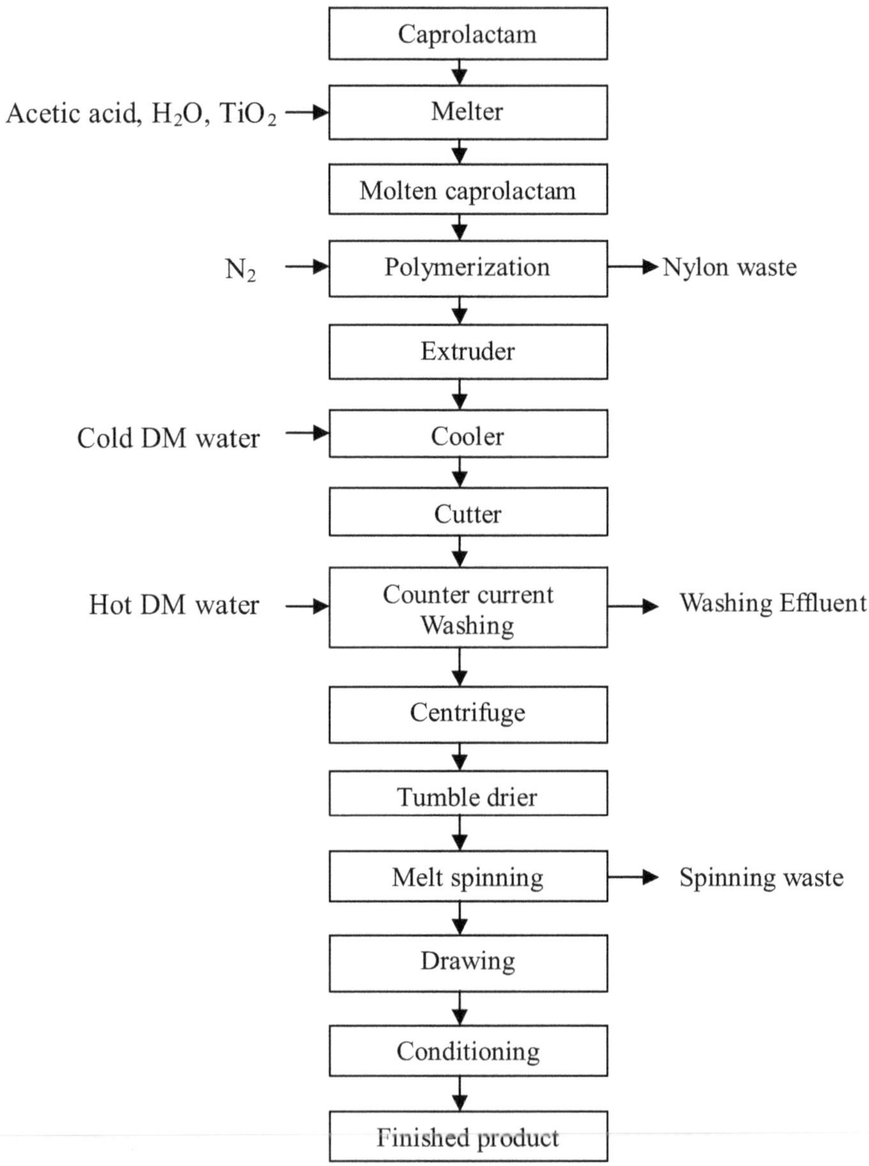

Figure 35.2: Process Flow Sheet for Nylon.

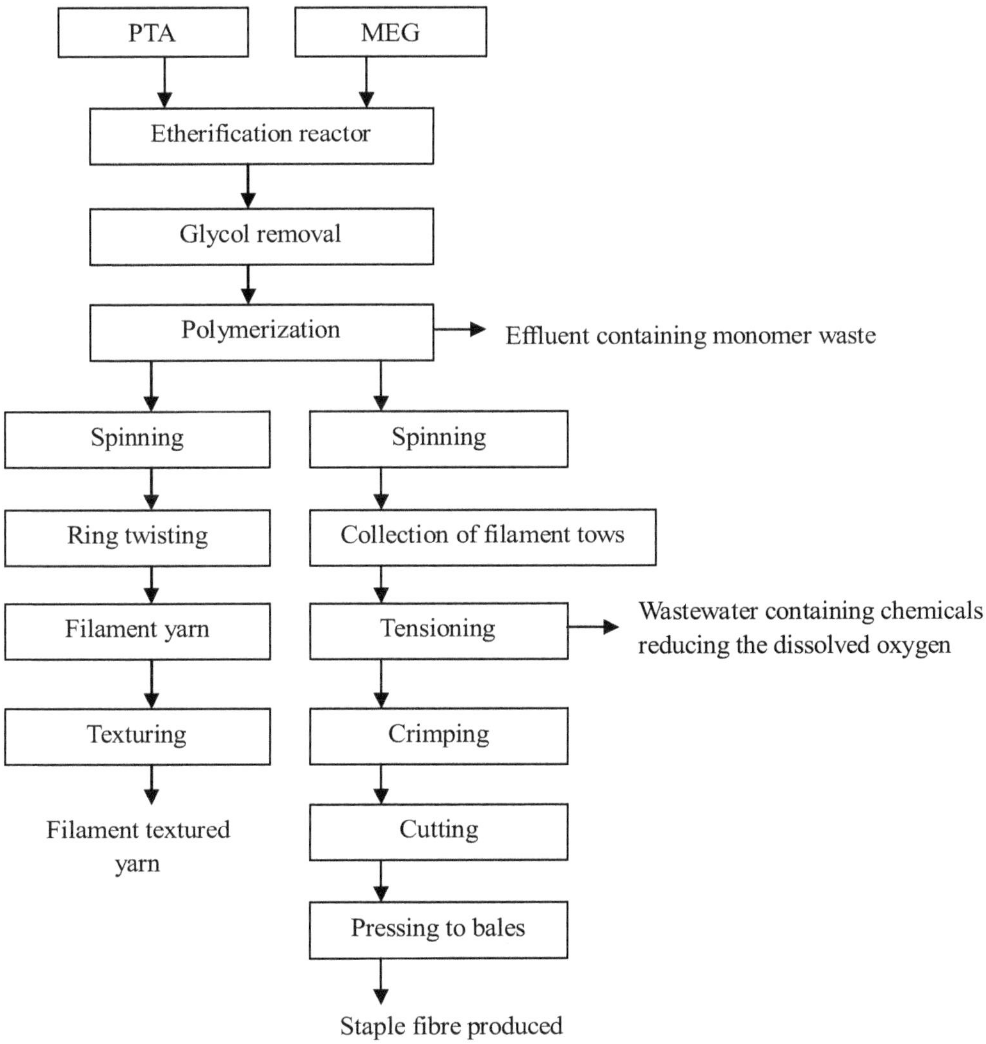

Figure 35.3: Process Flow Sheet for Polyester.

vapour separator to release the water molecule. The oligomer which is formed in the esterification reactor is fed to continuous polymerisation reaction vessel after filtration. At this stage delustrant titanium dioxide (TiO_2) and catalyst antimony trioxide (SB_2O_3) are added to oligomer which further reacts to produce polymer of desired viscosity under controlled temperature and vacuum. The filtered polymer is distributed to polyester filament yarn and staple fibre. The filtered polymer is distributed to manifold under pressure. The polymer melt is extruded through spinnerets and wound on a craft winders after a pass through quench chamber and finishing system for filament yarn. In PSF, the polymer is sent to manifold after filtration and is passed through spinnerets. The extruded filaments are passed through quench chamber and combined to form a tow. The tow is cut and collected in tow cans and processed through fibre lines from where the staple fibre produced is cut to the desired length and packed in bales. The manufacturing process with subsequent effluent generation is presented in Figure 35.3.

Synthetic Fibre–Acrylic Fibre

Acrylonitrile is the major raw material in the production of acrylic fibre. The reactions between acrylonitrile and solvent N,N, dimethylformamide (DMF) occurs in aqueous phase. The monomers solution is then conveyed to the polymerization reactor unit. A large amount of water is consumed in this section for washing of the resulted polymer. The purpose of this process is to obtain a homogenized polymer solution called dope suitable for spinning. The wash water contains higher concentrations of COD and TKN. The second stage is characterized by two processes such as solvent recovery and purification. Recovery of the solvent occurs in a wet solvent extraction process where some water is consumed and wastewater generated. The extracted solvent is purified through activated carbon adsorption. The activated carbon process generates some wastewater during bed-backwashing operation. Wet spinning is performed for the refined dope solution by extruding the dissolved polymer through stainless steel spinnerets immersed in large spinning baths. As the polymer emerges from the spinnerets, it coagulates to form fiber bundles. The excess solvent is recycled. Negligible amount of water is consumed in this stage and so hence little wastewater is generated. Various mechanical operations like washing for solvent extraction, stretching, drying and heating is performed in this section to give strength and stability to the fiber bundles, before product is sent for packaging. The manufacturing process and wastewater generation in acrylic fibre is presented in Figure 35.4.

Semi-Synthetic Fibre–Rayon Fibre

Wood pulp or cellulose is the basic raw material for manufacturing of rayon fibre. In steeping, cellulose is immersed in 19 per cent aqueous sodium hydroxide (NaOH) at a temperature of 18-25°C to convert cellulose to alkali cellulose. This conversion into alkali cellulose generates an effluent stream with high pH due to sodium hydroxide. In xanthation, the aged alkali cellulose crumbs are reacted with carbon disulphide under controlled temperature at 20-30°C to form cellulose xanthate. Then it dissolved in dilute sodium hydroxide at 15-20°C under high-shear mixing conditions to obtain a viscous caramel-orange colour solution called viscose. The redistribution and loss of xanthate occurs containing undissolved cellulose xanthate and hemicelluloses during ripening. The viscose solution is fed through spinnerets into a spin bath containing 10 per cent sulphuric acid to acidify the sodium cellulose followed by neutralization and acidification. Rapid coagulation of the rayon filaments occur which is followed by simultaneous stretching and decomposition of cellulose xanthate. Large quantity of wastewater is generated during the fiber spinning and finishing processes which comprises 80 per cent of the total quantity of concentrated effluent discharged from the washings. The waste from the spinning machine and acid bath contains sulphuric acid, sodium sulphate, zinc sulphate, waste fiber,

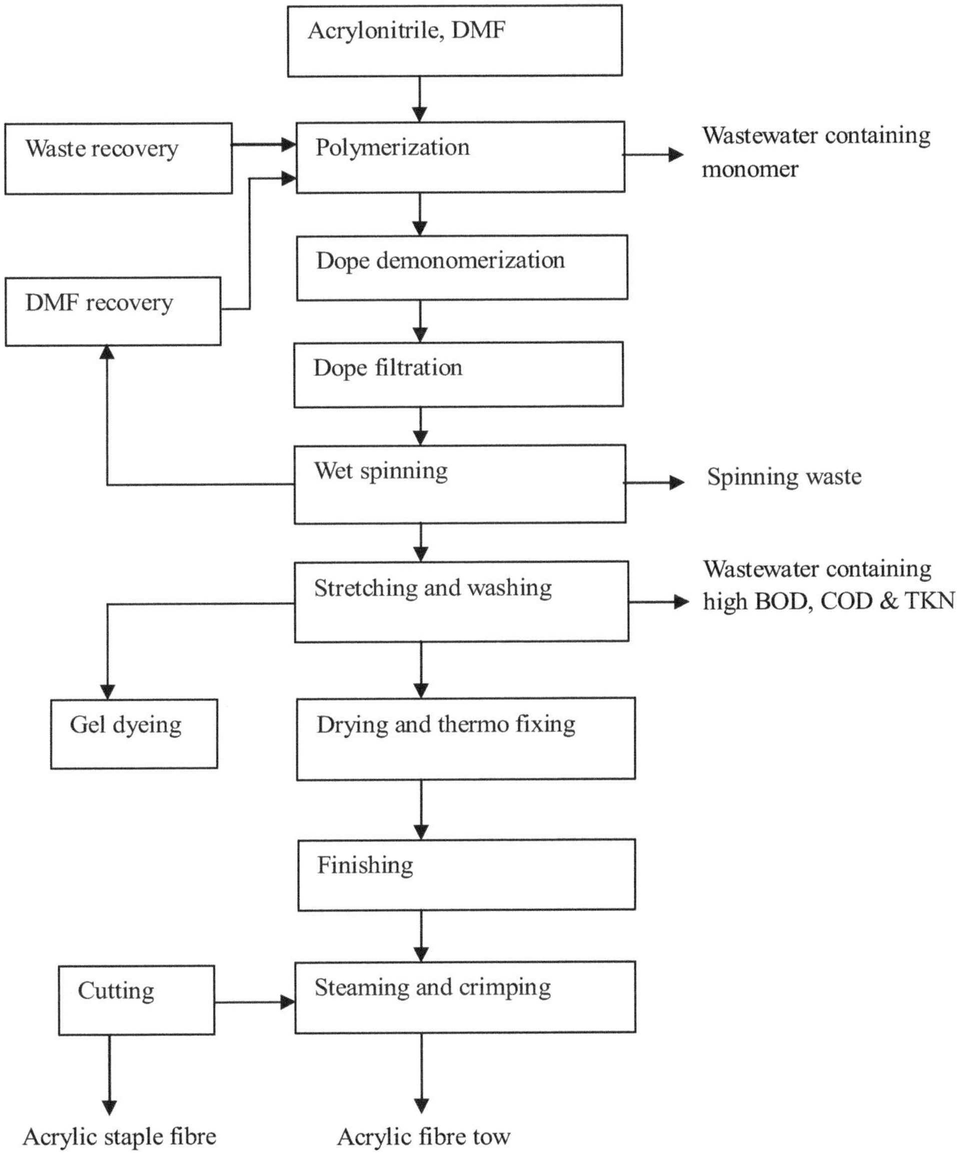

Figure 35.4: Process Flow Sheet for Manufacturing of Acrylic.

hydrogen sulphide, carbon disulphide and other organics. The process with effluent generation is mentioned in Figure 35.5.

Effluent Sources and Characteristics in MMF Industry

The effluents generated from MMF industry is due to use of large volumes of water either in the chemical process or during spinning, washing and finishing processes in both synthetic and semi-synthetic fibre. The quantity of wastewater

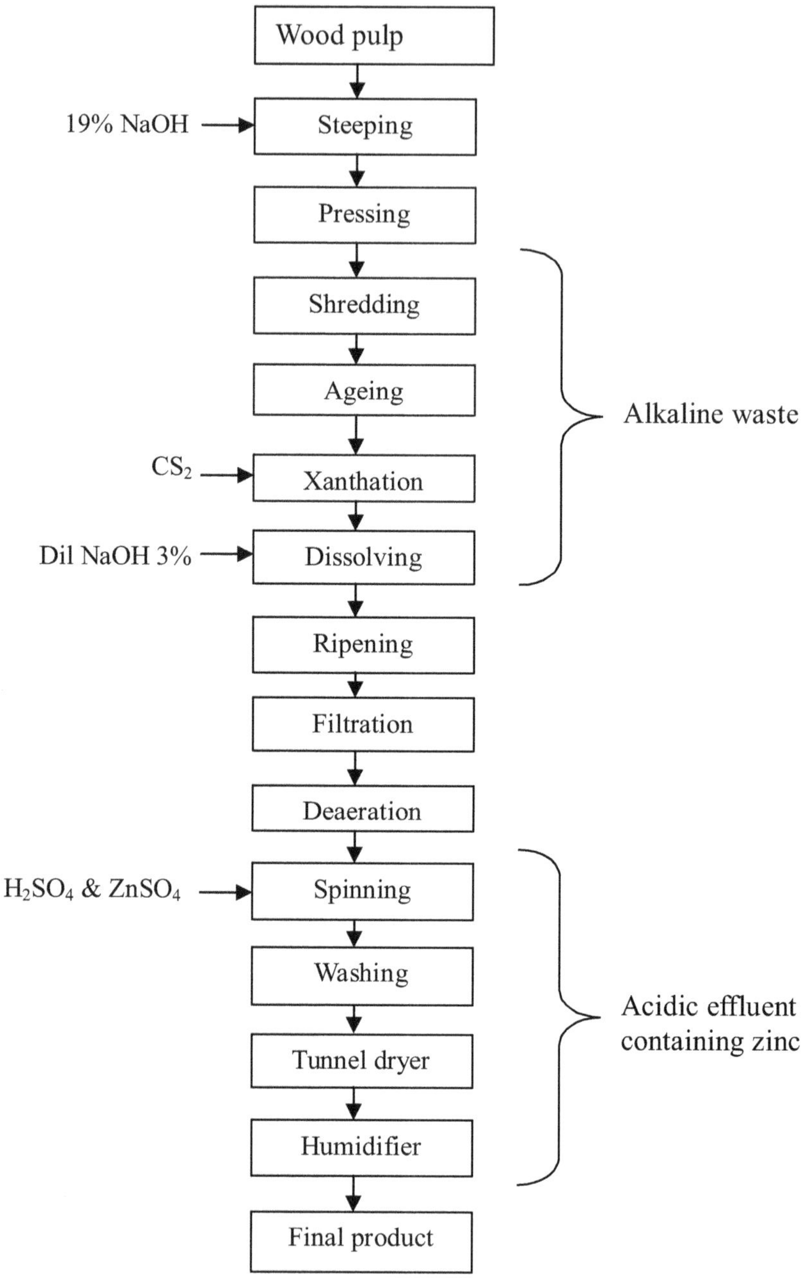

Figure 35.5: Process Flow Sheet for Manufacturing of Viscose.

discharged varies from one plant to another, depending upon the size of the unit, its production, availability of water and its usage. Wastewater quantification and characterisation in Indian MMF is elaborated further in the text (Tables 35.3 and 35.4).

Table 35.3: Effluent Generation in different MMF Industry [4, 6]

MMF Products	Production t/d	Effluent Discharge m^3/d	Waste Volumem^3/t of Product
Semi-synthetic fibre			
Viscose filament yarn (VFY)	27	13100	485
Viscose staple fibre (VSY)	165	28800	175
Viscose tyre cord (VTC)	12	4540	378
Synthetic fibre			
Nylon filament yarn (NFY)	7	330	47
Nylon tyre cord (NTC)	5	80	16
Polyester filament yarn (PFY)	16	1112	70
Polyester staple fibre (PSF)	27	2200	82
Acrylic staple fibre (ASF)	3	282	94

Table 35.4: Combined Effluent Characteristics from MMF Industry in India [6]

Parameters	Nylon	Polyester	Acrylic	Rayon	Minimum National Standards for MMF Units
Temperature	24-35	25-33	29-44	31-38	40
pH	2.3-7.2	3.4-7.3	4.0-6.9	2.7-6.9	5.5-9
SS	108-355	53-197	27-85	32-250	100
TDS	912-3341	948-3192	190-2435	594-2218	2100
BOD	51-938	68-864	18-590	22-87	30
COD	230-3600	223-3437	669-1002	214-2400	250
Oil and grease	5-20	0.1-12	0.5-4	4.6-18	10
Zinc Cyanide	0.1-0.6-	0.1-0.3-	–0.8-3	15-80-	50.2

All parameters are in mg/l except pH and temperature.

Synthetic Fibre – Nylon and Polyester

- ☆ Synthetic wastewater is generated during monomer and polymer recovery processes such as centrifuging, monomer stripping, and slurry tanks.
- ☆ Processes generate effluents with relatively high concentrations of contaminants; including equipment oil, spent solvent, raw material and drum residuals.
- ☆ Spinning and finishing section of nylon and polyester plant contain higher range of BOD, COD, condensation products ethylene oxide, fatty acids and aldehydes.
- ☆ Leaks and spills also generate waste and often occur at pumps, flanges, valves and agitator seals.

Semi-Synthetic Fibre Unit–Rayon

- Alkaline wastes are generated in dialysers during the viscose production process, effluents are generated from the washing of the alkali cellulose tanks, press and xanthating equipments. Highly alkaline effluents are generated from washing of filter clothes.
- Acid wastes from spinning machine, acid bath recycle, washing of the filters and first fibre rinse.
- Neutral waste from further washing of the fibre on the machine with bar conveyor and from ancillary water treatment and softening plant.
- The largest quantities of effluents are generated during fibre spinning and finishing processes. It comprises almost 80 per cent of the total quantity of concentrated effluent and is discharged from the washers. The remaining effluents are generated during the preparation of alkali cellulose and viscose. The alkaline wastes contain sodium hydroxide and residual viscose.
- The wastes from the spinning machine and acid bath contain sulphuric acid, sodium sulphate, zinc sulphate, waste fiber, hydrogen sulphide and carbon disulphide which are the contributors to the major pollution load.

Effluent Management in Manmade Fibre Industry

Wastewater management can be efficiently reduced by internal process modifications and proper housekeeping. It is the most economic way to reduce pollutant load in effluent, especially for old installations. In many instances the effluent quality obtainable by internal measures may not be sufficient to comply with regulations and may turn out to be expensive. External treatment methods or end of pipe (EOP) measures are the best alternative in such cases. EOP measures are used to reduce the pollutant load and discharge into the environment. Sometimes it is also necessary to adjust pH or temperature of the effluent[6]. The existing wastewater management practices in India MMF units depend on a combination of primary, secondary and tertiary treatments for wastewater management. Primary treatment includes equalisation, neutralization, chemical coagulation, flocculation and clarification. The secondary treatment uses activated sludge process for biological oxidation. The activated sludge process yields higher COD removal efficiency in the range of 70-90 per cent for treatment of large wastewater volumes[6].

Nylon Unit

In nylon fibre industry effluent from different sections of process are collected in the equalization tank. The equalised effluents are neutralised first (lime) followed by use of coagulants for precipitation at pH 8.0-8.5 (Floccol-120, Floccol-1010). Lime addition depends upon the pH of the raw effluent. The dissolved organic matter is biologically degraded by acclimatized microbes in the aeration tanks. The final effluent (COD: < 250ppm; pH: 7.5-8.0; TDS: < 2100 ppm) is polished by passing through pressure sand filters to reduced suspended solids (Figure 35.6a).

Polyester Unit

The effluent generated during the production of polyester fibre contains organic and inorganic impurities and small volumes of oil and grease. The wastewater is equalized followed by neutralization with caustic solution. The equalized wastewater is pumped to accelerated gravity separator for physico-chemical treatment. Poly Aluminum Chloride (PAC)/polyelectrolyte is added as coagulant/flocculent at adequate doses for chemical aided clarification. The chemically treated wastewater is further treated through biological oxidation unit (aeration tank) in which the wastewater is treated using aerobic bacteria and aerators provide mixing and aeration to the suspended biomass. Biological treatment is followed by tertiary treatment through sand filter, activated carbon and chlorination before reuse within the plant (Figure 35.6a).

Acrylic Fibre

The wastewater contains high concentrations of COD and BOD. The wastewater is equalized and neutralized using caustic. The neutralized effluent is treated through biological oxidation route where the dissolved organic matter is removed (Figure 35.6a).

Rayon Fibre

The viscose fiber industry requires large quantum of water of which only a small fraction is incorporated in the product whereas the balance finds their way as wastewater. Basically two effluent streams are generated in the viscose rayon plant *viz.*, acidic and alkaline. The acidic stream contains sulfuric acid, sodium sulphate and zinc sulphate as the inorganic salts. The alkali waste contains sodium hydroxide, hemi-cellulose of raw pulp residual viscose, sodium polysulphides and sodium thiosulphate. Effluent from acid plant and the boiler house is collected in the neutralization tank. This effluent contains soda, ash, charcoal and ash-dust. The effluent is neutralized with lime addition after equalization process. The neutralized effluent is further treated through physico-chemical treatment through coagulation-flocculation process. The sludge free effluent water is treated through bio-oxidation. The zinc rich waste from rayon plant is separated from rest of the effluent streams and treated through the precipitation with addition of lime and coagulants to precipitate zinc in the form of zinc-hydroxide. The precipitation is thickened in sludge thickener/centrifuge. The sludge acquired from clarifloculator of the normal effluent treatment plant and zinc recovery plants are transferred to the gravity thickener where sludge is concentrated and dewatered through centrifuge for solids. The supernatant from the zinc high stream is mixed with the normal stream for combined treatment if required through bio-oxidation (Figure 35.6b).

Specialized Effluent Treatment System

Specific treatment through tertiary/advanced treatment is relevant to MMF industry for removal of zinc and other heavy metals which includes chemical precipitation, solvent extraction, ion-exchange, adsorption, membrane filtration and electrochemical methods. The performance these advanced treatments are mentioned in Table 35.5.

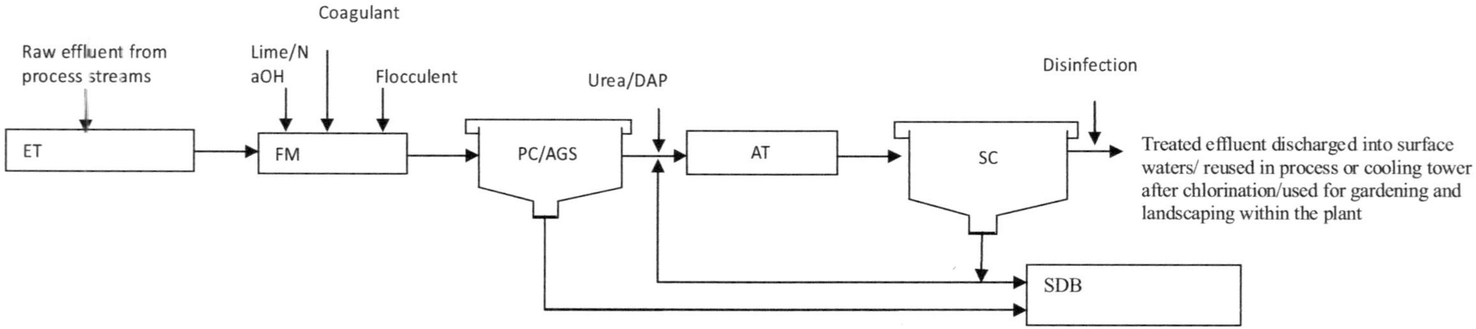

Figure 35.6a: Effluent Treatment Scheme in Nylon, Polyester and Acrylic Units.

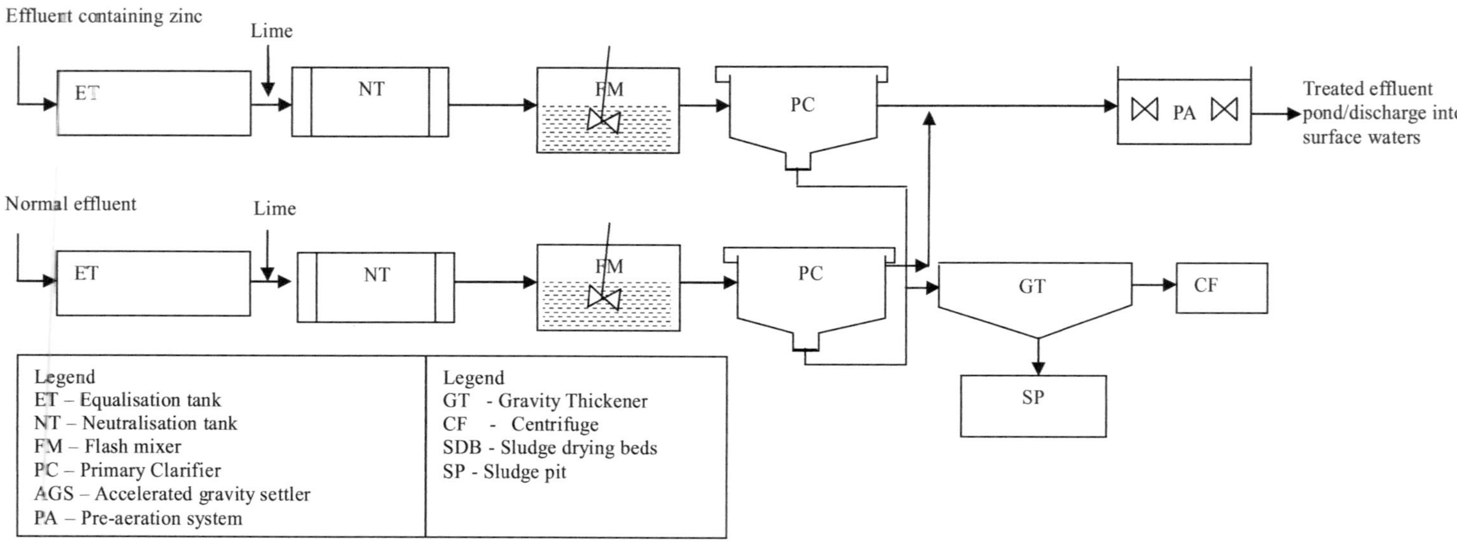

Figure 35.6b: Effluent Treatment Scheme in Viscose Rayon Unit.

Table 35.5: Chemical and Advanced Treatment Method for Zinc Removal from Effluents of MMF Industry

Methods	*Zinc Metal Concentration (mg/l)*	*Optimum pH Range*	*Precipitant (mg/l)*	*Removal Efficiency (%)*	*References*
Chemical precipitation	32	9-10	CaO	99-99.3	[15]
	100	7-11	CaO	99.4-99.6	[16]
	450	11.0	$Ca(OH)_2$	99.7	[18]
Solvent extraction	100	4.2	Cynex 272*	99.7	[9]
	100	2.9	Cynex 303*	96.5	
Ion exchange	100	5.0	Dinoptilolite	100	[21]
	100	–	NaPl zeolite	100	[22]
Electrochemical treatment					
– Electro-coagulation	217-236	6.0	–	100	[24]
– Electro-flocculation	20	7.0	–	96	[26]
Membrane filtration					
– Reverse osmosis	64-170	–	CDA	98.9	[31]
– Ultra-filtration	–	–	SDS	97-9	[29]

* Indicates that extractant in wastewater treatment process

Chemical Precipitation

Chemical precipitation is used in wastewater management systems because of simplicity and cost[14]. Chemical precipitation using hydroxide is common in rayon industry. The solubility of the metal hydroxides is minimized in the pH range of 8.0-11.0[15,16]. Lime is the preferred choice of base used in hydroxide precipitation at primary stage of treatment[17,18]. The addition of coagulants such as alum, iron salts, and organic polymers can enhance the removal of zinc from wastewater employed through chemical coagulation and precipitation to treat synthetic wastewater[19]. A zinc reduction of up to 0.45 mg/l could be achieved by chemical precipitation through zinc concentration as high as 100 mg/l.

The precipitation using sulfide is an effective process for the treatment of toxic heavy metals ions in semi-synthetic fibre industry[19]. The major advantage of using sulfides is solubility of the metal sulphide being lower than hydroxide. Sulfide reacts with divalent soluble metals to form insoluble metal sulfides of the zinc-bearing wastewater[16]. Sulphide precipitation technique is applied to effluents having sulfate, zinc and TOCs concentration up to 6000, 400 and 1500 mg/l of initial concentration.

Chemical precipitation is also successful in combination with other treatment methods. Electro-Fenton followed by lime precipitation is another route for treatment of wastewater containing high COD and zinc[13]. Effluents with characteristics of COD (2400 mg/L) and Zn^{2+} (32 mg/L) could be effectively treated through Electro-Fenton and chemical precipitation route. Electro-Fenton and chemical precipitation method with current density in the range of 0.45 A/dm^2 to 1.80 A/dm^2 with optimized

H_2O_2 dosages in the range of 765 -3825 mg/L obtained a COD reduction of 88 per cent of zinc removal of 99-99.3 per cent [14].

Solvent Extraction

Solvent extraction process is effective in separation of heavy metals from the complex low metal bearing effluents of MMF industry[9]. Thiophosphinic extractants Cynex 272 and 302 were used for recovery of zinc from rayon effluent by solvent extraction technique[9]. The extraction is pH dependent and their separation could be achieved by precise control of the pH[20]. The extractant Cyanex 302 has been found selective and recovery upto of 99.99 per cent of zinc have been obtained at pH 3.4 and organic aqueous (O/A) ratio of 1/30[9]. The zinc from the loaded Cyanex 302 can be stripped with 10 per cent sulphuric acid at an O/A ratio as low as 10. The stripped solution thus obtained could be recycled back to the spinning bath of the rayon plant and the raffinate could be disposed safely without affecting environment.

Ion Exchange

Ion exchange is used to remove heavy metals such as zinc using chelating ion exchange resin dinoptilolite[21]. Zn (II) ions are exchanged on weak acidic carboxylic cation exchangers[22]. Ion exchangers have great affinity for hydrogen ions hence they cannot be used for Zn (II) ions sorption in the acidic medium [23]. The regeneration of Zn (II) ions from the carboxylic cation exchange resin proceeds readily for application of acid baths as a regenerator in semi-synthetic fibre industry.

Electrochemical Methods

Electrochemical methods (Electrocoagulation-flocculation) have gained importance globally in the past two decades[24]. Electrocoagulation-flocculation (EC-EF) refers to the generation of coagulants dissolving electrically either aluminium or iron ions from aluminium or iron electrodes[25]. The metal ion generation takes place at the anode and hydrogen gas is released at the cathode which can help to float the flocculated particles out of the water[24, 26]. In the EC system aluminium electrodes are successful for removing heavy metals in some cases was greater than 99 per cent [25].

Adsorption

Adsorption is an effective and economic method for treatment of organic substances and divalent metals from wastewater in MMF industry. The adsorption process offers flexibility in operation and in many cases produces high-quality treated effluent. Activated carbon (AC) adsorbents are widely used in the removal of organic contaminants and heavy metals[27, 28]. The heavy metals removal proved highly effective as removal efficiency increased with increasing pH while it decreased with increasing metals concentration[16]. The removal efficiency was high for zinc ranging from 81-97 per cent[28]. Although efficient, adsorption process always accompanies with the problem of carbon regeneration and waste disposal.

Membrane Filtration

Membrane filtration techniques show great promise for their high efficiency and easy operation. A prominent advantage is reduced foot print for treatment. The choice of the membrane processes, whether it is ultra filtration (UF), nanofiltration (NF), reverses osmosis (RO) is dependent on purpose of recycling and reuse of effluents in MMF industry. UF membrane processes can be used effectively in synthetic fibre industry[29]. The Ultra filtration membrane process could be used as in MBR to achieve COD and TKN removal[30]. Reverse osmosis used as end-of-pipe treatment for effluent recycling[31]. RO is one of the techniques able to remove a wide range of dissolved salts from the effluents with efficiency up to 98 per cent, through certain specific polymeric membranes, such as cellulose acetate or nylon[32]. In semi-synthetic industry, the effluent from spinning bath, washing section, water and chemicals recovery is achievable through use of RO. NF could be used in recovery of sodium salts from the spinning section effluents. A water recovery is up to 94 per cent have been achieved using cellulose acetate membrane.

Resource Recovery in MMF Units

In recent years, due rising prices attention has been given to the recovery of raw materials and water in MMF industry. This is mainly for environmental pollution control and proper utilization for wastewater management in MMF industry[4].

Chemical Recovery

Chemicals like sodium sulphate, caustic soda, zinc and formaldehyde can be recovered by evaporation, membrane separation and through liquid-liquid extraction and specifically applied based on fibre type, requirement and spinning methods[8]. Dry spinning processes typically use condenser or scrubbers for recovering solvent from the spin cell. Efficient solvent recovery is particularly important in dry spinning since solvent is used at three to five times the quantity of polymer[4].Wet spinning processes typically use distillation to recover solvent from the spinning bath, drawing and washing operations[13].

Water Recovery

Water recovery is achieved in monomer recovery section or barometric legs on condensers in the multiple effect evaporators and is used for cooling purpose or dilutes acidic effluents [7,9]. This water could also be routed through finishing and washing section[12]. The implementation of this system can reduce the volume of effluent half[20].

Conclusion

MMF industry is an important industry as it has wide ranging applications and is the secondary largest producer of employment. However sustainable development of this industry is closely linked to environmental degradation to which the industry is dependent. Advances in process chemistry, in-house keeping and wastewater management has considerably improved the status of MMF units since last three decades. Being one of the largest water users, water conservation and reuse of treated effluent is very critical to this industry as it can reduce the environmental impact. Increased recovery of sodium and zinc from the process and gypsum from the effluent treatment plant reduces the impact on the receiving water bodies.

References

1. Indian Manmade Fibre Industry, Credit Analysis and Research Ltd., 2009-10.
2. Current Status and Future Prospects, Association of Synthetic Fibre Industry of India, *Indian Chem Fib Ind J*, 2010.
3. Petrochemical Directory, Monitoring and Evaluation Division, Dept. of Chem. and Petrochem., Ministry of Chemicals and Fertilizers Govt. of India, New Delhi, 2006
4. IL and FS Eco smart Ltd., *Technical EIA guidance manual for Manmade Fibre Industry*, 2010.
5. http: //www.bharatbook.com/detail.asp?id=104063 and rt=Indian-Man-Made-Fibre-Industry.html
6. Centre Board for the Prevention and Control of Water Pollution, Govt. Of India, New Delhi, *Comprehensive Industry Document Man-Made Fibre Industry*, 1979.
7. Matthew N, Rayon Fibres the Versatile Choice for Wipes, 2008, http: // www.kelheim-fibres.com/pdf/atlanta_1208.pdf.
8. U.S. Environmental Protection Agency, EPA-310-R-97-006, *Profile of the Plastic Resin and Manmade Fibre Industry*, Washington, DC, 1997.
9. Jha M K, Kumar V, Bagchi D, Singh R J, Lee J C, Processing of rayon waste effluent for the recovery of zinc and separation of calcium using thiophosphinic extractant, *J of Haz Mat*, 145 (2006) 221–226.
10. Ghosh P, Samanta A N, Ray S, Reduction of COD and removal of Zn^{2+} from rayon industry wastewater by combined electro-Fenton treatment and chemical precipitation, *Desal* (2011) 213-217.
11. Kurniawan T A, Chan G Y S, Lo W H, Babel S, Physico–chemical treatment techniques for wastewater laden with heavy metals. *Chem Eng J*, 118 (2006) 83–98.
12. Fischer G, *Climate changes a key issue for 21st century fibre makers 48th manmade fibre congress*, Industry News, www.fiberjournal.com, 12 December 2009.
13. Ray B, *Treatment and management of wastewater from an acrylic fibre industry*, Thesis, Asian Institute of Technology, Bangkok, 2006.
14. Fenglian F, Wang Q, Removal of heavy metal ions from wastewaters: A review, *J Env Man* 92 (2010) 407-418.
15. Andre Dupont, Lime Treatment of Liquid Waste Containing Heavy Metals, Radionuclides and Organics, *Nat Conf on Manag of Uncontrolled Haz Waste Sites*, 7th Edition, Prod Engg, 1997, 84-88.
16. Benefield L D, Morgan J M, *Chemical precipitation, Water Quality and Treatment*, Edited by R D Letterman, McGraw-Hill Inc., NY, 1999, 10.1–10.57.
17. Tunay O, Kabdasli N I, Hydroxide precipitation of complexed metals, *Wat Res*, 10: 28, (1994) 2117–2124.

18. Charerntanyarak L, Heavy metals removal by chemical coagulation and precipitation, *Wat Sci Technol*, 39: 10-11 (1999) 135-138.

19. Trivedi P R, Gurdeep R. Treatment of manmade fibre unit effluent for removal of zinc by double precipitation method, Vol. 21, Ency of Env Sci, Akashdeep Publication House, New Delhi, India, 1992, 1425.

20. Ali A M I, Ahmad I M, Daoud J A, CYANEX 272 for extraction and recovery of zinc from aqueous waste solution using a mixer settler unit, *Sep and Purif Tech*, 47 (2006) 135-140.

21. Athanasiadis K, Helmreich B, Influence of chemical conditioning on the ion exchange capacity and on kinetic of zinc uptake by clinoptilolite. *Wat Res*, 39 (2005) 1527-1532.

22. Landaburu A J, García V, Pongrácz E, Keiski R L, The removal of zinc from synthetic wastewaters by micellar-enhanced ultra filtration: statistical design of experiments. *Desal* 240 (2009) 262-269.

23. Wang L K, Vaccari D A, Li Y, Shammas N K, *Chemical precipitation, Physicochemical Treatment Processes*, Vol 3, Edited by Wang L K, Vaccari D A, Li Y, Shammas N K, Humana Press, New Jersey, 2004, 141–198.

24. Chen G H, Electrochemical technologies in wastewater treatment. *Sep Purif Tech* 38 (2004) 11-41.

25. Heidmann I, Calmano W, Removal of Zn II, Cu II, Ni II, Ag I and Cr VI present in aqueous solutions by aluminium electro-coagulation, *J Haz Mat* 152 (2008) 934-941.

26. Belkacem M, Khodir M, Abdelkrim S, Treatment characteristics of textile wastewater and removal of heavy metals using the electroflotation technique. *Desal* 228 (2008) 245-254.

27. Bose P, Bose M A, Kumar S, Critical evaluation of treatment strategies involving adsorption and chelation for wastewater containing copper, zinc, and cyanide, *Adv Env Res* 7 (2002) 179–195.

28. Kurniawan T A, Chan G Y S, Lo W H, Babel S, Comparisons of low-cost adsorbents for treating wastewaters laden with heavy metals, *Sci Tot Env* 366: 2-3 (2006), 409-426.

29. Huang J H, Zeng G M, Zhou C F, Li X, Shi L J, He S B, Adsorption of surfactant micelles and $Cd^{2þ}$ / $Zn^{2þ}$ in micellar-enhanced ultrafiltration, *J Haz Mat*, 183 (2010) 287-293.

30. Ray B, Lamasan, Annachhatre A P, Treatment and management of wastewater from an acrylic fibre industry, GASARN International Conference on Sustainable Development: Issues and Prospects for GMS.

31. Ujang Z, Anderson G K, Application of low-pressure reverse osmosis membrane for Zn^{2+} and Cu^{2+} removal from wastewater, *Wat Sci Tech* 34: 9 (1996) 247–253.

32. Lee K H, Kang B C, Lee J B, Acrylic wastewater treatment and long-term operation using a membrane separation system, *Desal*,191 (2006)169–177.

Chapter 36

Performance Parameters of Impingement Scrubber in Flue Gas Treatability System for Small and Medium Scale Industrial Emissions

Padma Rao, V.A. Mhaisalkar and S.R. Wate

CSIR-NEERI, Nagpur

ABSTRACT

The performance parameters such as liquid flux distribution and collection efficiency in a pilot impingement scrubber was tested successfully with experimental data. The liquid flux distribution is estimated for an entire range of simulated gas conditions. It is observed that the collection efficiency increases with uniform flux distribution formed due to increase in gas velocity. The design parameters (aspect ratio and nozzle diameter) have an optimum that offers the most uniform flux distribution and the maximum collection efficiency.

***Keywords**: Aspect ratio and Nozzle diameter.*

Introduction

Air pollution problems in developing countries have gained larger fraction in the last decade especially due to non functioning and non implementation of effective air pollution control devices in industries. In industrial wastewater management, adequate treatability studies are conducted to arrive at a techno-economic treatment option. However no such studies were done for reducing air pollution or emission from industries until now in India. Little information was available about such studies in other countries. Rao, B PS *et al.*, novel technique known as flue gas treatability studies was used to study and optimise air pollution control systems for select emission categories. This study is a tool for techno-economic selection of air pollution control systems specially for small/medium scale industrial emissions.

The impingement scrubber one of the control system in the treatability modules is a high-energy impaction atomizing system that is useful for controlling both particles and gases pollutants (Figure 36.1a-b). The particulate collection efficiency is complex and depends on many variables including aerodynamic particle size, liquid to gas (L/G) ratio, gas velocity, liquid droplet distribution, The performance study suggests that collection efficiency increases with uniform flux distribution formed due to increase in gas velocity.

Wet Scrubbers

These scrubber have a constant pressure drop (at constant volume), present no secondary dust problem in disposing of the collected dust, and can handle high-temperature or moisture-laden gases. These can also handle corrosive gases or aerosols, but corrosion-resistant construction may add materially to their cost. Space requirements are reasonably small. Disposal of the wastewater or its clarification for reuse may, however, be difficult or expensive.

The collection efficiency varies widely with different designs. Most collectors decline rapidly in efficiency for particles between 1 and 10 microns. Many investigators believe that collection efficiency is directly related to the total power expended in forcing the gases through the collector and in generating the water spray.

The process of contacting an air-contaminated gas with a scrubbing liquid results in dissipation of mechanical energy in fluid turbulence and, ultimately, in heat. The power dissipated is termed the contacting power. Semrau (1960) made an exhaustive literature survey to correlate scrubber efficiency with contacting power. He states that contacting power can be derived from (1) the kinetic energy or pressure of the gas stream, (2) the kinetic energy or pressure head of the liquid, or (3) energy supplied mechanically by the rotor. He concludes: "Efficiency is found to have little relation to scrubber design and geometry, but to be dependent on the properties of the aerosol and on the contacting power".

Theory of Collection

The principal mechanism by which liquids may be used to remove aerosols from gas streams are as follows:

1. Wetting of the particles by contact with a liquid droplet,
2. Impingement of wetted or unwetted particles on collecting surfaces followed by their removal from the surfaces by a flush with a liquid.

Mechanism for Wetting the Particle

The particles can be wetted by the following mechanisms:

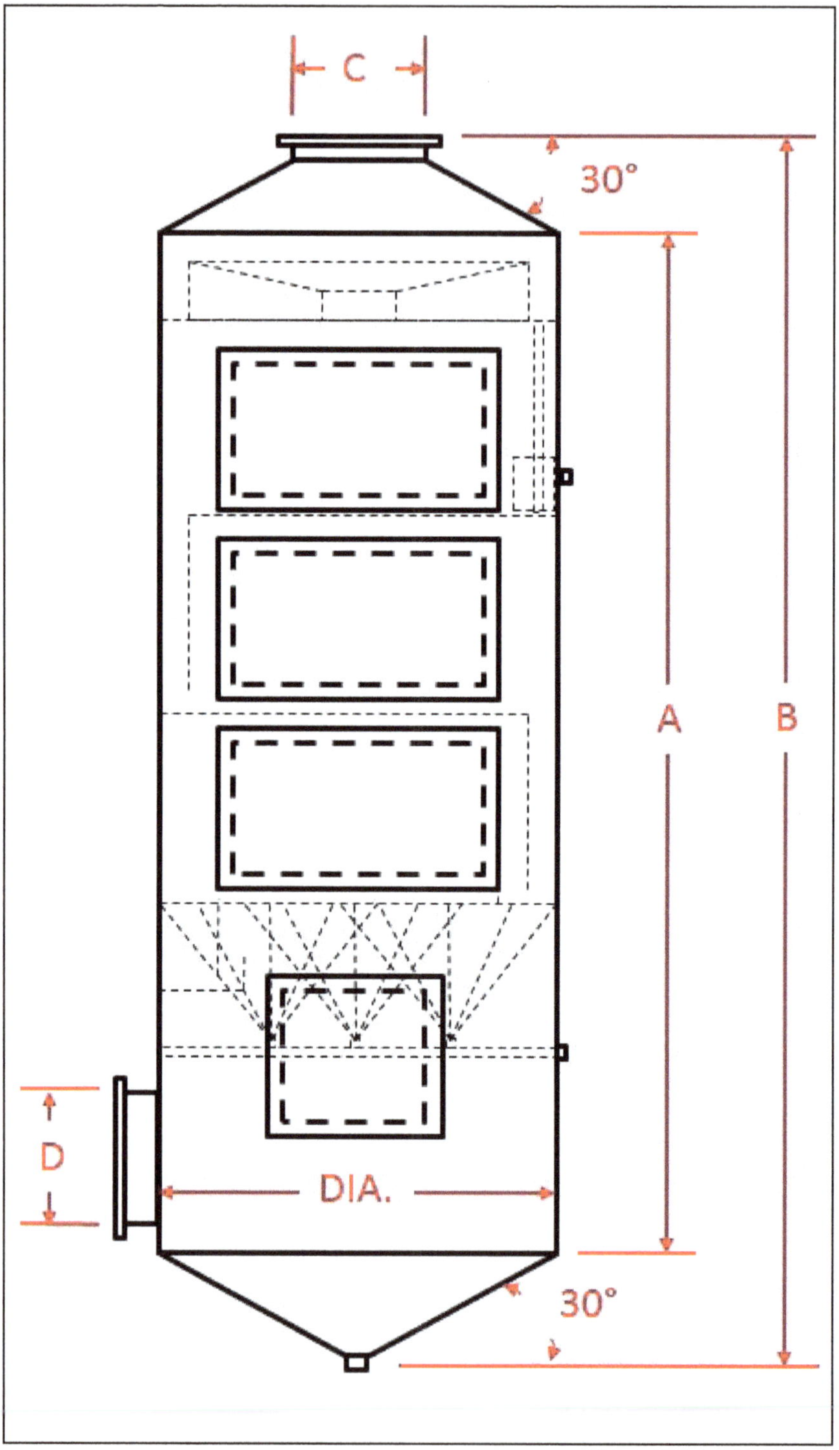

Figure 36.1(a): Impingement Scrubber.

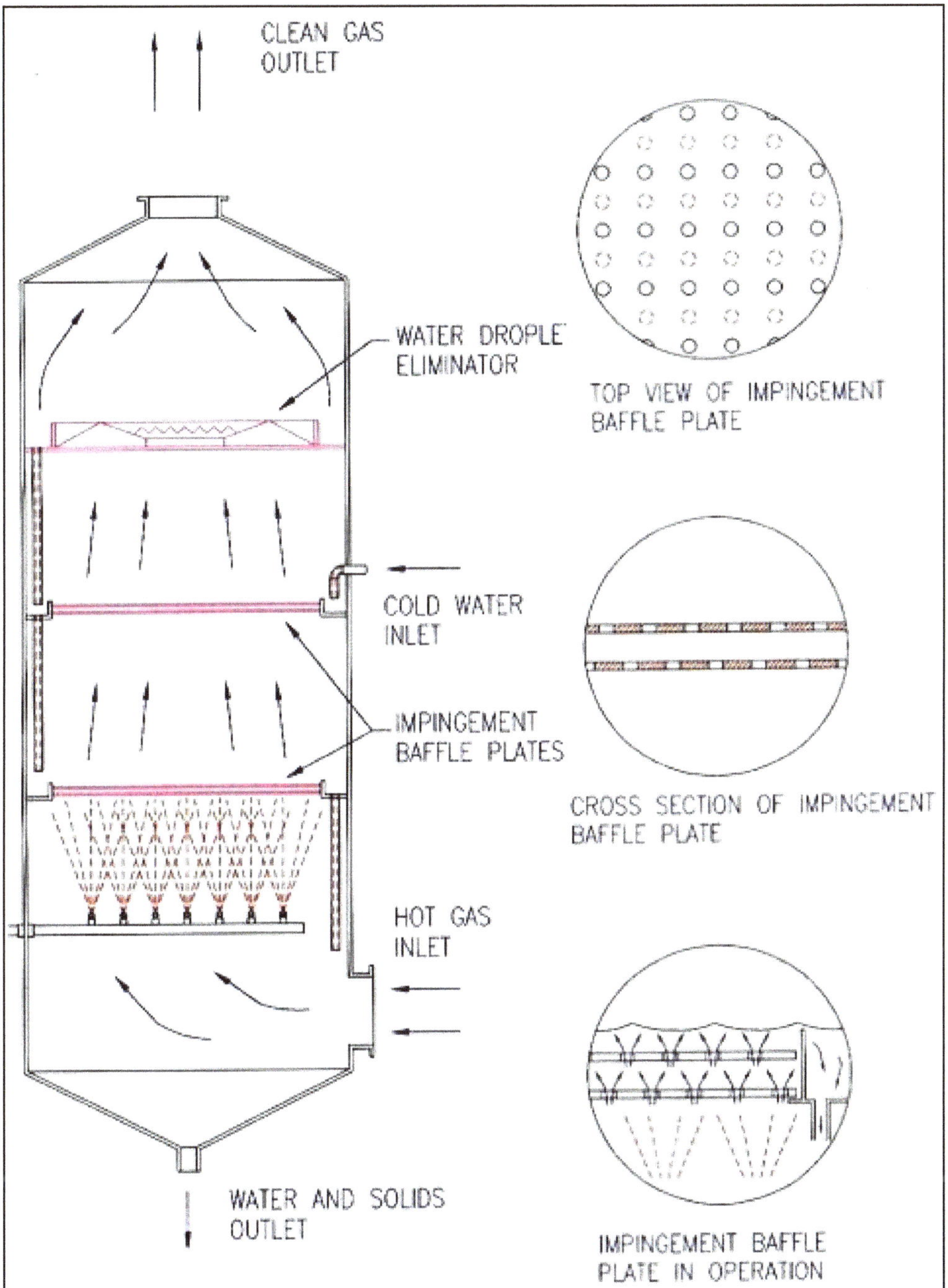

Figure 36.1(b): Impingement Scrubber.

1. Impingement by spray droplets. A spray directed across the path of the dust particles impinges upon them with an efficiency proportional to the number of droplets and to the force imparted to them. Johnstone and Roberts (1949) states that the optimum droplet particle size is about 100 microns. Above 100 microns, there are too few droplets, and below 100 microns, The droplets do not have sufficient force.
2. Diffusion. When liquid droplets are dispersed amongst dust particles, the dust particles are deposited on the droplets by Brownian movement or diffusion. This is the principal mechanism in the collection of submicron particles. Diffusion as the fluid turbulence may also be an appreciable mechanism in disposition of dust particles on spray droplets.
3. Condensation (Lapple, 1963). If a gas is cooled below the dewpoint in passing through a wet collector, then condensation of moisture occurs, the dust particles act as condensation nuclei. This effective increase in the particle size makes subsequent collection easier. Condensation alone can remove only relatively small amounts of dust, since the large concentration is greater than can be achieved.
4. Humidification and electrostatic precipitation have been suggested as mechanisms that facilitate collection of particles by causing them to agglomerate. These effects are not, however, well understood and cannot be relied upon to play any significant role in the collection mechanisms.

Several investigators have used wetting agents for scrubbing water in an effort to improve efficiency. In most cases, little or no improvement has been found (Friendlander *et al.*, 1952). In order to be wetted, a particle must either make contact with a spray droplet or impinge upon a wetted surface. When either of these occurs, the particles is apparently wetted as adequately without the use of wetting agents as it is with their use.

Particles that have been wetted must reach a collection surface if the collecting process is to be completed. They may be impinged against surfaces placed in the path of the gas flow; or centrifugal action may be used to throw them to the outer walls of the collector; or simple gravity settling may be employed.

In some devices impingement is the principal collection mechanism, the water sprays being used merely to remove the dust from the collection surfaces.

Centrifugal action may be provided by a vessel that is essentially the same as a dry cyclone separator. Helical vanes in the cylindrical vessel are extensively used to supply centrifugal action. In some devices, baffles are shaped and placed so that act both as impingement and collection surfaces, and as imparters of cyclonic motion to the gas stream.

Impingement Scrubber

In this, scrubber a chamber containing two/three impingement plates over which spray nozzles are placed. The gas stream velocity decreases as it enters the chamber, and the wetted particles settle and are collected at the bottom of the chamber. The outlet of the chamber is equipped with eliminator plates to help prevent the liquid from being discharged with the clean stream. Its efficiency as a dust collector is improved due baffle plates upon which particles are impinged. Water rates ranging from 3 to 8 gallons per minute (gpm) per 1,000 cfm.

This scrubber operate with pressure ranging from 100 to 150 psi. Increasing the pressure at the spray nozzles has been found to increase collection efficiency by creating more droplets and giving them more force. Very small nozzle orifices are used, and in most cases this precludes recirculation of water. Nozzles are so located that collision between water droplets are minimized, and hence design ensures maximum collision between water droplets and the dust particles. Water consumption ranges from 5 to 10 gallons per 1,000 cfm.

In this study, considering a nonuniform distribution concentration of droplets throughout the scrubber zone, a three-dimensional mathematical model was used to predict the SO_2 removal efficiency of a scrubber.

Mathematical Model

To obtain a suitable mathematical model for simulation of a impingement scrubber, it is necessary to have information on the mechanics of operation, as well as the gas-absorption rate in the scrubbing zone. The liquid spray from the nozzle penetrates the scrubber as fine droplets. As soon as droplets are formed they are dispersed by convection and diffusion. It was assumed that droplets have an average diameter that can be calculated by equations developed by Boll or Nukiyama and Tanasawa, depending on the conditions of liquid injection. The distribution of droplet concentration, Cd, is evaluated by a three-dimensional dispersion model. The main equation of that model, which expresses material balance for droplets in each increment of the scrubbing zone, is:

$$\frac{\partial (V_d C_d)}{\partial x} = E\left(\frac{\partial^2 C_d}{\partial y^2} + \frac{\partial^2 C_d}{\partial z^2}\right) + S \tag{1}$$

In the above equation, it is assumed that droplets are generated at a point source. The location of this point source is determined by the penetration length of the liquid spray. Eddy diffusion of droplets, E_d, is related to eddy diffusion of gas, E_g, by means of a force balance. The source strength, S (number of droplets generated per unit volume per unit time), is equal to the droplet flowrate through each nozzle per unit volume. In fact, this source strength is used instead of the liquid stream provided by each nozzle. Further details on evaluating the parameters needed for solving equation 1 are given by Fathikalajahi *et al.*, Distribution of droplet concentration throughout the scrubber is determined by solving the above equation. In the next step, the gas absorption rate and pollutant concentration distribution should be evaluated.

The following dispersion model was used to obtain the pollutant concentration distribution in the gas phase:

$$\frac{\partial (V_g C_g)}{\partial x} = E_g \left(\frac{\partial^2 C_g}{\partial y^2} + \frac{\partial^2 C_g}{\partial z^2} \right) - N_A \pi D_d^2 C_d \tag{2}$$

In the above equation, NA is the rate of mass transfer into moving droplets and is related to the pollutant concentration in the liquid phase. To find the pollutant concentration distribution in the liquid phase, material balance in a differential control volume in liquid phase was written as follows:

$$\frac{\pi}{6} D_d^3 \frac{\partial (V_d C_d C_l)}{\partial x} = \pi D_d^2 C_d N_A \tag{3}$$

To solve the above equations one needs to obtain the mass transfer rate of SO_2 into moving drops (N_A).

There have been many investigations into the SO_2 absorption of a moving drop. Among them, Baboolal *et al.*, have developed the most comprehensive absorption model for predicting the mass transfer rate of SO_2 into falling drops. In their model, the complete Navier-Strokes equation and mass-transfer equation have been solved, and comparison between experimental data and the results confirms the validity of their model. However, due to its complexity, this model cannot be applied here.

To find a simple model for evaluating the mass transfer rate, the following assumptions were made:

1. The mass transfer rate in the gas phase can be evaluated by the following equation:

$$N_A = K_g \left(C_g - C_{g_s} \right) \tag{4}$$

where, K_g is given by the following correlation:

$$N_{sh} = \frac{K_g D_d}{D_g} = 2 + 0.6 \left(N_{Re}^{0.5} N_{Sc}^{0.33} \right) \tag{5}$$

2. At the liquid surface, the concentration of the dissolved gas is C_{ls}, in equilibrium with the concentration of SO_2 in the gas interface, C_{gs}:

$$C_{gs} = f_{eq}(C_{ls}) \tag{6}$$

The equilibrium between gas and liquid phases is illustrated in Appendix A.

3. The laminar-jet model is applicable to evaluating the mass transfer rate in the liquid phase. In this model, the flow of droplets is considered as a laminar jet, for which the mass transfer rate in the liquid phase can be calculated by the following equation.

$$N_A = C_{ls} \sqrt{\frac{D_l}{\pi t}} \tag{7}$$

At equilibrium, the mass transfer rates in gas and liquid phases are equal. Thus, after time, by knowing gas concentration, C_g, the mass transfer rate, N_A can be evaluated by simultaneous solution of equations 4 – 7.

The validity of using a simple laminar-jet model is confirmed by comparing the results of that model to the model by Baboolal *et al.*, And experimental data of Walcek et. al. shows this comparison and reveals that, indeed, a laminar-jet model is an appropriate tool in this situation.

Furthermore, when pollutant gas undergoes a very fast chemical reaction, as in the case in absorption of SO_2 by a strongly alkaline solution, the surface concentration in the gas phase can be considered to be zero. Thus, N_A can be simply evaluated by the following equation:

$$N_A = K_g \; C_g \tag{8}$$

To solve equation 1, 2 and 3, an upwind control-volume method was used. This method is applicable to convection and diffusion equations when the Peclet number is greater than 2. For high Peclet numbers, the common numerical error of false diffusion can be neglected and, in spite of the simplicity of the upwind scheme, the numerical results are valid.

Results and Discussion

The pilot scale impingement scrubber (Figure 36.1) is tested under simulated emission to identify the effect of design and operating parameters on collection efficiency. The operating parameters used for simulation include L/G ratio of 0-100 gal/1000ft^3 and gas velocity ranging from 25-100 ft/s. The performance parameter include plate diameter particle L/G ratio, inlet concentration etc. as given in Table 36.1. The emission characteristics simulated for various small and medium industries are given in Table 36.2. The design data of the impingement scrubber is given in Table 36.2.

Effect of Particle Size and Gas Velocity

Figures 36.2 illustrates the effect of particle size and gas velocity on collection efficiency and it is observed that high collection efficiency is achieved at higher gas velocities for all conditions such as L/G ratio, nozzle diameter and scrubber geometry remaining constant. The narrow particle size range below 1u size is critical as the efficiency is low in this region, however it increases beyond 5 u size with increased gas velocities which produces finer droplets thus increasing the particle collection efficiency. This effect is in agreement with the observations reported by Viswanathan (1997). However, for gas velocity beyond 90 ft/s, there is no significant improvement in removal efficiency. This can be explained by the fact that although finer droplets are formed with increasing gas velocities, the flux distribution remains the same and the finer droplets are accelerated very quickly to the gas velocity giving the scrubber

Table 36.1: Performance Characteristics of Scurbber and their Wet Collectors

Device	*Test Aerosol*	*Inlet Concentration*	*Particle Size at Inlet, μ*	*Efficiency Per cent*	*Resistance H_2O*	*Velocity (Face of wet cell), fpm*	*Water Rate (per wet cell), gal/ 1,000 ft³*	*Remarks*
Wet Cell washer (a 9–element washer consisting of 3 units in series, each made up of 2 concurrent[a] wet filter cells followed by a dry eliminator pad)	Normal air	0.2 to 0.5 grain/ 1,000 ft³	Most particles between 0.3 and 0.5 micron (estimated)	57	0.19 to 0.28 (per wet cell)	216	3.3 to 5	Wet cells: A: 3-in cells of 150-μ fibers random packed 7.9 lb/ft³
				65	0.38 to 0.56 (per wet cell)	216		B: 6-in cells of 150μ fibers random packed 7.9 lb/ft³
				67	0.21 to 0.23 (per wet cell	216		C: 8-in cells of 255μ fibers partially oriented, packed 4.9 lb/ft³
	Dust composed of spheres of copper sulfate	1 to 2 grains/ 1,000 ft³	2.5 (mass median)	90	same as A	216	3.3 to 5	Same as A
				92	Same as B	216	3.3 to 5	Same as B
				94	Same as C	216	3.3 to 5	Same as CDry pads were 2-in thick and composed of 10-μ fibers packed 0.4 lb/ft³
				Cumulative efficiency stages 1 2 3				Wet Cells were 8-in thick and composed of 255-μ fibers partially oriented with densities of 3.3 lb/ft³
Wet Cells washer (an -8 element washer stage 1 has one counter-current[b] followed by one concurrent wet cell.	Dust composed of CO_3 spheres	0.01 to 0.06 grain/ 1,000 ft²	0.8 (mass median)	20 75 80	0.14 to 0.17 (concurrent) 0.22 to 0.23 (counter-current)	216	15.3	

Contd...

Table 36.1–*Contd...*

Device	*Test Aerosol*	*Inlet Concentration*	*Particle Size at Inlet, μ*	*Efficiency Per cent*	*Resistance H_2O*	*Velocity (Face of wet cell), fpm*	*Water Rate (per wet cell), gal/ 1,000 ft³*	*Remarks*
Stage 2 and 3 each have on e counter-current and one concurrent wt cell followed by a dry eliminator pad)				13 74 80	0.21 to 0.23 (concurrent) 0.32 to 0.36 (counter-current per wet cell)	216	15.3	4.9 lb/ft³ Dry pads were 2-in thick and composed of 10-μ fibers packed 0.46 lb/ft³
Ducon No. 5	Dust from stone and sand-drying kiln	5.8 grains/ft³	1.5	74 (weight)			1 (rated)	Flow rate was 9,300 cfm. Temperature was 230ºF in and 120ºF out. Aerosol was effluent from Sirocco No. 20 Cyclones
Venturi scrubber	H_2SO_4 mist from sulphuric acid plant	2.92 grains/ft³		99.7	9.7	10,260	4.3	Outlet concentration was 0.0092 grain/ft³
		3.98 grains/ft³		99.7	29.3	19, 200	3.9	Outlet concentration was 0.011 grain/ft³
	SiO_2 from silicon ore furnaces		0.01 to 0.35	86.7	11.0	15,060	4.7	
	Iron oxide fume from open hearth furnace (oxygen lanced)	1 to 5.99 grains/ft³	0.02 to 0.5	92 to 99 (weight)	14.0	15,840		Outlet concentration was 0.0092 to 0.069 grain/ft³

Contd...

Table 36.1–*Contd...*

Device	*Test Aerosol*	*Inlet Concentration*	*Particle Size at Inlet, μ*	*Efficiency Per cent*	*Resistance H_2O*	*Velocity (Face of wet cell), fpm*	*Water Rate (per wet cell), gal/ 1,000 ft³*	*Remarks*
Cyclone scrubber	Lime dust from lime kiln	9.2 grains/ft³	2.0 to 40.0	99	2 to 4 (rated)		5 to 10 (rated)	Outlet concentration was 0.08 grain/ft³
	iron ore and coke dust from blast furnace	3.0 to 24.0 grains/ft³	0.5 to 20.0	99 (weight)	2 to 4 (rated)		5 to 10 (rated)	Outlet concentration was 0.03 to 0.08 grain/ft³
Centri-merge	Na_2CO_3 fume	30 x 10^4 particles/ft³		96.2 (count)	5.5	2,000	40	
Multi-wash	Foundry dust	99.7 x 10^4 particles/ft³	<15	88.9 (count)	4	3,300		Laboratory test

a: Concurrent means that water is sprayed in the direction of the air flow.

b: Countercurrent means that water is sprayed against the air flow.

Source: Friendlander *et al.*, 1952.

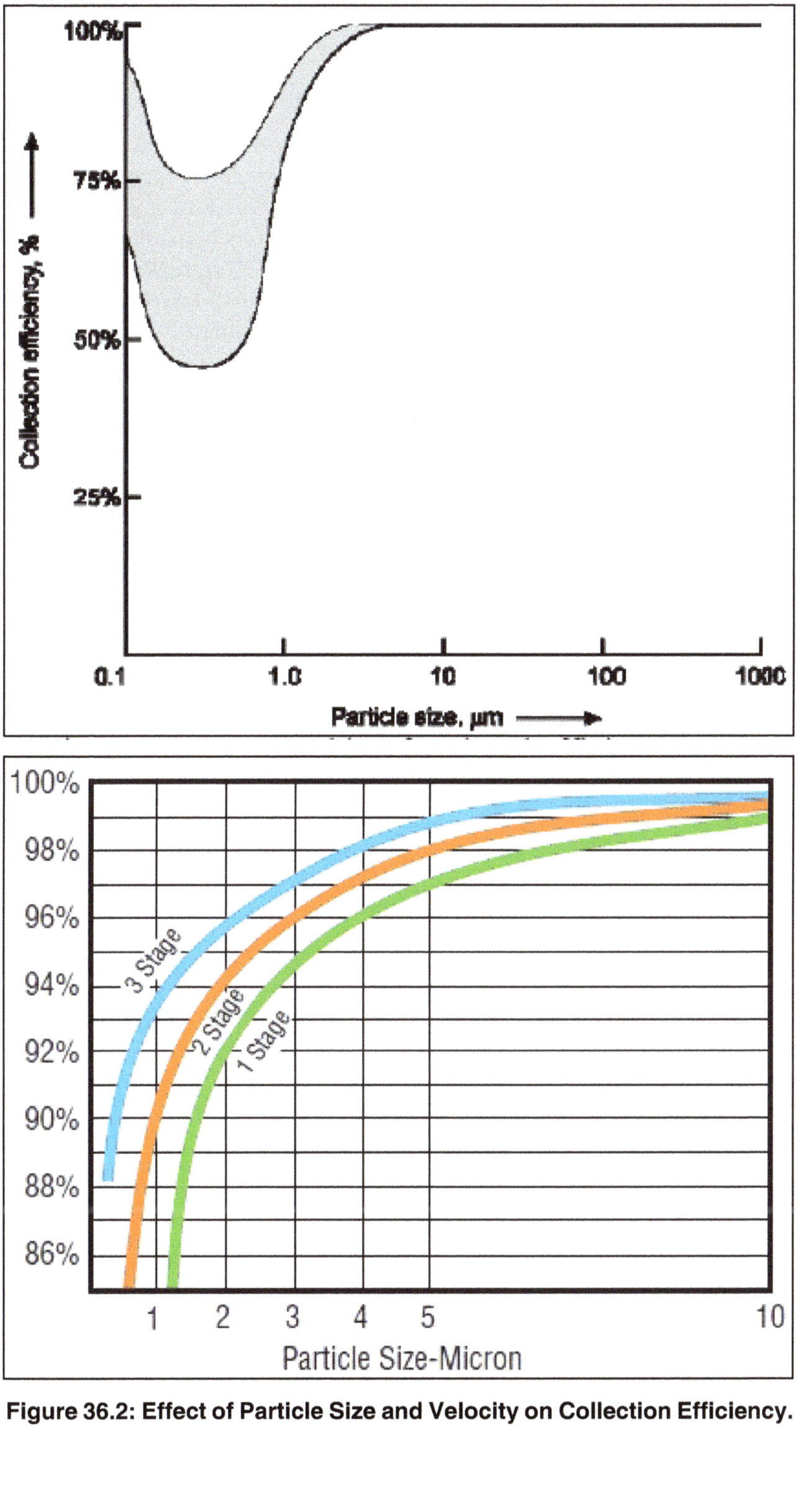

Figure 36.2: Effect of Particle Size and Velocity on Collection Efficiency.

very little interaction time thereby negating all other factor aiding the collection process.

Effect of Liquid to Gas Ratio on Removal Efficiency

It can be seen from Figures 36.3 that there is an optimum L/G ratio at which the droplets are uniformly distributed and the collection efficiency is maximum while all other parameters are fixed. Increasing L/G ratio at constant gas velocity, affects penetration due to changes in initial liquid momenta. This in turn affects the uniformity of flux distribution. The L/G ratio was varied from 0 to 100 gal/1000ft^3 of gas. It was reported by Fathikalajahi, 1996 that higher L/G provides more liquid available for scrubbing thereby providing bigger drops and non uniformity of flux distribution producing dominating negative effect on particle collection resulting in lower collection efficiencies. It is evident from the figure that maximum collection efficiency is obtained at L/G ratio of 80gal/1000ft^3. Hence it can be said that L//G and flux

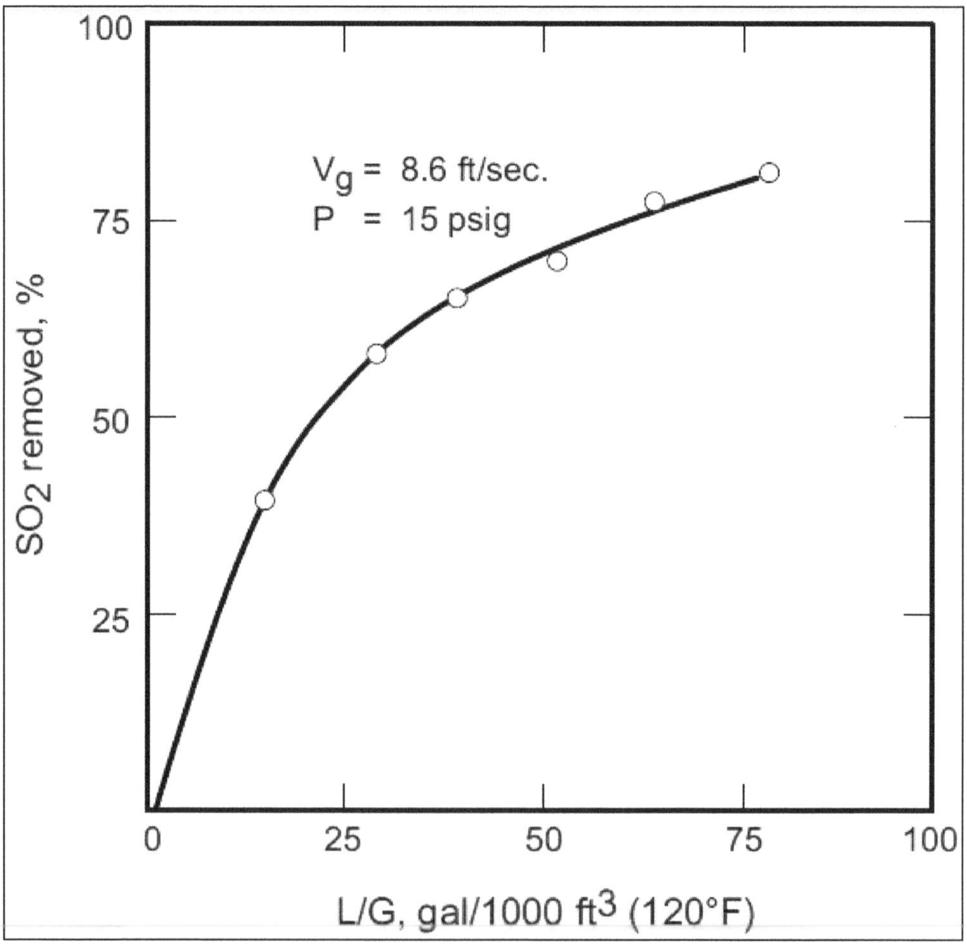

Figure 36.3: Effect of L/G Ratio on Collection Efficiency

Table 36.2: Emission Characteristics from Small Medium Scale Industries

Sl.No.	Small/Medium Scale Industries	Flow Range (m^3/hr)	Temp. (°C)	Velocity (m/s)	Gaseous Pollutants (mg/Nm^3)			Dust Pollution (mg/Nm^3)	
					SO_2	NOx	VOC	Conc.	Particle Size and its Per cent
1.	Lime Kiln	1000–2000	50–80	1–3	0.24–0.42	2–4	300	1000–2000	80–90 per cent are < 10μ size
2.	Brick Kiln	2500–4000	75–150	2–3	0.26–1.12	0.08 – 0.2	1000	300–2000	50–60 per cent are <10μ size
3.	Stone crusher	1000–4000	25–30	< 1	—	—	—	> 10000	50–60 per cent are >10μ size
4.	Bagasse fired boiler	1500–5000	100–250	19–29	—	—	400–500	400–1500	80–90 per cent are <10μ size
5.	Rice husk boiler	800–3500	90–220	10–13	—	—	300–500	300–2000	80–90 per cent are <9μ size
6.	Glass Industry	1500–4000	80–250	> 10	200	500	300	2000–4000	80–90 per cent are <10μ size
7.	Coke Ovens Bee Hive type	1300-1500	90-150	1-8	250	600	3000	2100 – 4500 as tarry matter	80–90 per cent are <10μ size

Table 36.3: Design Data of Impingement Scrubber

☆ Gas flow rate range	:	2083-2500 m^3/hr
☆ Gas Temperature	:	50º C
☆ Inlet dust concentration range	:	1.0-2.5 gm/m^3
☆ Inlet acid concentration range	:	0.9-1.5 gm/m^3
☆ Type of Scrubber	:	Impingement Plate
☆ Size	:	610 mm Dia x 3900 mm Ht.
☆ No. of stage	:	2 stages
☆ Capacity	:	1500–2500 m^3/h
☆ Design efficiency	:	99.5 per cent down to 3μ and above
☆ Pressure Drop thro' Scrubber	:	150 mm WC
☆ Water Recirculation rate range	:	1-10 m^3/hr
☆ Seal Pot	:	610 mm Dia x 1100 mm Ht
☆ Material of Construction	:	PP/FRP
☆ Fan		
• Type	:	Backward curved, Single inlet Centrifugal
• Capacity	:	1500-2500 m^3/hr
• Static Pressure	:	225 mm WC
• Impeller Speed	:	2800 RPM
• Absorbed Power	:	3 HP
• Motor	:	3 HP/2800 RPM, TEFC Squirrel cage Induction Motor, IP55, Class F, 415V, 50 Hz, 3 Ph, AC
• Drive Arrangement	:	Impeller mounted on Motor Shaft
• Accessories	:	a. Manual control Damper in fan inlet dust
		b. Flexible Canvas connector at inlet
☆ Recirculation Pump		
• Type	:	Vertical, glandless Pump
• Capacity		10 m^3/hr
• Head	:	25 m
• Material of Construction	:	PP
• Motor	:	3 HP/280 TEFC Squirrel cage Induction Motor, IP55, Class F, 415 V, 50 Hz, 3 Ph, AC
☆ Ducting		
• Size	:	170 mm Dia.
• Velocity thro' duct	:	18 m/sec
• Material of Construction	:	PP/FRP for intake Duct M.S., 3 mm for clean air Duct
☆ Skid		
• Size	:	1000 Width x 2000 Long
• Material of Construction	:	M.S. Channel

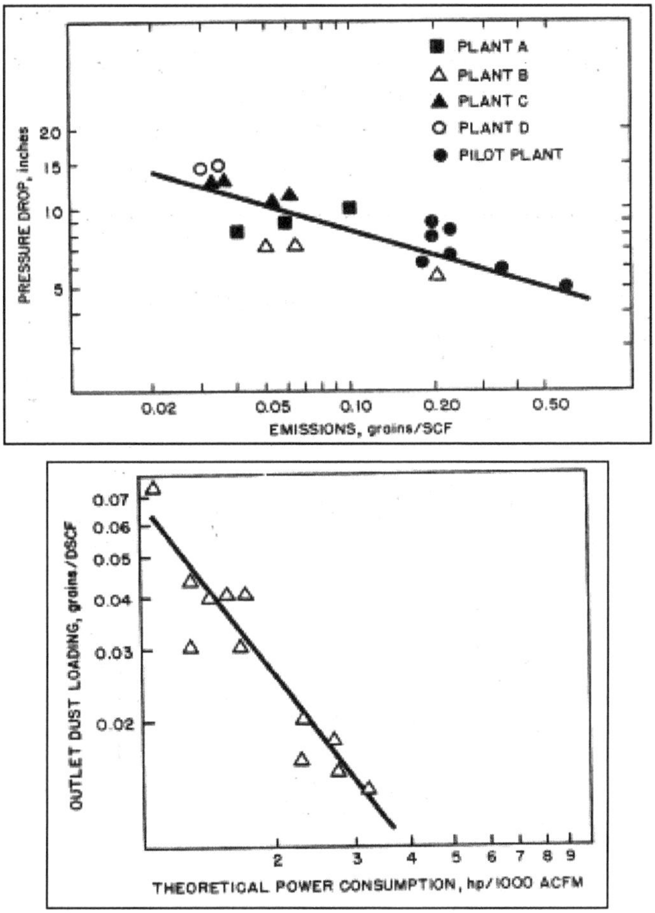

Figure 36.4: Effect of Emission Load on Pressure Drop and Power Consumption.

distribution are the two main operating parameters considered fundamental for any impingement scrubber.

Effect of Aspect Ratio

It is observed by varying aspect ratio from 0-5, different flux distribution is achieved keeping other parameters constant. Collection efficiency was found maximum at aspect ratios between 2 and 3.5. It was also observed that in this range of aspect ratio, the flux distribution was uniform

Effect of Plate diameter

Change in diameter of plate affect the liquid velocity. A decrease in plate dia. increases the liquid velocity and vice versa for a scrubber geometry at constant L/G and gas velocity. Higher liquid velocity through plates increases penetration length flow as film on the walls. However, very high penetration length causes non uniformity in droplet flux distribution and impedes the collection efficiency. It is observed that by increasing diameter from 1.0 to 2.0 mm, a 14 per cent rise in efficiency is obtained, however it drops down by 10 per cent for a further increase in diameter to 4 mm. Similar observations were observed by Fathikilajahi *et al.* (1996) using a 3D model approach for the Brine k and constant scrubber.

Conclusion

The experiment shows that the flux distribution is a function of gas velocity. L/G ratio, plate diameter and aspect ratio. There seems to be an optimum working range for all these parameters, which can be called as performance parameters wherein maximum collection efficiency can be obtained. In order to estimate the cost effectiveness of the scrubber a correlation between pressure drop and collection efficiency and power consumption (Figure 36.4) should be established.

References

1. Rao BP, Rao BS, Manthapurwar NS, Hasan MZ. "Flue gas treatability studies: a tool for techno-economic control of industrial air pollution", *Environ Monit Assess.* 2003 Feb; 82 (1): 75-81
2. Lapple, C. E., Kamach, H. J., Chemical Engineering, *Progr.* 51, 110 (1955)
3. Semrau, K. T., Correlation of dust scrubber efficiency, *J. Air Poll. Contr. Technol. Assoc.*, 10 (3), 200, 1960.
4. Viswanathan, S., Gnyp, A. W., and St. Pierre, C. C. (1985). Annular pressure drop for Pease–Anthony venturi scrubbers. *AIChE Journal*, 12 (31), 1947–1958.
5. Fernández Alonso, D., Gonçalves, J.A.S., Azzopardi, J.J., and Hills, J.H., 1999, Improvements to the modelling of venturi scrubbers, 2nd *European Congress of Chemical Engineering*, October 5-7, Montpellier.
6. Johnstone, H. F., and M. H. Roberts. 1949. Deposition of aerosol particles from moving gas streams. *Industrial and Engineering Chemistry*. 41: 2417-2423
7. X. Li, R. Tankin, Droplet size distribution: A derivation of a Nukiyama–Tanasawa type distribution function. *Combust. Sci. Technol.* 56 (1987)65–76.
8. Fathikalajahi, J., Taheri, M. and Talaie, M. R. (1995). Theoretical study of liquid droplets dispersion in a venture scrubber. *Journal of the Air and Waste Management Association*, Vol. 45, pp. 181-185.
9. Baboolal, L.B., Pruppacher, H.R., Topalian, J.H., 1981. A sensitivity study of a theoretical model of SO_2 scavenging by water drops in air. *Journal of the Atmospheric Sciences* 38, 856–870.
10. Walcek and H. R. PRUPPACHER Department of Atmospheric Science, University of California, Los Angeles, CA 90024, U.S.A.
11. "*Chemical Engineers Handbook*", Perry, R. H., Don W. Green, 7th ed., 1997, McGraw Hill Company, Inc.
12. "*Engineering Thermodynamics*" Francis F. Huang, 1976, Macmillan Publishing Co. Inc.

13. *"Handbook of Separation Techniques for Chemical Engineers"*, Philip A Schweitzer, 3rd ed., 1997 McGraw Hill Company, Inc.
14. *"Fundamentals of Fluid Mechanics"*, Bruce R. Munson, Donald F. Young, Theodore H. Okiishi, 4th ed., 2002 John Wiley and Sons. Inc.
15. *"Gas Engineers Handbook"*, 1st ed., 1965 Industrial Press Inc.
16. *"Petroleum Fluid Flow Systems"*, O. W. Boyd, 1st ed., 1983, Campbell Petroleum Series.
17. *"Separation of Handbook"*, E. J. Halter, 1st ed., 1966, Burgess Manning Company.
18. *"Applied Process Design"*, Ernest E. Ludwig, Volume 1, 3rd ed., Gulf Publishing Company.
19. *"Engineering Data Book"*, Gas Processors Suppliers Association, Volumes 1 and 2, 11th ed., 1998, Gas Processors Suppliers Association.
20. *"Flow of Fluids"*, Engineering Department, Crane Valves, 1988, Crane Co.
21. *NEERI Report*, Pilot Scale Flue Gas Treatability Studies for Stone Crusher Emissions, a research project of APC Div, Nagpur, India (2000-2002).

Chapter 37

Ecological Economics of Wetlands: An Indian Case Study

J.S. Pandey, V. Joseph and S.N. Kaul

CSIR-NEERI, Nagpur

ABSTRACT

Although wetlands occupy only about 2 per cent of the world's land area, they are estimated to contain 10 per cent -14 per cent of carbon. Moreover, the aerobic-anaerobic stratification of wetlands' sediments is important far out of proportion to their area because of their role in the global cycling of sulfur, nitrogen, phosphorus and carbon. Conversion of wetlands to other land-use classes releases large quantities of CO_2 to the atmosphere, thus contributing significantly to the greenhouse problem. Therefore, in view of their importance, we have carried out an exercise of their economic assessment in Indian context based on their costs as well as benefits. Karnataka (7896.5 million US$ yr^{-1}), Gujarat (7689.4 million US$ yr^{-1}) and Andhra Pradesh (7670.9 million US$ yr^{-1}) are found to be the richest states in terms of wetland-wealth, whereas Nagaland (3.1 million US$ yr^{-1}), Meghalaya (5.9 million US$ yr^{-1}) and Sikkim (15.9 million US$ yr^{-1}) turn out to be the poorest. Chandigarh (2.5 million US$ yr^{-1}) happens to have the lowest value amongst Union Territories.

Keywords: *Economic evaluation, Natural wetlands, Artificial wetlands, Ecological potential, Indian states.*

Introduction

Wetlands are transitional areas between aquatic and terrestrial ecosystems covered by shallow, freshwater for at least part of the annual cycle[1]. They include marshes, swamps, flood-plains, bogs, peatlands, shallow ponds, littoral zones of larger water bodies, tidal marshes etc.[2]. Inter alia, they offer excellent economically beneficial possibilities for the control of nutrients and sediments from point and non-point sources[3].

During last few decades, many experimental demonstrations have been carried out for exploring the possibilities of effective and purposeful application of wastewater to natural and artificial wetlands. Artificial wetlands have been designed and applied for flood control, water quality enhancement and land stabilization. Moreover, in view of the present crisis attributed to the accumulation of greenhouse gases in the atmosphere, there is an urgent need for controlling these gases either through appropriate measures at the place of their emissions or through enhancing their assimilation by designing and managing natural and artificial ecosystems such as wetlands, which offer enormous potential for the mitigation of greenhouse gases[4].

Wetlands, although occupy only about 2 per cent of the world's land area, they are estimated to contain 10 per cent -14 per cent of carbon[1]. Their soils contain up to 20 per cent carbon by weight. Moreover, the aerobic-anaerobic stratification of wetlands sediments is important far out of proportion to their area because of their role in the global cycling of sulfur, nitrogen, phosphorus and carbon. Conversion of wetlands to other land-use classes releases large quantities of CO_2 to the atmosphere, thus contributing significantly to the greenhouse problem[1,5–6].

India happens to be the second largest country in Asia and the second most populous country in the world, which supports one-seventh of human population on the earth. Moreover, the livelihood and the economy of the majority of this vast population depends very significantly on wetlands and the resources they generate[7]. Therefore, in view of the significant role the wetlands play in Indian economy, we have carried out an exercise of economic evaluation of wetlands existing in various Indian states based on their ecological potential. The exercise has mainly taken recourse to the methods currently being used in literature[8–12].

Indian Wetlands and their Economic Importance

India is very rich in wetland resources and exhibits significant ecological diversity, primarily because of variability in climatic conditions and changing topography[2]. Moreover, in India, wetlands cover approximately 17 million hectares. They range from high Himalayan lakes in the north to the salt flats of Kutch, the mangrove swamps of the Bay of Bengal and the innumerable small water reservoirs of the Deccan[4, 13]. This great diversity of wetlands supports an extremely rich biodiversity of flora and fauna. As national parks and wildlife sanctuaries, many important wetland zones have been well protected. The predominant wetland types in India's geographical zones are[7]:

- ✰ Tanks, Reservoirs, and other water bodies of the Deccan peninsula
- ✰ Backwaters and Estuaries of the west coast of the peninsula
- ✰ The Saline expanses of Rajasthan and Gujarat (Rann of Kachch)
- ✰ Freshwater Lakes and Reservoirs from Gujarat eastwards through Rajasthan and Madhya Pradesh
- ✰ Deltaic Wetlands, Lagoons and Salt Swamps of India's east coast

- ✰ Marshes, Jheels, Terai Swamps, and Chaur lands of the Indo-Gangetic plain
- ✰ Floodplain of the Brahmaputra and the Marshes and Swamps in the hills of northeastern India
- ✰ Lakes and Rivers of Kashmir and Ladakh

Estimates of the types of wetlands in India and the total area under them[7] are given in Table 37.1. As far as aquaculture (farming of fish and shell fish) production (Table 37.2) is considered, India happens to be the second largest producer after China[10]. World over, about 21 million metric tons are annually produced through aquaculture. So aquaculture accounts for about 20 per cent of the total global production of fish and shell fish (*i.e.* 112 million metric tons). In India, intercropped shrimp yields are nearly 200 to 500 kg/ha per crop[10, 14]. However, slightly more intensive farming techniques, in which, natural ponds are enclosed and mesh and weirs are used to control water levels and the influx and efflux of organisms, yield as much as 1 metric ton (*i.e.* 1000 kg)/ha per crop.

Table 37.1: Wetlands Types in India and their Area (in million ha) [a]

Sl.No.	*Wetland Type*	*Area*
1.	Area under paddy cultivation	40.9
2.	Area suitable for fish culture	
	a) freshwater (1.6)	
	b) brackish water (2.0)	3.6
3.	Area under capture fisheries	2.9
4.	Estuaries	3.9
5.	Backwaters	3.5
6.	Man-made Impoundments	3.0
	Total (excluding mangroves)	57.8

a: *Source*: DIW[7] (p 6).

Table 37.2: Distribution of Global Aquacultural Production [b]

Sl.No.	*Country*	*Global Production (per cent)*
1.	China	57
2.	India	9
3.	Japan	4
4.	Indonesia	4
5.	Thailand	3
6.	United States	2
7.	Phillippines	2
8.	Republic of Korea	2
9.	Other Countries	17

b: *Source*: Mitsch and Gosselink[10] (p 631).

Table 37.3: Unit Values of Different Ecosystems [c]

Sl.No.	*Ecosystem*	*Unit Value (US \$ ha^{-1} yr^{-1})*
1.	Estuaries	22832
2.	Wetlands	14785
3.	Lakes/Rivers	8498
4.	Forests	969
5.	Grasslands	232

c: *Sources*: Costanza *et al.*[8] and Mitsch and Gosselink[10] (p 608).

In the most intensive aquacultural techniques, wetlands are dredged to form ponds in which water levels are controlled by pumps. "Seed" organisms, the young post larvae, are raised in separate hatcheries. The young organisms are fed in the ponds on synthetic diets, often composed of fish bycatch from commercial fisheries. Water quality is regularly monitored and maintained, the ponds are aerated and the wastes are treated. The worldwide boom in aquaculture, fueled by the high demand for fishery products, has led countries such as India to offer large incentives to initiate new fish farms and has drawn large corporations to invest in this industry[10].

Economics of Wetlands

In ecological economics, generally two different techniques are used for natural resource accounting and wetland valuation[9–10]. They are: willingness-to-pay (WTP)-valuation and energy-analysis (EA)-valuation.

In WTP-valuation, economic values for the benefits are determined by society's willingness to pay for the benefits. Hence, under this scheme, a comprehensive list of benefits is first prepared, and then the WTP for each one of them is determined. For example, for the wetlands, these benefits would include: a) commercial fishing and harvesting; b) recreational benefits; and c) storm protection. Issue of water quality is inherently inter-linked with all these benefits. The valuation of commercial fisheries is straight forward and, in any case, it is in economic terms only. The recreational benefits are normally determined through travel cost method, the details of which are already available in Costanza *et al.*[9]. WTP, in general, utilizes the concepts of opportunity cost, utilitarian value, option value, existence value and contributory value, while making the evaluation.

An alternative chosen will always be at the expense of some opportunity foregone, and this theme of *opportunity cost* is central to economic decision making. For example, the cost of providing a scenic view can be directly derived from the net value of outputs and uses foregone such as timber harvest and dispersed grazing. This is the essence of *opportunity cost* quantification[9]. Utilitarian (or use) value refers to the value of using an ecosystem's products and amenities to derive present and future benefits, which include timber harvest, recreation and aesthetics.

Option value refers to the values which the individuals may be willing to pay now, for the option of using a resource in the future. *Existence value* is the value which a person may be willing to pay simply to know that a resource exists even when there is no intention of using it. *Contributory value* is the value assigned to a resource not because of its direct use value, but because of its indirect role in maintaining and accentuating the ecosystem processes, which support these direct benefits. These indirect benefits include: maintenance of environmental quality (air, water and soil), control and regulation of climate, flood control, the maintenance of a genetic library and supporting various ecological food-webs, trophic levels and nutrient cycling[15].

As far as evaluation based on the Energy Analysis[9,11,16] is concerned, it looks at the total amount of energy captured by natural ecosystems as an estimate of their potential to do such useful work which is economically beneficial. In this methodology, the total biological productivity of wetland is converted into an equivalent economic value on the basis of the cost, which the society will be required to pay so as to replace this energy source with fossil fuel. In other words, this technique uses the Gross Primary Production (GPP) of the ecosystem as an Index of the solar energy captured by the system, and converts this energy value into dollars using energy-dollar conversion factor. GPP for an ecosystem is analogous to GNP for an economic system. Although both of these are crude indices, they are at the same time quite essential measures of overall system performance. In short, the EA-technique basically involves following three steps[9]:

1. Determination of the GPP (Kcal m^{-2} $year^{-1}$) of the natural area in question either by field measurements or by laboratory experiments;
2. Conversion of this estimate into fossil fuel equivalents (FFE) by taking into account the fuel efficiency of each source;
3. Conversion of the FFE-value into dollars by using the ratio of economic value per unit of energy[12].

Results and Discussion

The estimation of the economic value of natural wetlands is a difficult and complex task. However, it is highly essential and unavoidable from environmental management point of view. For the economic evaluation of natural wetlands the values given in Costanza *et al.*[8–9] have been mainly used. As far as costs of artificial (man-made) wetlands were concerned they were computed on the basis of the model given in Mitsch and Gosselink[10] (*i.e.* Cost = \$ 196336 $A^{-0.511}$, where A = Wetland area in hectares), while for the benefit-valuation for even the artificial wetlands, the values (Table 37.3) by Costanza *et al.*[8] have only been used. The wetland data for Indian states was taken from the Directory of Wetlands of India [2,7].

Indian states for which the present wetland-economic-valuation exercise has been carried out. These states have been subsequently divided into five zones: Northern, Southern, Eastern, Western and Central. While the wealth of the wetlands in these states has been estimated individually, their zone-wise wetland-wealth has been plotted. Economic values of wetlands range from 3.1 million US\$ yr^{-1} for Nagaland (Eastern Zone) to 7896.5 million US\$ yr^{-1} for Karnataka (Southern Zone), which is a very large variation showing clearly a very uneven distribution and presence of wetlands in different zones. So the wetlands have been grouped under four classes according to their economic values: (A) those between 2.5 to 50.0 million US$ yr^{-1}; (B) those between 50.0 to 1000.0 million US$ yr^{-1}; (C) those between 1000.0 to 4000.0 million US$ yr^{-1}; and (D) those between 4000.0 to 8000.0 million US$ yr^{-1}. Selects the wetlands which happen to be having highest (A) and lowest (B) economic values. Karnataka (7896.5 million US$ yr^{-1}), Gujarat (7689.4 million US$ yr^{-1}) and Andhra Pradesh (7670.9 million US$ yr^{-1}) are found to be the richest states in terms of wetland-wealth, whereas Nagaland (3.1 million US$ yr^{-1}), Meghalaya (5.9 million US$ yr^{-1}) and Sikkim (15.9 million US$ yr^{-1}) turn out to be the poorest. Chandigarh (2.5 million US$ yr^{-1}) happens to have the lowest value amongst Union Territories.

Mangroves have not been included in the present analysis because they are designated as a distinct category by themselves according to the Directory of wetlands prepared by the Ministry of Environment and Forests, India. Even by Costanza *et al.*[8], they have been evaluated separately. In our present analysis, the classification under "natural wetlands" includes lakes, bogs, marshes etc., and the term "artificial wetlands" includes tanks, reservoirs etc. Although according to Costanza *et al.*[8] wetlands and lakes have been put under separate categories and given different values (lakes and rivers, have the same values according to them), we have not made a distinction between natural wetlands and lakes, primarily because of the data limitations and constraints as the Directory of Indian wetlands has included and reported lakes as a part of natural wetlands themselves.

Environmental decision making cannot do without economic quantification of ecological benefits and costs. However, this process of economic evaluation can carry significant bias depending on the fact whether the valuation is carried out by an individual, societal or regional perspective. Ecosystems' economic value should include the value of their explicit physical, chemical and biological role in the over all human system. And, depending upon their importance, it can focus and concentrate on the roles which are relatively more important. This particular aspect calls for the need of utilizing additional mathematical and computational techniques for carrying out sensitivity analysis with respect to various relevant ecological parameters.

Referances

1. Odum, E.P. (1983), *Basic Ecology*. CBS College Publishing, Saunders College Publishing, Holt-Saunders Japan.
2. WID. (1990), *Wetlands of India–A Directory*. Government of India, Ministry of Environment and Forests, New Delhi.
3. Nichols, D.S. (1983), Capacity of Natural Wetlands to Remove Nutrients from Wastewater, *Journal of Water Pollution Control Federation*, 55, 495-505.
4. Pandey, J.S., S.C. Deb, and P. Khanna. (1997), Issues Related to Greenhouse Effect, Productivity Modeling, and Nutrient Cycling: A Case Study of Indian Wetlands, *Environmental Management*, 21 (2), 219-224.
5. Pandey, J.S., and P. Khanna. (1994), Development of Plant Function Types for Studying Impact of Greenhouse Gases on Terrestrial Ecosystems, *Journal of Environmental Systems*, 23 (1), 67-82.

6. IGBP. (1992), Global Change: Reducing Uncertainty (International Geosphere Biosphere Programme). The Royal Swedish Academy of Sciences, Stockholm, Sweden, pp.11-13.

7. DIW. (1993), Directory of Indian Wetlands. Compiled by the World Wide Fund for Nature (WWF), India in collaboration with the Asian Wetland Bureau.

8. Costanza, R., R. Groot d'Arge, S. Farber, M. Grasso, B. Hannon, K. Limburg, S. Naeem, R.V. O'Neill, J. Paruelo, R.J. Raskin, P. Sutton, and M. van den Belt. (1997), The value of the world's ecosystem services and natural capital, *Nature,* 387, 253-260.

9. Costanza, R., S.C. Farber, and J. Maxwell. (1989), Valuation and management of wetland ecosystems, *Ecological Economics,* 1, 335-361.

10. Mitsch, W.J., and J.G. Gosselink. (2000), *Wetlands.* John Wiley and Sons, Inc., New York.

11. Turner, M.G., R. Costanza, T.M. Springer, and E.P. Odum. (1984), Market and Nonmarket Values of the Georgia Landscape, *Environmental Management,* 12, 209-217.

12. Cleveland, C.J., R. Costanza, C.A.S. Hall, and R. Kaufmann. (1984), Energy and the United States Economy: A Biophysical Perspective, *Science* 255, 890-897.

13. AWB. (1989), A Status Overview of Asian Wetlands. Compiled by D.A. Scott and C.M. Poole for the Asian Wetland Bureau (AWB) and World Wide Fund for Nature (WWF). Published by the Asian Wetland Bureau.

14. James, P.S.B.R. 1999. Shrimp farming development in India–an overview of environmental, socio-economic, legal and other implications. *Aquaculture Magazine, Online Edition. www.aquaculturemag.com*

15. Costanza, R., and H.F. Daly. (1987), Toward an ecological economics, *Ecological Modeling,* 38, 1-7.

16. Costanza, R. (1980), Embodied energy and economic valuation, *Science,* 210, 1219-1224.

Chapter 38

Protection of Ecological Sensitive Areas: Origin of Rivers and Upper Catchment Area

T.K. Ghosh, S.B. Merchant and S.N. Kaul

CSIR-NEERI, Nagpur

ABSTRACT

The economy of India depends upon the major and medium rivers originated from different parts of the country. Relatively minor disturbances due to anthropogenic activities near the origin and upper catchment area of a river may result in major changes on the river system. As many as 12 and 48 rivers in India are classified as major and medium rivers respectively covering a total catchment area of 278million ha. Quarrying upstream affects activities further downstream, sometimes quite drastically. It has been evidenced that deforestation in upper catchment areas of the Godavari, Krishna and Cauvery rivers resulted in severe water scarcity in downstream. The sources and upper catchment areas of major rivers need special attention by developing a strategy which includes social, economic and environmental issues. The aspects to cover management strategy include water quality and quantity, weeds, fire management, grazing management, feral animals, mining, land degradation, nature conservation, agriculture, fisheries management, tourism and recreation, and cultural heritage.

Keywords: *Origin of river, Upper catchment, Ecological sensitivity.*

Introduction

While historically human settlements have tended to flourish mainly in the plains, Indian civilization recognized the central role of mountains in ensuring survival in the densely populated river basins and valleys. Mountain watersheds, thus, have often been treated as sacred and have been conserved. The sacred Himalaya is the source of the major rivers of North India-the Ganga, the Yamuna, the Brahmaputra and their many tributaries. The Vindhya and Satpura Ranges feed the Tapti, the Narmada, the Sone, the Mahanadi etc. The Western Ghats are the origin of the major rivers of peninsular India like the Godavari, the Krishna and the Kaveri. These rivers are the lifeline of the economy, and the mountains from which they renew their flow are the foundation for a stable economy.

The catchment of any region acquires importance because it determines the annual water-yield of the river draining it. The physical characteristics of the catchment, *viz.* the sub-streams, the land-use pattern and the changes therein, the nature and the extent of the vegetative cover, the permeability of the soil strata, and finally the presence of man-made structures for harnessing water in the sub-catchments/micro-catchments, are the factors which determine the rate of surface water run-off.

To harness the water resources of different rivers and their tributaries, a large number of dams have been (or are going to be) constructed in India. One of the major threats to the life of a dam is the problem of siltation of the reservoir. The rate of siltation, in turn, is primarily dependent on the condition of the catchment area of the project, *i.e.* the healthier the catchment the smaller will be the volume of silt flowing into the reservoir, and so, the longer will be the life of the dam. If the catchment area is degraded, the silt load increases in the reservoir reducing the life of the dam.

Since the economy of India depends upon the major rivers originated from different parts of the country, relatively minor disturbances due to anthropogenic activities near the origin and upper catchment area of a river may result in major changes on the river system, including erosion rates in the mountains, fluctuations in the hydrologic regimes, silt accumulation rates, flooding water in the low lands and natural system of recharge.

Importance of the Criteria

Origin of Rivers

A glacier, mountain or spring from where a water stream originates is referred to as origin of river. The Himalayan rivers, after origin, run parallel to the mountain ranges in many reaches. Across these mountain ranges there are many rapids and cascades due to hard resistant rock bands outcropping in the midst of softer bands. Since the Himalayan rivers traverse through poorly consolidated sedimentary rocks affected by folds, faults and thrusts, there is greater erosion and removal of silt. Landslide debris also add to the river sediment. Due to high gradient and tremendous velocity, these rivers are highly erosive. The eroded material (pebbles, cobbles and boulders) act as excellent cutting tools to carve out deep gorges, which have been steepened by periodic uplifts. Headward erosion has also been active and this has resulted in what is known as river piracy.

The southern part of the Indian peninsula has been subjected to at least three periods of peneplaination and uplift. A longitudinal profile of a river in the shield area would show the beginning stage, when the streams started as rills and springs from the Western Ghats, meandering on the older peneplains, and cascading down the active erosion slope, and again flowing on the youngest peneplain in a meandering fashion. The course of the rivers also tends to follow joints and faults which are zones of relatively greater weathering and erosion. The western coastline is a straight line giving further evidence of a tear and subsequent foundering of the land. The little marginal land came up due to subsequent uplift.

Upper Catchment Area

Catchment area, also referred to as drainage area, is a basin like structure for collecting and draining water. Upper catchment area is typically referred to indicate a basin collecting precipitation mostly in the mountainous region/upper reaches of a river following its origin. The collected water is absorbed by the soils or drains into the river.

The Upper catchment area from which water is collected into the upper stretch of a river varies widely from river to river. It is dependent on various factors, *viz.* location of origin of the river, slopes of the basin, tributaries, annual discharge, geology, soil characteristics, forest coverage and river development plan in downstream.

The health of a catchment area is indicated by the land pattern and extent of its forest cover. The forests play a valuable role in conserving water. Acting like sponges, the forests absorb the water during the rains and continue to release the stored water into streams, rivers and wells in the dry seasons. In most of the rain-fed river basins in India, where the catchment area is denuded of forests, floods occur after the heavy monsoon showers. The river basins then run dry for several months before the onset of the next monsoon. The opposite is true for rivers where catchment area possesses thick vegetation. These rivers yield water till the end of summer.

On barren land, moisture from the top layers of the exposed soil quickly evaporates under the scorching sun. If there is a forest cover, the moisture is retained for several months. If soil is under forest cover, it takes about 1,74,200 years to loose six inches of soil by way of erosion. The same soil can be washed away within 17 years, if it is uncovered.

The upper parts of the streams have an impact on the lower parts, and quarrying upstream affects activities further downstream, sometimes quite drastically. In general, the high mountain systems are very sensitive environments which offer unique challenges to and opportunities for global change research. Under UNCED – (United Nations Commission of Environment and Development) Agenda 21 entitled 'Managing Fragile Ecosystems – Sustainable Mountain Development' the so-called mountain agenda, special relevance is attached to: a) generating and strengthening the knowledge of ecology and sustaining development of mountain ecosystems. b) promoting integrated watershed development and alternative livelihood opportunities. The available estimates show that the water yield from a high Himalayan catchments is roughly double that from an equivalent one located in peninsular India and this is mainly due to additional inputs from snow and ice melt contributions from high altitudes.

Specific Areas of Concern in India

India is a land of many rivers and mountains. Its geographical area of about 329 million hectare (m ha) is criss crossed by a large number of small and big rivers, some of them figuring amongst the mighty rivers of the world. The rivers and mountains have a greater significance in the history of Indian cultural development, religious and spiritual life. It may not be an exaggeration to say that the rivers are the heart and soul of Indian life. The total amount of water flowing from the Himalayas to the plains is estimated to be 8634 km^3 yr^{-1} [1,2].

These rivers can be categorized into four groups:

- ☆ Rivers that flow down from the Himalayas and are supplied by melting snow and glaciers. This is why these are perennial, that is, they never dry up during the year.
- ☆ The Deccan Plateau rivers, which depend on rainfall for their water.
- ☆ The coastal rivers, especially those on the west coast, which are short and do not retain water throughout the year.
- ☆ The rivers in the inland drainage basin of west Rajasthan, which depend on the rains. These rivers normally drain towards silt lakes of flow into the sand.

Table 38.1: Major River Basins in India

Sl.No.	*Name of the River*	*Origin*	*Length (Km.)*	*Catchment Area (Km^2)*
1.	Indus	Mansarovar (Tibet)	1114	321289
2.	The Ganga system			
	(a) Ganga	Gangotri (Uttar Kashi)	2525	861452
	(b) Brahmaputra	Kailash Range (Tibet)	916	194413
	(c) Barak and other rivers flowing into Meghna, like Gomti, Muhari, Fenny etc.			41723
3.	Sabarmati	Aravalli Hills (Rajasthan)	371	21674
4.	Mahi	Dhar (Madhya Pradesh)	583	34842
5.	Narmada	Amarkantak (Madhya Pradesh)	1312	98796
6.	Tapi	Betul (Madhya Pradesh)	724	65145
7.	Brahmani	Ranchi (Bihar)	799	39033
8.	Mahanadi	Nazri Town (Madhya Pradesh)	851	141589
9.	Godavari	Nasik (Maharashtra)	1465	312812
10.	Krishna	Mahabaleshwar (Maharashtra)	1401	258948
11.	Pennar	Kolar (Karnataka)	597	55213
12.	Cauvery	Coorg (Karnataka)	800	81155
	Total	**2528084**		

Source: Anonymous[3].

As many as 12 rivers, in India, are classified as major rivers whose total catchment area is 252.8 m ha. Of the major rivers, the Ganga–Brahmaputra Meghana system is the biggest with catchment area of about 110 m ha which is more than 43 per cent of the catchment area of all the major rivers in the country (Table 38.1). The other major rivers with catchment area more than 10 m ha are Indus (32.1 m ha), Godavari (31.3 m ha), Krishna (25.9 m ha) and Mahanadi (14.2 m ha). The catchment area of medium rivers is about 25 m ha, of which Subernarekha with 1.9 m ha catchment area is the largest river among the medium rivers in the country (Table 38.2).

Table 38.2: Medium River Basins in India

Sl.No.	*Name of the River*	*Village/Distt. (Origin)*	*State*	*Length (km)*	*Catchment Area (km^2)*
1.	Ozat	Kathiawar	Gujarat	128	3189
2.	Shetrunji	Dalkania	Gujarat	182	5514
3.	Bhadar	Rajkot	Gujarat	198	7094
4.	Aji	Rajkot	Gujarat	106	2139
5.	Dhadhar	Panchmahal	Gujarat	135	2770
6.	Purna	Dhosa	Maharashtra	142	2431
7.	Ambika	Dangs	Maharashtra	142	2715
8.	Vaitarna	Nasik	Maharashtra	171	3637
9.	Dammanganga	Nasik	Maharashtra	143	2357
10.	Ulhas	Raigarh	Maharashtra	145	3864
11.	Savitri	Pune	Maharashtra	99	2899
12.	Sastri	Ratnagiri	Maharashtra	64	2174
13.	Washishthi	Ratnagiri	Maharashtra	48	2239
14.	Mandvi	Belgaum	Karnataka	87	2032
15.	Kalinadi	Belgaum	Karnataka	153	5179
16.	Gangavati or Bedti (in upper reaches)	Dharwar	Karnataka	152	3902
17.	Sharavati	Shimoga	Karnataka	122	2209
18.	Netravati	Dakshina Kannada	Karnataka	103	3657
19.	Chaliar or Baypore	Elamtalvi Hills	Kerala	169	2788
20.	Bharathapuzha (known as Ponnani)	Annamalai Hills	Tamil Nadu	209	6186
21.	Periyar	Sivajini Hills	Kerala	244	5398
22.	Pamba	Devarmalai	Kerala	176	2235
23.	Burhabalang	Mayurbahanj	Orissa	164	4837
24.	Baitarni	Keonjhar	Orissa	365	12789
25.	Rushikulya	Phulbani	Orissa	146	7753
26.	Bahuda	Ramgirivillage	Orissa	73	1248
27.	Vamsadhara	Kalahandi	Orissa	221	10830

Contd...

Table 38.2–*Contd...*

Sl.No.	*Name of the River*	*Village/Distt. (Origin)*	*State*	*Length (km)*	*Catchment Area (km^2)*
28.	Nagavali	Kalahandi	Orissa	217	9410
29.	Sarda	Vishakhapatnam	Andhra Pradesh	104	2725
30.	Eleru	Vishakhapatnam	Andhra Pradesh	125	3809
31.	Vogarivagu	Guntur	Andhra Pradesh	102	1348
32.	Gundlakamma	Kurnool	Andhra Pradesh	220	8494
33.	Musi	Nellore	Andhra Pradesh	112	2219
34.	Paleru	Nellore	Andhra Pradesh	104	2483
35.	Muneru	Nellore	Andhra Pradesh	122	3734
36.	Swarnamukhi	Koraput	Orissa	130	3225
37.	Kandleru	Vinukonda	Andhra Pradesh	73	3534
38.	Kortalaiyar	Chinglepet	Tamil Nadu	131	3521
39.	Palar (including tributary Cheyyar)	Kolar	Karnataka	348	17871
40.	Varahandi	North Arcot	Tamil Nadu	94	3044
41.	Ponnaiyar	Kolar	Karnataka	396	14130
42.	Vellar	Chithri Hills	Tamil Nadu	193	8558
43.	Vaigai	Madurai	Tamil Nadu	258	7031
44.	Pambar	Madurai	Tamil Nadu	125	3104
45.	Gundar	Madurai	Tamil Nadu	146	5647
46.	Vaippar	Tirunolvolli	Tamil Nadu	130	5288
47.	Tambraparni	Tirunolvolli	Tamil Nadu	130	5969
48.	Subarnarekha	Nagri/Ranchi	Bihar	395	19296
	Total				**248505**

Source: Anonymous[3].

The water resources of India are unevenly distributed. Brahmaputra and Barak basin with 7.3 per cent of geographical area and 4.2 per cent of population of the country has 31 per cent of the annual water resources. Water scarcity (below 1000 m^3 $capita^{-1}$ yr^{-1}) is faced in Cauvery, Pennar and Sabarmati basins. In Sabarmati basin the per capita annual availability of water is as low as 360 m^3. Other water scarce river basins include the east and west flowing rivers. It may be mentioned that the upper catchments of the Godavari, Krishna and Cauvery rivers were thickly forested in the pre-British times and many of the early British Gazetteers give accounts of the destruction of these forests between 1800-1850, and the consequent siltation of many ports on the West Coast, especially in the Ratnagiri district of Maharashtra[4]. The locations of origin and upper catchment areas of select major rivers, which are of great concern, have been delineated here under:

Ganga

This is considered the holiest of all the great rivers of India. The Ganga is not known by this name in the beginning or at its end. It has its source at the Gangotri glacier at an elevation of 7010 m, where it flows from the cave Gomukh, as the Bhagirathi, which then joins the River Alaknanda (originated near the sacred Badrinath) as it flows towards Devprayag. The total catchment area of the northern tributaries of the Ganga is about 420,000 km^2 [(5)]. While the soil in the hill is brown, terai soils prevail in the area just at the foot of the hills.

Brahmaputra

This mighty river rises in the great glacier in the northernmost chain of the Himalayas in the Kailas range at an elevation of about 5,150 m (30°31'N; 82°10'E), just south of the lake Konggu Tsho. The river, under its Tibetan name Tsang Po, flows through southern Tibet for about 1800 km eastwards through the Himalayas, curves back across Arunachal Pradesh and Assam, then turns south to join the Padma and Ganga in Bangladesh and finally enters the Bay of Bengal. The water, in the upper reach (Tibetan portion) is clear and there is not much silt carried by the water. But on entry into India the silt content is heavy. It is longer and more voluminous than the Ganga, and gushes down with enormous force for most of its course through the mountain regions and the forests of north-east India. A unique feature of this river is that it is navigable even at a height of 10,000 feet. Petroleum refining units contribute most of the industrial pollution load into the basin along with other medium and small industries. The main problem facing the river basin is that of constant flooding. Floods have been occurring more often in recent years with deforestation, and other human activities being the major causes.

Indus

It has its source near Manasarovar close to southwestern Tibet (32°N, 81°E), at an altitude of 5,182 m and flows north-westward, through the Himalayas until it is joined, about 257 km from its source, by the Ghar. A short distance lower down, it enters the SE corner of Kashmir at an elevation of 4,206 m. It skirts Leh at 3,200 m, and is joined by the Zanskar river near the crossing of the great trade route into central Asia via the Karakoram pass.

Narmada

This is the largest west-flowing river in India and originates from Amarkantak at an elevation of 900 m in the Mekhala range of Shahdol district, Madhya Pradesh. The river rises there from a small tank, surrounded by a group of temples. The river Narmada drains the catchment between the Vindhyan mountains to the north of the river stretching east-west in general, and the Satpura mountain ranges to the south. It flows through the undulating plains of Hoshangabad about 300m high, dotted with occasional low buttes (a conspicuous isolated flat-topped hill with steep slopes or precipitous cliffs, often capped with a resistant layer of rocks and borders). The valley south of the Narmada is narrow and intersected by the ravines and deep streams which leap through the steep slopes of the Vindhyan ranges. This part is equally fertile and plain. Its northern margin is damp, dissected and covered with forests. The 'water-divide' separates the Narmada basin from the Chambal system in the north. In the south, the Satpuras separate it from the Tapti basin. It flows 1300 km west through the states of Madhya Pradesh and Gujarat. In terms of its catchment area, it is the seventh largest among the fourteen major river basins in the country.

Mahanadi

This river is considered the lifeline of Orissa, through which it flows. It originates from a pond near Parasiya village in south-eastern Chattisgarh near Raipur. It drains an area of 141,000 km^2 of which 53 per cent is in Chattisgarh. The basin has excellent surface and groundwater resources. Coastal alluvial tracts have freshwater formations near the surface and depths below 150 m.

Godavari

It is the largest and longest river of South India. This river originates in the Sahyadri range of western ghat in the northeast of Mumbai in Maharashtra. A large part of the basin lies over ancient crystallic and metamorphic rocks which are poor water bearing strata. It flows through that state and Andhra Pradesh before entering the Bay of Bengal. Its drainage basin is one of the largest in the country, second in size only to that of the Ganga.

Kaveri

The Kaveri is worshipped as a jeeva-nadi or perennial river. It has its origin from a spring at Talakaveri, which is a very small pond at Brahmagiri hills in the Sahyadri range of Western Ghats in Coorg district (12°25'N 75°34'E), Karnataka state. The origin is almost at the edge of Western Ghats at an elevation of 1341 m and within sight of Arabian sea, but it flows eastward crossing the mountain barrier. The upper course is in mountain territory covering a small jungle area of 19 km^2 in the Coorg district. The Kaveri is believed to flow underground for a distance of about 45 m in the Brahmagiris below its source. It surfaces at Bhagamandala to be joined by its first tributary Kannige. Thereafter the river traverses nearly 770 km.

Krishna

The Krishna is a great river of south India which, like the Godavari and the Kaveri, flows almost across the Peninsula from west to east. The Krishna rises about 64 km from the Arabian Sea in the Western Ghats at the hill station of Mahabaleshwar, and flows southwards, skirting the eastern spurs of the hills past Karad (Satara district), where it receives, on the right bank, the Koyna from the western side of the Mahabaleshwar hills, and Sangli where it receives the waters of the Varna, also from the west, until it reaches Kurundvad when the Panchganga joins it, again on the right bank.

Stepts to Protect the Sensitive Areas

The sources and upper catchment areas of major rivers need special attention by developing a strategy, which includes social, economic and environmental issues.

The aspects to cover management strategy include water quality and quantity, weeds, fire management, grazing management, feral animals, mining, land degradation, nature conservation, agriculture, fisheries management, tourism and recreation, and cultural heritage. A few select issues of special interest are delineated hereunder:

- *Restricted mining:* Open cast mining on upper catchment slopes has drastically reduced the water resources of the country. Mining increases surface run-off and decreases infiltration. The increased run-off combined with the choking of water courses with overburdens and fines are causing floods and droughts in regions-which had stable and perennial supplies of water. In the context of the unprecedented water scarcity facing the country, the role of mining in the hydrological destabilization of mountain watersheds can no longer be ignored, and leasing of the sensitive areas for mining should be avoided.
- *Water management:* Very few studies are undertaken on the Himalayan lake ecosystems and the water management programmes are either completely lacking or grossly inadequate. The inflow of high silt load from glaciers is rendering the lake waters turbid and unfit for biological activity and are gradually filling these lakes. The other impact is from pollution from agricultural, industrial, human and cattle wastes. Restoration plans for the lake systems should be undertaken on ecological considerations following their geophysical environment and annual rhythm in chemical and biological compositions.
- *Scientific monitoring and training:* Long-term monitoring programmes on snow and ice resources as well as on changing climatic conditions based on already existing programmes, taking account of local conditions and utilizing local expertise, be actively promoted. Models and other techniques for forecasting water supply, sediment load and flood hazards be further developed and applied with particular attention to the mitigation of glacier lake outburst floods. Data bases on snow and glacier hydrology should be developed for practical end uses and should be made freely available so that rational water management decisions may be made.
- *Conservation of biodiversity:* Biodiversity, categorised at different levels owes its origin to a variety of macromolecules, most notably DNA and proteins. The conservation of biodiversity refers to management of human interaction with the variety of life forms and ecosystems so as to maximise the benefits, they provide today and maintain their potential needs for future generation's needs and aspiration. For this, the hydrology and water management play a crucial role while dealing with ecosystems in changing moisture and temperature conditions on the ground depending on altitude and location.
- *Prevention of sediment erosion:* Sediment erosion is primarily caused by landslides, landslips, movement of debris and dumping of spoil. In order to prevent erosion, the plausible remedial measures are plantation on threatened slopes, construction of benches and restricted spoil dumping. The lateral root system of trees acts as long fibrous binders within the soil mass by interlocking the zones of weakness increasing their stability.

☆ *Public participation:* The conservation and management of the upper catchment area must be undertaken with the participation of the local people, so that it is sustainable.

References

1. Gupta, R.K. (1993). *The living Himalayas: Aspects of environmental and resources ecology of Garhwal*. Today and Tomorrow Printers and Publishers, New Delhi.
2. Kawosa, M.A. (1988). *Remote Sensing of Himalayas*, Natraj Publishers, Dehradun
3. Anonymous (2004). http: / / wrmin.nic.in / resource / wresource1.htm
4. Gadgil, M. (1985). The western ghats of India: an ecodevelopment approach. p. 309-336 in T.V. Singh and J. Kaur ed *Integrated Mountain Development*. Himalayan Books, New Delhi
5. Rao, K.L. (1975). *India's water wealth: its assessment, uses and projections*. Orient Longman, New Delhi.

Chapter 39

Treatment of Biologically Treated Distillery Effluent: A Case Study

R.A. Pandey, S. Malhotra, A. Tankhiwale, S. Pande, P.P. Pathe and S.N. Kaul

NEERI-CSIR, Nagpur – 440 020

ABSTRACT

The paper addresses physico-chemical treatment of biologically treated distillery effluent using conventional and non-conventional coagulants. Conventional coagulants, *viz.* ferrous sulphate, ferric sulphate, alum were used during the study. Ferrous sulphate in combination with lime was found to be optimal which could remove 70 to 84.6 per cent COD with efficient removal of colour. Further, commercially available polyelectrolyte aid, Percol 47 was found to be efficient which resulted in minimization of requirement of ferrous sulphate coagulant from 20 g/l tqwxcx o 10 g/l producing the same quality of effluent without affecting the efficiency of the treatment. Non-conventional coagulants, *viz.* wastewater from a Pickling industry and Titanium ore processing industry were also evaluated. The pickling wastewater also removed 87 per cent and 95.7 per cent residual COD and BOD respectively, from the biologically treated distillery spent wash. Further, commercial activated carbon was assessed for polishing the effluent after treatment with coagulants. A dose of 8 g/L of carbon with contact time of 45 minute was found to be optimal. The effluent generated after treatment with activated carbon had COD and BOD concentration of less than 100 mg/L and 10 mg/L, respectively meeting the discharge standards for inland surface water.

Keywords: *Distillery, Wastewater, Biologically, Coagulants, Activated carbon, Sludge.*

Introduction

Distillery wastewater originating from sugarcane molasses based alcohol industry is low in pH, has intense brown colour, high biochemical oxygen demand (BOD) and chemical oxygen demand (COD), and also high concentration of dissolved solids. However, the pollution load of the distillery wastewater depends on quality of molasses, unit operations for processing of molasses and process for recovery of alcohols.

Biological treatment is practiced in most of the Indian distilleries resulting in reduction of BOD and COD. The biological treatment system includes anaerobic treatment with recovery of biogas, followed by aerobic treatment for removal of residual BOD and COD. However, most of the biologically treated distillery wastewaters are dark brown in colour and contain high COD due to the presence of recalcitrant compounds, *viz.* caramel, melanoidins, variety of sugar decomposition products, anthocyanins and tannins, and other xenobiotic compounds formed during yeast growth and processing of alcohol (Gaikwad and Naik, 1999; Chandra, R. and Singh, H., 1999; White and Munns, 1947; Sellier, 1947). It is difficult to remove these constituents from the distillery effluent through conventional biological processes. Several workers have attempted to remove these recalcitrant compounds by physico-chemical methods using diluted distillery wastewater and biologically treated distillery effluent (Hashia and Bardia, 2000; Gadre and Godbole, 1990; Rao and Viraraghavan, 1984, and Sheehan and Greenfield, 1980). In the present study, wastewater from a biological treatment plant of a distillery unit was collected and studies carried out to evaluate different conventional and non-conventional coagulants with and without co-agulant aids for the removal of recalcitrant compounds from the biologically treated distillery effluent and characterize the chemical sludges generated from the treatment and evaluate the toxicity of sludges through leaching experiments. In addition, investigations were also carried out to assess the performance of activated carbon for removal of residual COD from physico-chemically treated effluent.

Materials and Methods

A distillery with an installed capacity of 45 KLd^{-1} of industrial alcohol production generates on an average 24 $m^3 hr^{-1}$ of spentwash. The generated effluent is biologically treated in an effluent treatment plant (ETP) comprising of screens, spentwash storage basin, acid phase reactor, hybrid methane reactor (Upflow Anaerobic Suspended Blanket Reactor), clariflocculator, anaerobic filter (I stage), anaerobic filter (Stage II), and aeration basins (I, II and III stages). Five litre samples of the biologically treated effluent from the ETP were collected on two hourly basis, and composited for a period of twelve hours for conducting the treatability studies.

The chemical coagulants, *viz.* ferrous sulphate, ferric sulphate and alum, used for the study were of commercial grade. The chemical reagents used for analysis were of Analytical Reagent (AR) grade. Wastewaters from Pickling and Titanium processing industries were used as non-conventional coagulants. Several commercially available polyelectrolytes were also used alongwith chemical coagulants as coagulant aid for treatment. Phipps and Baird Multiple Stirring Device (Jar Tester) equipped with stirring paddles and provision for controlled mixing by adopting the standard procedures was used for laboratory studies. Toxicity characteristics of the generated sludges were studied using Zero Head Extraction Vessel by adopting toxicity characteristics leaching procedure (TCLP). The colorimetric estimations were made using 'Bausch and Lomb' Spectronic-21 spectrophotometer. CHNS-O elemental analyser, Carlo Erba Model 1108 attached with data processor 200 was used for elemental analysis of sludges generated during the treatability studies.

All samples were analyzed for various physico-chemical parameters as per Standard Methods (APHA, 1976). Measured volumes of samples (500 ml) were flocculated using jar test apparatus with 1000 mL capacity beakers. The motor of stirrer paddles was started after the addition of coagulant into each beaker simultaneously. Rapid mixing was maintained at 90±10 rpm for 1 minute followed by flocculation for 30 minutes. The effluent obtained after sedimentation was analyzed for various parameters. The generated sludges were air dried and analyzed for various constituents. Fixed weights of various sludges obtained after coagulation were analyzed during standard TCLP in which extraction tests were performed on untreated sludges following 24 hours leaching contact period. The leachates were collected for further analysis. Further, activated carbon was evaluated as an adsorbent material for removing the residual COD from the treated effluent after chemical coagulation. Different doses of activated carbon were added in 500 mL of the effluent samples, and stirred for different periods of contact time. Samples were drawn from each flask at regular interval of time and analyzed for COD and colour parameters.

Results and Discussion

Characteristics of Biologically Treated Distillery Spentwash

Cane molasses based distillery generates two streams of wastewater. The first stream of the wastewater is generated from heavy yeast residual from fermentation tanks. The yeast sludge from fermenter tank is about 2 to 5 per cent of the total volume of spentwash discharged from the distillery. The second stream of the wastewater (spentwash) is generated from the distillation column. About 13-15 litres of spentwash is generated per litre of industrial alcohol produced at the distillery unit. The cooling and condenser water is about six times the volume of the spentwash. The water from cooling and condenser sections is recycled, while sludge from fermenter is treated separately. The spentwash is treated biologically comprising anaerobic followed by aerobic unit processes.

The physico-chemical characteristics of the biologically treated spentwash are presented in Table 39.1. BOD and COD values of the biologically treated effluent ranged between 400 to 420 mg/L and 7000 to 7800 mg/L, respectively and BOD/COD ratio is 0.056 to 0.057.

Evaluation of different conventional coagulants with and without coagulant aids for the removal of recalcitrant compounds from the biologically treated distillery spentwash.

Table 39.1: Characteristics of Biologically Treated Distillery Spentwash

Parameters	*Concentration (Range)*		
Colour	Intense Brown	Intense Brown	Intense Brown
pH	8.2	8.5	8.1
Temperature, ºC	27	26	28
Alkalinity (as $CaCO_3$)	2200	2185	2310
Ammonia-N	BDL	BDL	BDL
Total Nitrogen	12	18	9
Oil and Grease	BDL	BDL	BDL
COD	7800	7400	7000
BOD	425	420	400
Heavy Metals			
Cd	0.5	0.5	0.51
Ni	2.8	3.1	2.81
Fe	0.02	0.05	0.01
Pb	1.5	0.75	1.0
Cu	0.9	0.75	1.0
Mn	0.9	0.81	1.1
Cr	BDL	BDL	BDL
Zn	0.7	0.7	0.7

All values except pH and temperature are in mg/L

BDL: Below Detection Limit.

Table 39.2: Determination of Optimum Dose of Lime with Ferrous Sulphate (Ferrous Sulphate 20 g/L)

Sl.No.	*Dose of Lime (g/L)*	*pH*	*Colour (Pt-Co Unit)*		*COD (mg/L)*		*Per cent Reduction*	
			Untreated	*Treated*	*Untreated*	*Treated*	*Colour*	*COD*
1	15	8.1	60000	4875	7421	3528	91.8	52.4
	20	8.2	60000	3052	7421	2650	94.9	64.2
	25	8.2	60000	2528	7421	2014	95.7	72.8
	30	8.3	60000	1200	7421	1700	98.0	77.0
2	27	8.3	60000	2018	7421	1920	96.6	74.4
	28	8.3	60000	1498	7421	1852	97.50	75.0
	29	8.3	60000	1200	7421	1706	98.0	77.0

* PW: Pickling wastewater; **TW: Titanium processing wastewater.

All values represent an average of three sets of experiments.

Coagulation of the wastewater containing recalcitrant pollutants is practised through use of conventional salts of aluminium and iron as coagulants. When these salts of coagulants are added to biologically treated effluent, they react with inherent alkalinity and produce insoluble precipitates, *viz.* Fe $(OH)_3$ or Al $(OH)_3$. It is necessary to maintain optimum pH through addition of lime especially in case of $FeSO_4$ in biologically treated distillery spentwash for acquiring critical zeta potential for formation of maximum insoluble precipitates of metal ions. It was essential to determine the dosage of lime as well as coagulants. Table 39.2 presents the result on assessment of dose of lime by keeping the ferrous sulphate constant for the treatment of biologically treated distillery effluent. A dose of 29 gL^{-1} of lime was found to be optimal which resulted in 98 per cent removal of COD with 77 per cent removal of colour. Similarly, lime requirement for other coagulants were also assessed. A series of jar experiments were conducted with varying doses of each coagulant. The results of the studied coagulants. The optimum doses for $FeSO_4$, $Fe_2\,(SO_4)_3$ and alum were found to be 20 g/L, 12 g/L and 12 g/L, respectively with addition of lime for maximum COD removal. The results indicate that these coagulants removed COD in the range of 64.4 to 77.0 per cent, BOD 71.4 to 79.3 per cent, and 94.8 to 98 per cent of colour from the biologically treated distillery spentwash. Still, while comparing the costs of chemical coagulants, it was found that the ferrous sulphate is cost-effective and has a potential for large scale applications. Thus, ferrous sulphate was selected as suitable conventional coagulant.

Table 39.3: Characteristics of Industrial Wastewaters Used as Non-Conventional Coagulants

Parameter	*Source of Industrial Wastewater*	
	*PW**	*TW***
pH	0.63	0.16
Acidity, per cent	2.10	3.90
Chlorides	120 x 10^3	92
Sulphates	BDL	114.9 x 10^3
Total dissolved solids	122 x 10^3	133.7 x 10^3
Iron	73 x 10^3	18.6 x 10^3
Zinc	BDL	13.1
Cadmium	0.02	0.01
Nickel	3.1	0.77
Lead	0.24	1.5
Chromium	1.46	12.0
Copper	0.92	0.18
Manganese	BDL	54.0

All values except pH and acidity are expressed in mg/L.

Further, different commercially available organic polyelectrolytes with trade names, *viz.* Percol 47, Percol 173, Meg LT27 and Zetag alongwith ferrous sulphate were evaluated for achieving effective removal of residual COD and BOD from the biologically treated distillery effluent. The characteristics of treated effluent alongwith doses of conventional and non-conventional coagulants and polyelectrolyte.

Polyelectrolytes help in achieving better coagulation for effective removal of colloidal pollutants present in the biologically treated distillery effluent.

Among the studied polyelectrolytes, Percol 47 was found to be most effective as compared to others. The use of polyelectrolyte minimizes the requirement of ferrous sulphate. Table 39.6 indicates that each of the coagulant aid reduces the dose of coagulant required to the same level. COD concentration of the effluent generated after treatment using coagulant in presence of coagulant aid ranged from 958 to 1764 mg/L, and is very much comparable to the treated effluent generated without coagulant aids. However, the use of coagulant aids help the process of coagulation by decreasing the amount of coagulant required, making the flocs heavier, and thereby improving the sludge settleability.

Characteristics of Non-Conventional Coagulants

Wastewaters from pickling and titanium ore processing industries were collected, characterized and used separately as non-conventional coagulants for further treatment of biologically treated distillery spentwash. The pickling wastewater is generated from the processing of hot rolled steel sheets for production of cold rolled sheets in wide ranges of thickness, *viz.* drawing grade, deep drawing grade and galvanized sheets etc. Prior to cold rolling operations, hot rolled sheets are treated with hydrochloric acid, in a process known as pickling, in order to remove oxide scales and to impart better surface properties to the steel sheets. The pickling operations followed by rinsing with water generates pickling wastewater which contains appreciable concentration of mixture of ferrous and ferric chloride. Besides iron, the pickling wastewater contains other metal ions such as nickel, chromium, lead and copper. Titanium (Ti) is extracted from ilmenite ($TiFeO_3$) ore which is a mixture of TiO_2, Fe^{2+} ions and Fe^{3+} ions, through sulfate process. During this process, the powdered ore is treated with conc. H_2SO_4 thereby generating sulfates of iron. Thus, pickling wastewater was rich in iron and chloride ions, whereas the wastewater from titanium industry was rich in iron and sulphate ions. Hence, these were found to be the best substitutes for conventionally used coagulants, *viz.* iron chloride and iron sulphate. The titanium wastewater used in this study contained both ferrous and ferric ions (National Environmental Engineering Research Institute, NEERI, 1996). However the ferrous ions occur in excess. Table 39.3 presents the characteristics of pickling wastewater and titanium processing wastewater.

The characteristics of other metal ions alongwith anions in the pickling wastewater provide effectiveness for coagulation as the mineral content in the liquid environment changes resulting in minimization of flocculation time. Further, complex formation in liquid environment of biologically treated distillery effluent bridges the polynuclear network of cations and anions for effective coagulation (Bulusu and Pathak, 1973). The wastewater streams used for coagulation in the present study were strongly acidic and with high concentration of dissolved metal ions.

Evaluation of Non-Conventional Coagulants for the Treatment of Biologically Treated Distillery Spentwash

Chemical coagulation depends on coagulant dose and hydrogen ion concentration which determine the metal speciation and the net charge on the particles

(Bulusu and Pathamk, 1973). Hence, coagulation using acidic wastewater was carried out in presence of selected dose of lime to change the pH of the mixture. The coloured pigments, as already discussed, are mostly colloidal and are known to be hydrophilic in nature. At high concentration of added coagulant and at certain optimum pH, the stability of these colloids in solution becomes minimum, thereby causing their precipitation. The results for the optimum conditions where such colloids were precipitated using acidic wastewater. The results indicate that pickling wastewater removed 87.1 per cent COD, 92.9 per cent BOD and 98.9 per cent colour, whereas titanium processing wastewater removed 67.1 per cent COD, 80.9 per cent BOD and 98.5 per cent colour.

Among the non-conventional coagulants, the pickling wastewater showed the maximum COD removal efficiency. The coagulating effect of the acidic wastewater streams in presence of lime was found to be enhanced by addition of certain polymeric coagulant aids. Presence of such coagulant aids during the process of colloid precipitation decreases the stability of colloids by altering the hydration shells of hydrophilic colloids in solution. As the stability of a colloid depends on the hydration shells, the colloids get precipitated, once the hydration shells are disturbed. Accordingly, it was found that lesser doses of coagulants are required if coagulant aids are added. The results for the same are presented in Table 39.4.

Pickling wastewater when used alongwith coagulant aid/Percol 47 could result in 78.8 per cent COD removal, 94.28 per cent BOD removal and about 98.8 per cent colour removal. Meg LT-27 coagulant aid with pickling wastewater removed 87.1 per cent COD, 95.7 per cent BOD, and 98.9 per cent colour. Thus Meg LT-27 along pickling wastewater, the biologically treated effluent could be treated marginally better than Percol 47 (Table 39.4).

Table 39.4: Performance of $FeSO_4$ and Pickling Wastewater with Coagulant Aids

Coagulant	*Dose*	*Coagulant Aid*	*Dose, mg/L*	*pH (±0.2)*	*Colour*		*COD*		*BOD*	
					Untre-ated	*Tre-ated*	*Untre-ated*	*Tre-ated*	*Untre-ated*	*Trea-ated*
$FeSO_4$	10 g/L	Percol 173	0.1	8.8	60000	1050	7421	1764	420	40
$FeSO_4$	10 g/L	Meg LT 27	0.1	8.8	60000	1072	7421	1378	420	24
$FeSO_4$	10 g/L	Percol 47	0.1	8.8	60000	625	7421	958	420	30
$FeSO_4$	10 g/L	Zetag 63	0.15	8.9	60000	1105	7421	1408	420	24
PW waste	40 ml/L	Percol 173	0.1	8.8	60000	600	7421	1344	420	21
PW waste	40 ml/L	Meg LT 27	0.1	8.9	60000	640	7421	958	420	18
PW waste	40 ml/L	Percol 47	0.1	8.7	60000	700	7421	1568	420	24
PW waste	40 ml/L	Zetag 63	0.1	8.8	60000	725	7421	1568	420	30

(Lime dose–29 g/L).

All values represent an average of three sets.

Characterization of Chemical Sludges Generated from the Treatment and Evaluation of Toxicity of the Sludges

The characteristics of sludges generated during the coagulation of biologically treated distillery effluent using various coagulants are presented in Table 39.5. The empirical formulae for each of the chemical sludges were evaluated and these are found to be $C_{16}H_{80}SO_{190}$, $C_{18}H_{55}NS_2O_{82}$, $C_{19}H_{49}NS_2O_{70}$, $C_{15}H_{42}NO_{49}$ and $C_9H_{28}SO_{148}$ for $FeSO_4$/lime, $Fe_2(SO_4)_3$/lime Alum/lime, pickling wastewater/lime, and Titanium wastewater/lime systems, respectively. The weight of sludges ranged from 22 to 27 g/l with a pH value ranging from 10.4 to 12.1. The volatile matter content in these sludges ranged between 15.10 and 17.64 per cent wt/wt. Further these sludges generated from the coagulation contained metals, *viz.* Mn, Ni, Cu, Cr, Cd, and Fe. The concentration of Fe in the sludges was in the range of 5000 to 11,540 mg/L.

Table 39.5: Characteristics of Sludge Generated during Physico-Chemical Treatment of Biologically Treated Distillery Effluent

Parameters	*Coagulants*		
	$FeSO_4$/Lime	*$Fe_2(SO_4)_3$/Lime*	*PW*/Lime*
Wt. of sludges (g/L)	22	27	25
pH	12.2	10.4	12.1
Moisture Content (per cent)	1.02	1.40	1.3
Volatile Matter (per cent)	12.60	17.64	15.1

*PW: Pickling Wastewater.

All values represent an average of three sets of experiments.

The leachates were collected for further analysis. The leachates contained COD ranging from 4320 mg/l to 7488 mg/l, while metals, *viz.* Mn, Ni, Cu, Cr, Cd, Pb and Fe were in appreciable concentration as presented in Table 39.6. These sludges require proper disposal in a secured manner, otherwise it may pollute the surrounding environment through leaching of the pollutants.

Evaluation of Effectiveness of Activated Carbon for Removal of Residual COD from Biologically Treated Distillery Effluent after Chemical Coagulation

Since biologically treated spent wash after chemical coagulation generated an effluent having COD in the range of 958-1764 mg/L which warrants further treatment before discharge into inland surface water (Central Pollution Control Board, Ministry of Environment and Forests, MoEF, 1996). Thus, after chemical coagulation the effluent was subjected to adsorption by activated carbon.

Characterization of Activated Carbon

The efficiency of adsorption on activated carbon is directly related to its granular size and carbon content. Accordingly, the activated carbon used in the present study was subjected to characterization in terms of its general, proximate and ultimate characteristics. The results indicate that the activated carbon used in the study had in per cent (w/w) volatile matter 12.5, ash 12.6, fixed carbon 70.4, and moisture 4.5, while ultimate analysis indicated that the carbon, hydrogen, nitrogen and sulphur percentage was 79.4, 4.8, 0.8 and 2.1, respectively.

Table 39.6: Characteristics of Leachate Generated through TCLP

Parameters	*Coagulants*		
	$FeSO_4$/Lime	*$Fe_2(SO_4)_3$/Lime*	*PW*/Lime*
Heavy Metals (mg/L)			
Mn	5.59	0.22	0.87
Ni	ND	0.91	1.42
Cu	2.19	0.57	0.64
Cr	1.57	0.32	ND
Cd	0.27	0.1	0.12
Pb	ND	ND	ND
Fe	21.5	18.34	6.57
COD (mg/L)	4438	7488	4320

*PW: Pickling Wastewater.

All values represent an average of three sets of experiments.

Kinetics of Adsorption on Activated Carbon

Kinetics of COD Removal on Activated Carbon for Different Contact Times

The removal of COD by adsorption on activated carbon follows the equation as given below:

$$As = k_t.t^m \quad (1)$$

$$\text{or } \log As = m\log t + \log k_t \quad (2)$$

where,

t = Contact time

As = Per cent COD removal

k_t = Rate factor

Equation 2 gives a straight line on log-log plot with slope 'm' varying from 0.172 to 0.094 with an intercept k_t ranging from 60.830 to 75.610 for the ferrous sulphate-lime treated effluent. The same plot for pickling wastewater and lime treated effluent gave slope varying from 0.148 to 0.030 and K_t from 48.500 to 81.959. In both the cases the dose of activated carbon was varied from 4 g/L to 16 g/L. The respective values of m and k_t for both systems are presented in Table 39.7.

Table 39.7: Variation of per cent COD Reduction with different Contact Time for Activated Carbon

Coagulants	*Initial COD (mg/L)*	*Wt. of Activated Carbon (g/L)*	*Slope*	*Intercept*
$FeSO_4$ - lime treated system				
	2970	4	0.172	60.83
	2970	8	0.125	70.55
	2970	12	0.134	75.26
	2970	20	0.094	75.61
Pickling wastewater - lime treated system				
	950	4	0.147	48.5
	950	8	0.056	72.63
	950	12	0.067	76.6
	950	16	0.03	81.95

Contact time varied from 15 min to 60 min.

All values represent an average of three sets of experiments.

Increasing values of rate factor k_t with increasing activated carbon concentration indicates enhancement in the rate of COD removal. However, major proportion of total organic compounds are removed due to immediate solute adsorption on activated carbon with subsequent lesser removal of remaining COD. The larger values of 'm' indicate better adsorption mechanism for lower concentrations of activated carbon.

Kinetics of COD Removal on Different Weights of Activated Carbon at Fixed Contact Times

Removal of residual COD as a function of activated carbon concentration is written:

$$As = K.C^n \quad (3)$$

$$\text{or } \log As = n\log C + \log K \quad (4)$$

where,

As = Per cent COD removal

C = Concentration of activated carbon

K = Constant denoting extent of COD removal

The slope 'n' is representative of the rate of COD removal at a particular contact time. The representative log-log plot of per cent COD removal versus activated carbon concentration for ferrous sulphate-lime and pickling wastewater-lime systems, respectively. The values of slope 'n' and intercept are given in Table 39.8 for both systems. Value of slope 'n' reduces upto a contact time of 45 min. and then remains almost constant. This implies that rate of improvement in effluent quality becomes independent of contact time after 45 min.

Based on the above observations, a dose of 8 g/L of activated carbon with a contact time of 45 min. is considered as optimum for the given system. After activated carbon adsorption, COD and BOD of the treated effluent conform to the prescribed limits for surface water discharge.

Table 39.8: Variation of per cent COD with different Weights of Activated Carbon

Coagulants	*Initial COD (mg/L)*	*Wt. of Activated Carbon (g/L)*	*Slope*	*Intercept*
$FeSO_4$ - lime treated system				
	2970	15	1.437	58.67
	2970	30	1.292	61.1
	2970	45	1.182	62.79
	2970	60	1.242	61.53
Pickling wastewater - lime treated system				
	950	15	0.147	61.41
	950	30	0.497	76.73
	950	45	0.307	80.81
	950	60	0.415	82.54

Dose of carbon varied from 4 g/L to 16g/L.

All values represent an average of three sets of experiments.

Activated Carbon Adsorption Isotherm

Adsorption isotherm for the given effluent that defines the distribution of organic materials in liquid and solid phase at a constant temperature has been plotted. The equilibrium data has been processed in accordance with Freundlich Adsorption Isotherm defined by equation given below:

$$qe = K_f C^{1/n} \text{ or} \quad (5)$$

$$\log qe = \log K_f + 1/n \log C \quad (6)$$

where,

qe = Amount of COD removal per unit weight of activated carbon

K_f = Constant denoting adsorption capacity

C = Equilibrium COD value

The value of K_f as calculated was 1.44, whereas that of n was found to be 2.08. As the values of K and n are greater than one, it could be concluded that there is high adsorption throughout studied the concentration range. The performance of conventional and non-conventional coagulants for the treatment of Biologically Treated Distillery Spentwash are presented in Table 39.9.

Table 39.9: Performance of Conventional and Non-Conventional Coagulants for the Treatment of Biologically Treated Distillery Spentwash

Coagulant	*Dose*	*pH (±0.2)*	*Colour*		*COD*		*BOD*	
			Untreated	*Treated*	*Untreated*	*Treated*	*Untreated*	*Treated*
$FeSO_4$	20 g/L	8.3	60000	1200	7421	1706	420	87
$Fe_2(SO_4)_3$	12 g/L	8.1	60000	2200	7421	1703	420	93
Alum	12 g/L	8.1	60000	3100	7421	2640	420	120
PW Waste-water	80 ml/L	8.9	60000	625	7421	958	420	30
Ti Wastewater	150 ml/L	8.8	60000	890	7421	2438	420	80

(Lime dose–29 g/L).

All values represent an average of three sets of experiments.

Conclusion

- ☆ Molasses based distillery effluent after treatment in conventional biological treatment units generated effluent containing intense brown colour with a COD value ranging from 7000-7800 mg/L.
- ☆ The physico-chemical treatment of biologically treated distillery effluent using conventional coagulant–ferrous sulphate and non-conventional coagulant–pickling wastewater supplemented with coagulant aid generated effluent of similar quality. Thus the wastewater from one industry could be utilized for the treatment as a substitute for conventional coagulant.
- ☆ The effluent generated from coagulation could be treated further efficiently by using activated carbon to reduce COD and BOD to permissible level for discharge into the inland surface water.

Chapter 40

Removal of Refractory Organics from RO Reject using Advanced Oxidation Processes

K. Sri Bala Kameswari and Chitra Kalyanaraman

Environmental Technology Division,
CSIR- Central Leather Research Institute
Adyar, Chennai – 600 020, India

ABSTRACT

Recovery of water from wastewater is mandatory in Tamil Nadu for the leather sector and the recovered water is reused in leather processing. Secondary treated tannery wastewater contains high concentrations of Total Dissolved Solids (TDS) and other residual organic impurities and it is not suitable for reuse in the leather processing. To recover water from treated wastewater, reverse osmosis process is employed. About 75 per cent of water can be recovered for re-use in the industry and the remaining reject is evaporated. The reject generated from reverse osmosis process will contain both organic and inorganic impurities or pollutants. The pollutants are in the form of recalcitrant organic pollutants, residual dyes, calcium, magnesium, chloride, sulphate etc. While salinity in the recovered water is acceptable in some processes like soaking and pickling, the organic matter has to be removed completely in order to reuse the water in the process. When reject is evaporated in solar evaporation pans, the presence of organic pollutants will reduce the rate of evaporation and the recovered salt will be impure. Hence, removal of organic pollutants becomes essential in the RO reject, prior to the evaporation. Studies were carried out for treatment of reject to remove refectory organic pollutants by adopting various advanced oxidation processes such as UV, ozonation, hydrogen peroxide, Fenton's reagent alone and in combination with UV radiation. The findings of the study are presented in the chapter.

Keywords: *Advanced oxidation process, Biochemical oxygen demand, Chemical oxygen demand and Reject.*

Introduction

Leather processing is a traditional activity in India and nearly 30 to 50 m^3 of wastewater is generated in processing one tonne of raw hides/skins into finished leather. In view of its pollution problems and increasing demand for good quality water required both for domestic and other industrial purposes, it has become essential to treat the wastewater to an acceptable level prior to its disposal. Hence, tannery effluents undergo various levels of treatment known as primary, secondary and tertiary treatment. The treatment processes generally adopted in effluent treatment plants is the conventional treatment process consisting of physico-chemical treatment followed by biological treatment. It is difficult to degrade all the complex molecules present in the wastewater during biological treatment within the designed hydraulic retention time and solids retention time. Moreover, treated tannery wastewater contains high amount of total dissolved solids (TDS), which cannot be removed by conventional treatment process. The constituents that of TDS are sodium, chloride, sulphate, calcium, magnesium etc. The treated effluent from the effluent treatment plants is not suitable for reuse in the process straightaway due to its high TDS content. Also, discharge of these salts-bearing treated wastewater for agriculture purpose will affect the fertility of soil and reduce crop yield.

The State of Tamil Nadu in India has limited sources of water available for leather processing and Tanners are forced to buy water for their operations. Water has to be brought from faraway places and the cost of water is very high. Tamil Nadu Pollution Control Board (TNPCB), a regulatory body of Tamil Nadu State, India in order to meet the water requirements of tanning industry and to comply with the zero liquid discharge (ZLD) norms mandated in the state, has been enforcing implementation of reverse osmosis (RO) process by the tanning industry for recovery of water from treated tannery wastewater. At present, physico-chemical treatment followed by a two-stage biological treatment process are adopted for treatment of tannery wastewater. Tannery sector in Tamil Nadu state has already taken initiatives to recover and reuse water using RO process as per the Honorable High Court's Order and hence the existing treatment systems have been up-gradated to suit RO feed water. The process flow diagram is presented in Figure 40.1.

Advantages of ZLD

- Conservation of natural resources like water
- Demand for water will be reduced
- Recovered water can be reused in the process
- About 70 to 75 per cent of recovery of water is possible from treated wastewater.

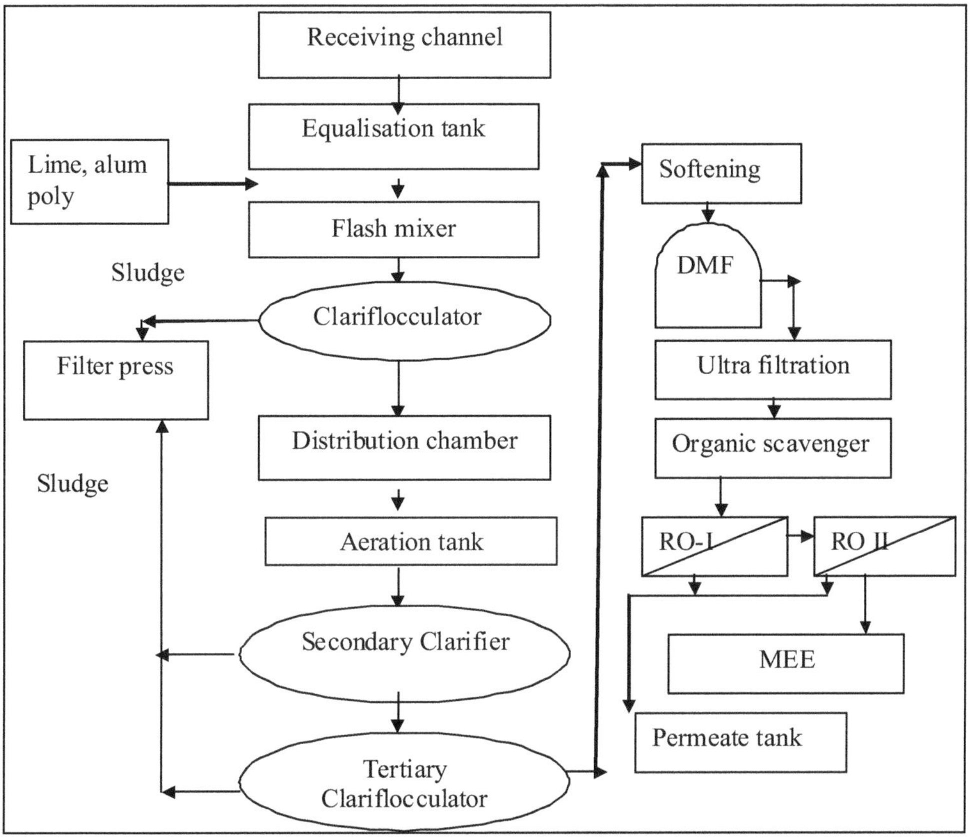

Figure 40.1: Process Flow Diagram for Recovery of Water from Treated Tannery Wastewater

RO: Reverse osmosis; DMF: Dual media filter; MEE: Multiple effect evaporator for evaporation of reject.

Disadvantages of ZLD

- About 25 to 30 per cent reject volume is to be evaporated
- No feasible method for disposal of reject
- High capital and operation cost for treatment, recovery and reuse
- High energy intensive process
- Advanced treatment units like MBR prior RO has to be included for efficient treatment and recovery of water
- Efficiency of RO unit depends on the feed water quality
- Life of RO membranes depends on various factors like quality of feed water, presence of silica, dyes etc.
- High recurring expenses.

Reject represents approximately 25 to 30 per cent of inflow depending on the recovery ratio and contains salts, dissolved nutrients, recalcitrant chemicals and persistent organic pollutants (POPs) such as pesticides and/or synthetic dyes which cannot be removed in biological wastewater treatment (Al-Rifai *et al.*, 2007; Pehlivanoglu-Mantas and Sedlak 2006; Westerhoff *et al.*, 2009; Bagastyo *et al.*, 2011).

Issues on Disposal of Reject

Reverse osmosis is the ultimate step to recover water through membrane filtration. Dispoosal of the reject containing inorganic matter and residual organic impurities is a major concern. Very little information is available on reject characterization and treatment of reject generated from RO plants, especially from wastewater treatment plants (Lee and Lueptow, 2001). Due to presence of residual organic impurities, the rate of evaporation will reduce and makes the recovered salt impure. Advanced oxidation processes (AOPs) generate hydroxyl radicals (•OH) as non-selective oxidants, and have been established to be highly efficient for the degradation of various refractory pollutants. Several processes including ozonation (Pophali *et al.*, 2011) heterogeneous photocatalysis (UV/TiO_2) (Westerhoff *et al.*, 2009), sonolysis and their combination processes (Zhou *et al.*, 2011) have been investigated for treatment of RO reject. The removal of organics from RO concentrate was carried out by Electro-Fenton process using a graphite-felt cathode and 62 per cent of COD removal was reported (Zhou *et al.*, 2012).

In the present study, in order to reduce or to remove the refractory organic pollutants from RO rejects generated from tannery wastewater, various advanced oxidation processes such as UV radiation, H_2O_2, Ozonation, Fenton's reagent, both alone and in combination, were investigated.

Primary Role of Advanced Oxidation Processes

Advanced oxidation process is a promising technology that has received increasing interest for treatment of water, wastewater, leachate and rejects generated from RO plants. AOPs predominantly involve the generation of a very powerful and non-selective oxidizing species, the hydroxyl radicals (OH^-) for the destruction of refractory and hazardous pollutants. AOP involves two stages of oxidation, namely, the formation of strong oxidants (*e.g.* hydroxyl radicals) and the reaction of these oxidants with organic contaminants present in the reject/water/wastewater (Marco *et al.*, 1997; Hang *et al.*, 1993). However, the term AOP refers specially to processes in which oxidation of organic contaminants occurs primarily through reactions with hydroxyl radicals. The ability of hydroxyl radical in breaking down the molecular structure of chemical compounds is advantageous for partial oxidation of high molecular weight organics. AOP treatment uses strong oxidizing agents (ozone, hydrogen peroxide) and/or catalyst (Fe, Mn, TiO_2) sometimes supported in activity by high energy radiation, *e.g.*, UV light. All of these processes can produce hydroxyl radicals, which can react and destroy a wide range of organic contaminants.

Experimental Methodology

One of the commercial tanneries in Chennai, Tamil Nadu processing wet-blue to finished leather process was operating an individual effluent treatment plant with

RO system for recovery of water from treated tannery wastewater. The reject from the RO plant was collected and was analyzed in triplicate and the average values were considered as reject characteristics.

Chemical Oxidation/Advanced Oxidation Processes Adopted

For removal of organics from RO reject, Chemical oxidation using Hydrogen peroxide and AOPs such as UV radiation, Ozonation, Fenton's reagent alone and in combination were investigated.

UV Radiation

UV oxidation is a destructive process that oxidizes organic constituents by the addition of strong oxidizers and irradiation with UV light. Oxidative degradation can occur either by direct reaction with the applied oxidant, or via the production of highly reactive secondary species, most commonly, hydroxyl radicals, which is one of the most powerful oxidants known (Metcalf and Eddy, 2007). UV radiation is commonly used to promote the formation of hydroxyl radicals. This can be achieved by a number of methods including photo-catalysis with titanium dioxide (TiO_2) (Egerton *et al.*, 2006; Murray and Parsons 2006) or by direct reaction of hydrogen peroxide (H_2O_2) (Rosenfeldt and Linden, 2004). Other systems, such as Ozone/H_2O_2 and UV/Ozone are also considered as AOPs since they too promote the formation of hydroxyl radicals. The duration of operation and maintenance of UV radiation depends on the influent water turbidity, contaminant, metal concentration, existence of scavenger and the required maintenance intervals on UV reactor. A wide variety of organic contaminants are susceptible for destruction of UV radiation, including petroleum hydrocarbons, chlorinated hydrocarbons used as industrial solvents and cleaners.

In the present study, approximately one liter of RO reject was taken and filled through the funnel of the UV apparatus fitted with medium pressure lamp. The experiments were conducted for 6 hours duration and every one hour the samples were collected and analyzed for COD and BOD.

Hydrogen Peroxide

Hydrogen Peroxide (H_2O_2) is a powerful oxidizing agent ($E^{\%}$=1.8V) and it decomposes spontaneously into water and oxygen. In acidic solutions, H_2O_2 is one of the most powerful oxidizers known stronger than chlorine, chlorine dioxide, and potassium permanganate. Also, through catalysis, H_2O_2 can be converted into hydroxyl radicals ($^{\bullet}$OH), which are highly reactive. The strong oxidizing power of hydrogen peroxide makes it suitable for the destruction of a variety of pollutants. The reject samples pH was adjusted to 3.0 using H_2SO_4 and buffer was added to maintain the required pH. Hydrogen peroxide dose was varied from 10, 15, 20, 25, 30 and 35 mL of 30 per cent H_2O_2 in one liter of reject and a contact time of 30 minutes was maintained. After 30 minutes of contact time, the samples were analyzed for COD and BOD removal efficiency. The interference due to the residual peroxide was eliminated before the estimation of COD and BOD (Talinli *et al.*, 1992).

Ozonation

Ozone is an extremely strong oxidant ($E^{\%}$= 2.07 V) and reacts rapidly with most organic compounds. It is now widely accepted that ozone reacts in aqueous solution with various organic and inorganic compounds, either by a direct reaction of molecular ozone in acidic pH *i.e.*, < 6 or through the hydroxyl radical produced by the ozone decomposition in water at alkaline pH (Olivia *et al.*, 2006). Ozonation experiments were carried out in a bench scale glass reactor of diameter 60 mm and height 1800 mm. Ozone was supplied through a ceramic diffuser fitted at the bottom. The laboratory ozone generator Model No. L6G from Faraday Instruments, India was used with pure oxygen as the feed inlet gas. Ozone concentration before and after reject treatment was measured as per the procedure (Part 4500-O_3 B) given in standard methods 20th edition (APHA 1998). At regular intervals of time *i.e.*, 10, 20, 30, 40, 50 and 60 minutes, the samples were collected, analyzed for COD and BOD removal efficiency.

Fenton's Reagent

The Fenton's process has attracted great interest in view of its high efficiency to generate hydroxyl radicals through decomposition of hydrogen peroxide (H_2O_2) by Fe^{2+} in acidic condition. The oxidation process utilizing activation of H_2O_2 by iron salts, classically referred to as Fenton's reagent, is known to be very effective in the destruction of many hazardous organic pollutants (Neyens *et al.*, 2002). There are two factors affecting the rate of Fenton reaction: the H_2O_2 dose and the Fe^{2+} concentration. The H_2O_2 dose is important to obtain better degradation efficiency, while the iron concentration is important for the reaction kinetics. When the amount of Fe^{2+} employed exceeds that of H_2O_2, the treatment tends to have the effect of chemical coagulation. When the two amounts are reversed, the treatment tends to have the effect of chemical oxidation (Neyens and Baeyens 2003). Hence, for the purpose of chemical oxidation in the present study, the ratio of Fe^{2+} to H_2O_2 was kept less than 2.0 as reported by Neyens and Baeyens (2003). From the results obtained from H_2O_2 treatment, for fixed concentration of H_2O_2, the Fe^{2+} concentration was varied from 18, 27, 36, 45 and 54 mL of 1M $FeSO_4$ to find out the maximum removal efficiency of organic contaminants from RO reject. A contact time of 30 minutes was maintained.

H_2O_2 Coupled with UV Radiation

UV/H_2O_2 oxidation proved to be the most effective treatment (Bagastyo *et al.*, 2011). In the previous section reject was treated using UV radiation and H_2O_2 separately. However in order to improve the efficiency, H_2O_2 coupled with UV radiation was tried for treatment of RO reject. From the previous studies optimum H_2O_2, dose was selected and simultaneously RO reject was subjected to UV radiation. The experiments were conducted for 6 hours duration and every one hour the samples were collected and analyzed for COD and BOD removal efficiency.

Fenton's Reagent Coupled with UV Radiation

In the previous section reject was treated using UV radiation and Fenton's reagent separately. However in order to improve the efficiency, Fenton's reagent coupled with UV radiation was tried for treatment of RO reject. From the previous studies optimum dosage of Fenton's reagent was selected and simultaneously RO reject was subjected to UV radiation. The experiments were conducted for 6 hours duration and every one hour the samples were collected and analyzed for COD and BOD removal efficiency.

Results and Discussion

Characteristics of RO Reject

The RO reject samples collected from a commercial tannery operating an individual effluent treatment plant with RO system for recovery of water from treated tannery wastewater. The characteristics of RO reject are presented in Table 40.1.

Table 40.1: Characteristics of RO Reject

Sl.No.	*Parameter*	*Average Value*
1.	pH	6.5
2.	BOD (mg/L)	85
3.	COD (mg/L)	1955
4.	Chlorides as Cl^- (mg/L)	3400
5.	TDS (mg/L)	20760
6.	Sulphates as SO_4^{2-} (mg/L)	12630

It was observed from the reject characteristics that a considerable amount of refractory organics were present in the RO reject requiring their removal prior to evaporation in order to improve the rate of evaporation as well as to recover organics-free salts from the evaporator.

Reject Treatment using UV Radiation

After UV radiation, the samples were taken and analyzed for COD and BOD removal efficiency and the results are presented in Figure 40.2.

After 3 hours of UV radiation, only 8.7 and 13.09 per cent removal of COD and BOD respectively were observed. There was no phenomenal increase in COD and BOD removals observed even after UV radiation of reject samples for 6 hours. The poor removal efficiency is due to high TDS which lowers the oxidation rate. Carbonate and bicarbonate ions are hydroxyl radical scavengers, affects the reaction rate of hydroxyl radicals with organic compounds (Wang *et al.*, 2000).

Reject Treatment using Hydrogen Peroxide

After addition of required dosages of H_2O_2, the reject samples allowed for reaction. In order to avoid the interferences due to un-reacted H_2O_2, the samples were analyzed after 24 hours and the results are presented in Figure 40.3.

By increasing the H_2O_2 dose from 10 to 25 mL/L, increase in COD and BOD removal was observed. At H_2O_2 dose of 25 mL/L, the COD and BOD removal of 38.04

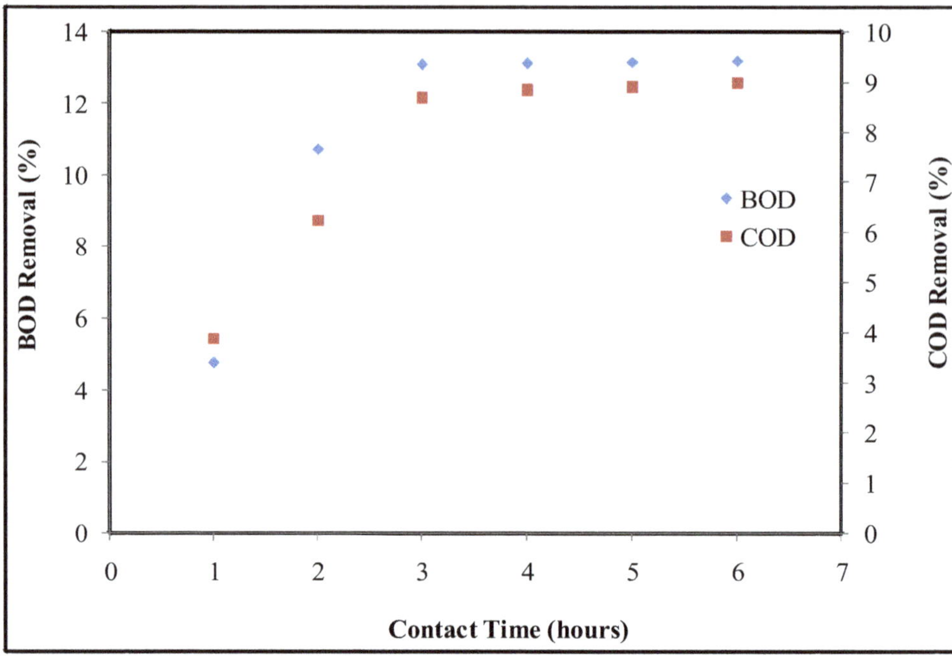

Figure 40.2: Reject Treatment using UV Radiation.

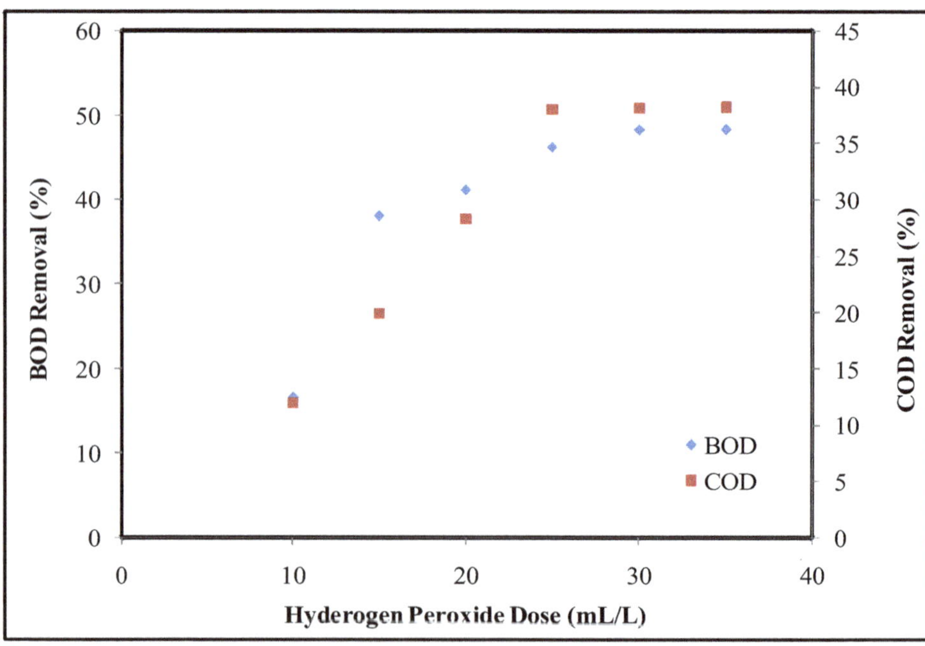

Figure 40.3: Reject Treatment using Hydrogen Peroxide.

and 46.28 per cent respectively were observed. Further increasing the dose, the increase in removal of COD and BOD was marginal. Also it was evident from the results that H_2O_2 has oxidation potential next to ozonation.

Reject Treatment using Ozonation

After Ozonation, the samples were taken and analyzed for COD and BOD removal efficiency and the results are presented in Figure 40.4.

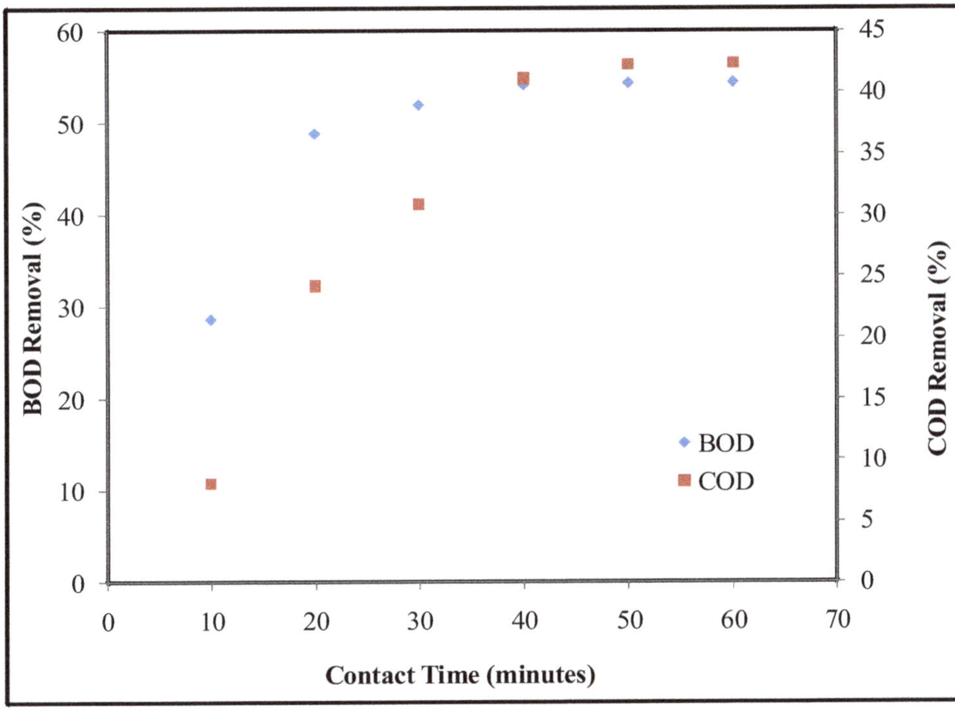

Figure 40.4: Reject Treatment using Ozone.

It was observed that, for the ozonation time of 40 minutes, 41.11 per cent and 54.04 per cent removal of COD and BOD was observed. Increasing the ozonation time further upto 60 minutes, resulted in only 1.2 and 0.3 per cent increase in COD and BOD removal. Considering the economics, 40 minutes of ozonation time is considered as optimum. The results of the present study are comparable with the findings reported by Pophali *et al*. (2011). High salinity levels in landfill leachates do not significantly affect ozone reactions and similar observations were noticed in the present study also (Monje-Ramirez and Orta de Velasquez 2004).

Reject Treatment using Fenton's Reagent

The results of treatment of RO reject using Fenton's reagent is presented in Figure 40.5.

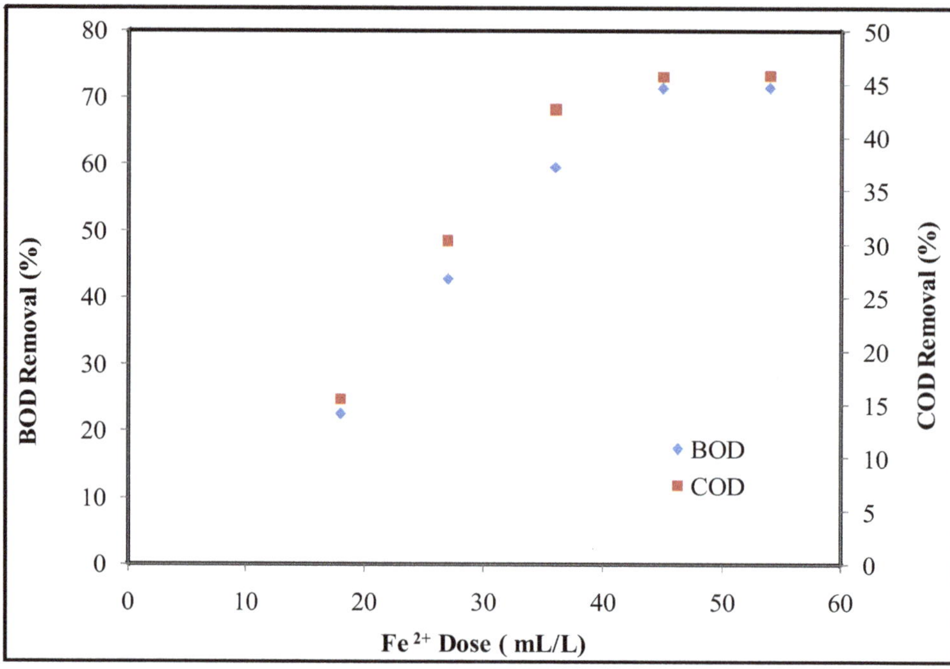

Figure 40.5: Reject Treatment using Fenton's Reagent.

It was observed that by increasing the Fe^{2+} dose for the fixed dosage of H_2O_2 *i.e.*, 25 mL/L the increase in COD and BOD removals was observed. A maximum of 45.8 and 71.52 per cent of COD and BOD removal was observed with the Fe^{2+} dose of 45 mL/L. Addition of catalyst enhanced the reaction process, thereby increase in removal of COD and BOD was observed using Fenton's reagent in comparison with H_2O_2 alone. Removal of COD and BOD was considerably improved by H_2O_2 coupled with UV radiation due to generation of additional hydroxyl radicals.

Reject Treatment using H_2O_2 coupled with UV Radiation

The results of reject treated using H_2O_2 coupled with UV radiation are presented in Figure 40.6.

H_2O_2 concentration was kept constant *i.e.*, 25 mL/L and then subjected to UV radiation. It can be seen from the Figure 40.6 that at 4 hours of UV radiation time, 50.33 and 76.19 per cent of COD and BOD removal was observed. By further increasing the contact time, the increase in COD and BOD removal was only marginal. Hence for the reject treated with the combination of UV/H_2O_2, peroxide dose of 25 mL for UV radiation period of 4 hours yielded a maximum of 76 and 50 per cent removal of COD and BOD in comparison with H_2O_2 and UV radiation alone. It is clear from the studies that UV radiation alone is not effective for treatment of reject. However H_2O_2 coupled with UV radiation enhanced the removal of COD and BOD from rejects by 13 and 30 per cent respectively.

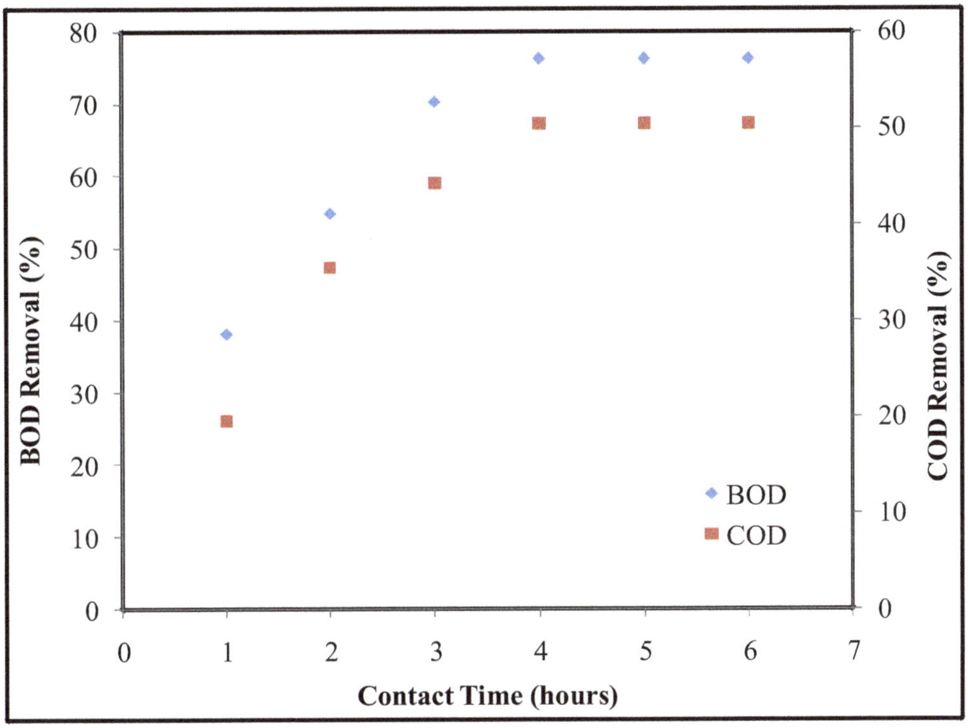

Figure 40.6: Reject Treatment using Hydrogen Peroxide Coupled with UV Radiation.

Reject Treatment using UV/Fenton's Reagent

The results of reject treated using Fenton's reagent coupled with UV radiation are presented in Figure 40.7.

It can be observed from Figure 40.7 that at 4 hours of UV ration time, 59.54 and 84.52 per cent of COD and BOD removal was observed. By further increasing the contact time, the increase in COD and BOD removal was only marginal. Hence for the reject treated with the combination of UV/Fenton's reagent 25 mL H_2O_2 dose, Fe^{2+} dose of 45 mL and UV radiation period of 4 hours yielded a maximum of 59.54 and 84.52 per cent COD and BOD removal. It is clear from the studies that UV radiation alone is not effective for treatment of reject. However Fenton's reagent coupled with UV radiation enhanced the removal of COD and BOD from rejects by 13 and 14 per cent respectively. Combining Fenton's reagent with UV radiation produces more hydroxyl radicals in comparison to the conventional Fenton's reagent, thus promoting the degradation of organic pollutants. The higher production of hydroxyl radicals is due to the combination of oxidant compounds and metallic catalysts in the presence of UV radiation (Pignatello *et al.*, 2006). It was reported that the photo-fenton's reagent is most efficient process for treatment of landfill leachate (Primo *et al.*, 2008). Similar findings were observed in the present study also for treatment of RO rejects.

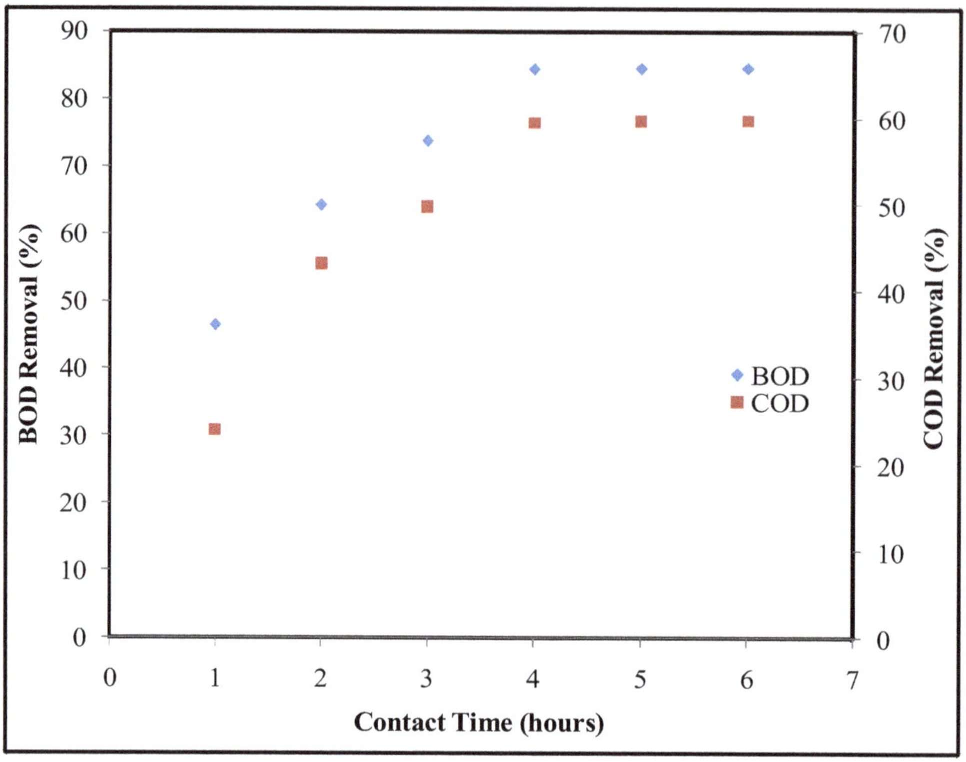

Figure 40.7: Reject Treatment using Fenton's Reagent Coupled with UV Radiation.

Conclusion

From the characteristics of rejects, it is evident that the rejects contain refractory organics in addition to inorganic impurities. The presence of organic impurities will affect the rate of evaporation thereby increasing the energy costs for evaporation. Also, the presence of refractory impurities hinders the quality of recovered salt. Considering these problems, the reject was subjected to UV radiation, H_2O_2, Ozonation, Fenton's reagent alone and in combination for removal of COD and BOD.

From the results obtained from various AOP's, UV radiation alone is not effective. For optimum H_2O_2 dose of 25 mL/L the COD and BOD removal of 38.04 and 46.28 per cent respectively were observed. Whereas, for ozonation time of 40 minutes, 41.11 and 54.04 per cent removal of COD and BOD were observed. It is clear that, H_2O_2 has oxidation potential next to ozonation. A maximum of 45.8 and 71.52 per cent of COD and BOD removal was observed with Fenton's reagent and comparing the COD and BOD removal using H_2O_2 and Fenton's reagent, addition of catalyst in Fenton's reagent enhanced the COD and BOD removal efficiencies. By coupling H_2O_2 and Fenton's reagent with UV radiation further enhancement in COD and BOD removal was observed during reject treatment. However a significant removal in COD and BOD was observed with Fenton's reagent alone or by coupling the H_2O_2, Fenton's reagent with UV radiation. Based on the results obtained, Fenton's reagent for treatment of RO rejects is a better option from a field application point of view.

Acknowledgement

Authors are thankful to the Director, CSIR- Central Leather Research Institute (CLRI), Adyar, Chennai, India for permitting to publish this work.

References

1. Al-Rifai, J.H., Gabelish, C.L., Schafer, A.I. (2007). Occurrence of pharmaceutically active and non-steroidal estrogenic compounds in three different wastewater recycling schemes in Australia. *Chemosphere* 69, 803- 815.
2. APHA, AWWA, WEF. *Standard methods for the examination of water and wastewater*, 20th edn. American Public Health Association/American Water Works Association/Water Environment Federation: Washington, DC, USA, 1998.
3. Bagastyo, A.Y., Keller,J., Poussade, Y., Batstone, D.J. (2011). Characterisation and removal of recalcitrant's in reverse osmosis concentrates from water reclamation plants. *Water Research* 4 5, 2415-2427.
4. Egerton, T.A., Christensen, P.A., Kosa, S.A.M., Onoka, B., Harper, J.C., Tinlin, J.R. (2006). Photoelectrocatalysis by titanium dioxide for water treatment. *International Journal of Environmental Pollution* 27 (1-3), 2 -19.
5. Huang, C.P., Dong, C., Tang, Z. (1993). Advanced chemical oxidation: Its present role and potential future in hazardous waste treatment. *Waste Management* 13: 361-377.
6. Lee, S. and Lueptow, R.M. (2001). Membrane rejection of nitrogen compounds. *Environmental Science and Technology* 35, 3008- 3018.
7. Marco, A., Esplugas, S., Saum, G. (1997). How and why combine chemical and biological processes for wastewater treatment. *Water Sci Technol.*, 35 (4): 321-327.
8. Metcalf and Eddy (2007). *Water Reuse: Issues, Technologies, and Applications.* McGraw Hill, New York.
9. Monje-Ramirez, I., Orta de Velasquez, M.T. (2004). Removal and transformation of recalcitrant organic matter from stabilized saline landfill leachates by coagulation–ozonation coupling processes. *Water Research* 38, 2359–2367.
10. Murray, C.A., Parsons, S.A., 2006. Preliminary laboratory investigation of disinfection by-product precursor removal using an advanced oxidation process. *Water and Environment Journal* 20 (3), 123-129.
11. Neyens, E., Baeyens, J., Weemaes, M.B, De heyder (2002). Advanced biosolids treatment using H_2O_2-oxidation. *Environ. Eng. Sci.* 19: 27–35.
12. Neyens, E., Baeyens, J. (2003). A review of classic Fenton's peroxidation as an advanced oxidation technique. *Journal of Hazardous Materials* 98: 33–50.

13. Olivia, S., Soares, G.P., Jose, J.M.O., Dionisia, P., Antonio, V., Pereira, R.M.F. (2006), Ozonation of textile effluents and dye solutions under continuous operation: Influence of operating parameters, *Journal of Hazardous Materials* B137, 1664-1673.

14. Pehlivanoglu-Mantas, E., Sedlak, D.L. (2006). Wastewater-derived dissolved organic nitrogen: analytical methods, characterization, and effects: a review. *Critical Reviews in Environmental Science and Technology* 36 (3), 261- 285.

15. Pignatello, J., Oliveros, E., MacKay, A. (2006). Advanced oxidation processes for organic contaminant destruction based on the Fenton reaction and related chemistry, *Crit. Rev. Environ. Sci. Tech.* 36, 1–84.

16. Pophali, G.R., Hedau, S., Gedam, N., Rao, N.N., Nandy, T. (2011). Treatment of refractory organics from membrane rejects using ozonation, *J. Hazard. Mater.* 189, 273–277.

17. Primo, O., Rivero, M.J., Ortiz, I (2008). Photo-Fenton process as an efficient alternative to the treatment of landfill leachates. *Journal of Hazardous Materials* 153, 834–842.

18. Rosenfeldt, E.J., Linden, K.G. (2004). Degradation of endocrine disrupting chemicals bisphenol A, ethinyl estradiol, and estradiol during UV photolysis and advanced oxidation processes. *Environmental Science and Technology* 38 (20), 5476-5483.

19. Talinli, I.; Anderson, G. K. Interference of hydrogen peroxide on the standard COD test. *Water Res.* 1992, 26, 107"110.

20. Westerhoff, P., Moon, H., Minakata, D., Crittenden, J. (2009). Oxidation of organics in retentates from reverse osmosis wastewater, *Water Research* 43 (16), 3992- 3998.

21. Zhou, T., Lim, T-T., Chin, S-S., Fane, A.G. (2011). Treatment of organics in reverse osmosis concentrate from a municipal wastewater reclamation plant: feasibility test of advanced oxidation processes with/without pretreatment, *Chem. Eng. J.* 166, 932–939.

22. Zhou, M., Tan, Q., Wang, Q., Jiao, Y., Oturan, N., Oturan, M.A. (2012). Degradation of organics in reverse osmosis concentrate by electro-Fenton process, *Journal of Hazardous Materials* 215–216, 287– 293.

Chapter 41

Applications of SAR in Environmental Monitoring

Gayatri Mirajkar
Shivaji University, Kolhapur

1. Remote Sensing

Remote sensing is the acquisition of information about an object or phenomenon, without making physical contact with the object [1]. In modern usage, the term generally refers to the use of aerial sensor technologies to detect and classify objects on the earth (both on the surface, in the atmosphere, and oceans) by means of propagated signals (*e.g.*, electromagnetic radiation emitted from aircraft or satellites). There are two main types of remote sensing: passive remote sensing and active remote sensing [2]. Passive sensors detect natural radiation that is emitted or reflected by the object or surrounding area being observed. Reflected sunlight is the most common source of radiation measured by passive sensors. Examples of passive remote sensors include film photography, infrared, charge-coupled devices, and radiometers. Active collection on the other hand, emits energy in order to scan objects and areas whereupon a sensor then detects and measures the radiation that is reflected or backscattered from the target. RADAR and LiDAR are examples of active remote sensing where the time delay between emission and return is measured, establishing the location, height, speed, and direction of an object.

Remote sensing makes it possible to collect data on dangerous or inaccessible areas. Applications of remote sensing include monitoring deforestation in areas such as the Amazon basin, glacial features in Arctic and Antarctic regions, and depth sounding of coastal and ocean depths. Remote sensing also replaces costly and slow data collection on the ground, ensuring in the process that areas or objects are not disturbed. Orbital platforms collect and transmit data from different parts of the electromagnetic spectrum, which in conjunction with larger scale aerial or ground-based sensing and analysis provides researchers with enough information to monitor trends such as *El Niño* and other long and short term phenomena. One of the biggest challenges nowadays is the precise understanding of the earth environment, with respect to land use, land cover changes, and exploration and conservation of the natural resources. Remote sensing plays a key role in such areas of earth sciences such as natural resource management, agricultural fields such as land usage and conservation, national security and overhead, and ground-based and stand-off collection on border areas. The process of remote sensing is also helpful for city planning, archeological investigations, military observations, and geomorphological surveying.

In the last decades, there has been an increased interest in sensors operating in the microwave region of the spectrum, such as synthetic aperture radar (SAR). This is due to the fact that these sensors have advantages, which can be enumerated as follows [3]:

1. SAR images are related to the dielectric properties of the target returning, thus, information that may not be visible to optical sensors. In other words the information retrieved by SAR and optical sensors is very difficult.
2. SAR systems can operate at different frequencies and polarizations, and each combination extracts different kinds of information from the same target.
3. They are sensitive to microtextures as, for instance, differences between calm and rippled water surfaces. In this way, it is possible to infer the presence or absence of wind over lakes, the sea, etc.
4. Their spatial resolution is related to the power of the emitted signal and to the kind of processing. Fine spatial resolutions can be attained, even with orbital platforms.
5. Good quality digital elevation models can be generated.
6. Clouds and other atmospheric constituents do not interfere with detection at the wavelengths used for radar remote sensing. Also since the platform carries its own primary energy source, radar imaging can be carried out at any time of the day and under any weather conditions.
7. Since the wavelengths used for radar are about four or five orders of magnitude longer than those employed in optical remote sensing, quite different properties of earth cover types can be detected at microwaves. Radar scattering is determined largely by geometric properties, such as shape and surface roughness, and by moisture content. Also, depending on the wavelength employed and the moisture content of the near earth surface being imaged, it is sometimes, possible to image beneath the surface. It is certainly possible to image below vegetation canopies at longer wavelengths. As a result of all these effects, radar image data can capture a different set of properties of the region being merged than is the case for optical data.

8. With radar we have control over the properties of the incident energy. This allows a wide variety of data types to be recorded and enables innovative applications such as topographic mapping, landscape change detection, and to a limited extent, three dimensional modeling of the volume detail of a resolution element.

The SAR image is formed by sending a microwave signal towards the target, and by recording and processing the reflected echo.

2. InSAR

Interferometric synthetic aperture radar (InSAR) is a rapidly evolving remote sensing technology that directly measures the phase change between two phase measurements of the ground pixel of the Earth's surface. Two coherent SAR phase images of the same portion of the earth's surface are required to form a phase difference image that is called a interferogram, in which a fringe pattern may appear. The two coherent SAR images used to form an interferogram can be acquired either from two antennas on same space platform and separated perpendicularly to the flight direction (azimuth direction), or from different passes of the same SAR antenna at different times, known as repeat-pass interferometry [4]. Any factor that can affect the phase of the backscattered radar signal can affect the fringe pattern and number of fringes in the interferogram, and thus can potentially be measured by the InSAR technology. These measurements include surface displacements, land topography, land changes, land subsidence/uplift, water levels, soil moisture, snow accumulation, stem volume of forest, etc. Therefore InSAR has found very broad applications in the field of earth and environmental sciences.

Repeat-pass interferometry allows the detection and mapping of the earth surface by using the temporal and spatial coherence characteristics, which can be successfully used for land cover classification, mapping of flooded areas, monitoring of geophysical parameters. The basic measurement of interferogram is the changes of spatial and/or dielectric properties using two images. The space deformation and/or dielectric property change can be due to various forcings: earthquakes, landslides, lake or river surface water flow, oceanic water motion, movement of glaciers and ice sheet, accumulation of snow, forest canopy height, sand dune movement, dielectric constant change resulted from soil moisture change, freezing, or thawing, land subsidence due to groundwater withdrawal, underground mining, hydrocarbon extraction, and permafrost melting, etc. The ground surface "deformation" can be anything the ground pixel is displaced between the two image acquisitions such as crustal deformation due to earthquakes, volcanoes, surface subsidence due to groundwater exhaust or underground mining activities, river water level changes, etc. Therefore, application of InSAR can be very broad – in seismology: land surface deformation due to earthquakes; in natural disaster monitoring and assessment: volcanoes and landslide movement; in surface water hydrology: water level monitoring, snow accumulation; in groundwater hydrology: land subsidence due to excess water pumping or uplift due to groundwater recharge; in glaciology: ice sheet motion and rheology, glacier flow, postglacial rebound of the lithosphere; in mining: land subsidence due to mining;

in forestry: forest canopy height, forest mapping and monitoring; in environmental and structural engineering: ground subsidence and structural stability.

3. Applications of InSAR in Seismology

Remote sensing has been used for earthquake research from the '70s, with the first appearance of satellite images. First of all it was used in structural geological and geomorphological research. Active faults and structures were mapped on the base of satellite images. This method is very limited in time series analysis. It was simply an extrapolation of airphoto geological interpretation methods into space. The modern version of this method is active tectonic analysis with the application of alignment analysis. Time series of alignment distributions on the earth's surface are investigated before and after an earthquake.

The current situation of remote sensing application for earthquake research indicates a few phenomena, related with earthquakes, particularly the earth's surface deformation, surface temperature and humidity, atmospheric temperature and humidity, gas and aerosol content. Both horizontal and vertical deformations scaled from tens of centimeters to meters are recorded after the shock. Such deformations are recorded by the InSAR technique with confidence. Pre-earthquake deformations are rather small, on the order of centimeters. A few cases of deformation mapping before the shock using satellite data are known at present time. Future developments lay in precision longwave SAR systems with medium spatial resolution and combined with the GPS technique. There are numerous observations of surface and near surface temperature increases of 3-5° prior to earth crust earthquakes. Methods of earthquake prediction are being developed using thermal infrared (TIR) surveys. Multiple evidence of gas and aerosol content changes before earthquakes are reported for ground observations. Satellite methods allow one to measure the concentrations of gases in atmosphere: O_3, CH_4, CO_2, CO, H_2S, SO_2, HCl and aerosols. However, the spatial resolution and sensitivity of modern systems restricts the application of satellite gas observation in seismology and the first promising results have been obtained only for ozone, aerosol, and air humidity.

3.1 Deformations

One of the main directions of remote sensing applications for seismology is deformation mapping. Surface deformations in seismic cycles can be divided into three phases: pre-seismic or inter-seismic, co-seismic and post-seismic ones. Co-seismic deformations are evaluated up to meters and tens of meters while pre-seismic movements amount to centimeters. Post-seismic deformations are also measured in centimeters, but subsequent landslides can increase deformations to meters. Most current research is focused on co-seismic and post-seismic (landslide) deformations.

3.2 Applications

The InSAR technique is used to examine small-scale features in the deformation field associated with earthquakes. The InSAR results show significant deformation signatures associated with faults, fractures, and subsidences. The interferogram also clearly indicates surface deformation related to earthquakes.

The first application of satellite interferometry for earthquake research was demonstrated in the '90s by Massonnet *et al.,* [5]. The well-known "butterfly" image (Figure 41.1) of the Landers earthquake (M=7.3, 28 June 1992) was compiled on the base of pre-seismic image of April 24, 1992 and post-seismic scenes: August 7, 1992, July 3 1992, and June 18 1993 [6]. The comparison of the observed interferometric picture and modeled scheme of earthquake deformations shows many similar details (Figure 41.2). Similar images have been obtained for all the significant earthquakes of the last decade: Neftegorsk, 1995, Russia [7], Bhuj, 2001, India [8], Bam, 2003, Iran [9], Sumatra, 2004, Indonesia [10], etc. All these cases demonstrate co-seismic and post-seismic deformations.

The Tokai region is a well-known seismic gap located about 200 km southwest of Tokyo, at the conjunction of the Eurasian plate, the Philippine sea plate and the North-American plate. The vertical deformation recorded on the base of InSAR data coincides with ground leveling and GPS observation. The subsidence continuous to this moment, but an earthquake has still not happened (Figure 41.3).

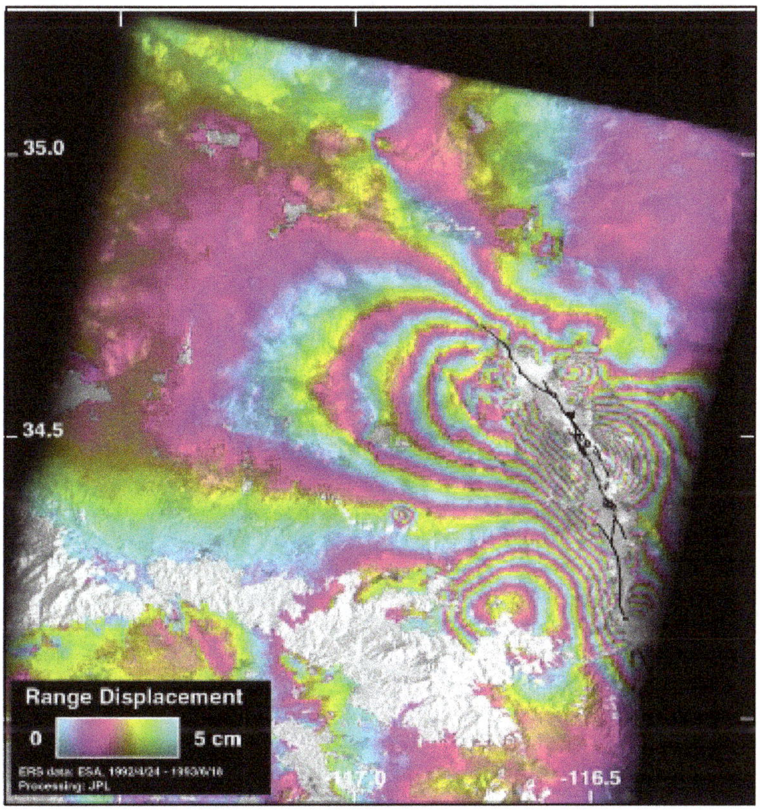

Figure 41.1: Observed Co-seismic Interferogram for the Landers Earthquake (M=7.3, 28 June 1992, CA, USA). One full color cycle represents here 5 cm of range displacement. Solid lines depict the fault geometry as mapped in the field. Gray areas are zones of low phase coherence that have been masked before phase unwrapping [6].

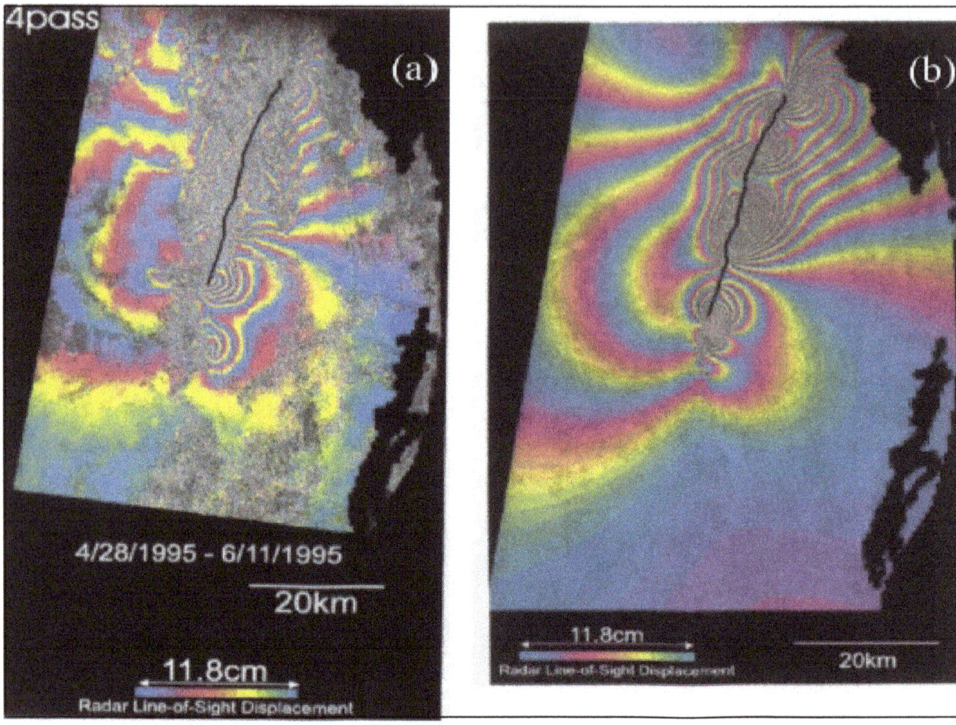

Figure 41.2: Ground Deformations Induced by the Neftegorsk Earthquake (M = 7.6, 28 May 1995, Sakhalin, Russia), (a) Radar Interferogram, (b) Deformation model prediction [7].

The following research has shown surface deformations three year before the Chi-chi earthquake (1999, Taiwan [11]). ERS-2 radar scenes were used to identify

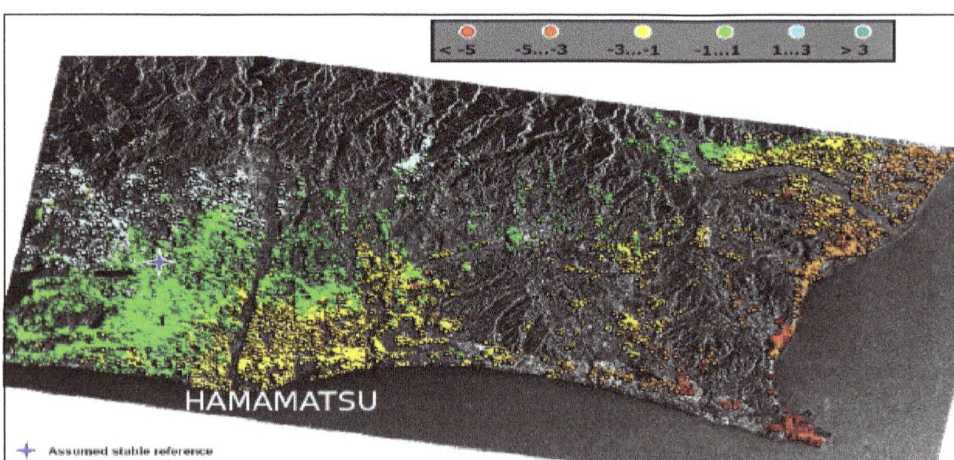

Figure 41.3: Average Displacement Rate in Tokai area, Japan, 1992 – 2000 years, Vertical Deformation, mm/year [12].

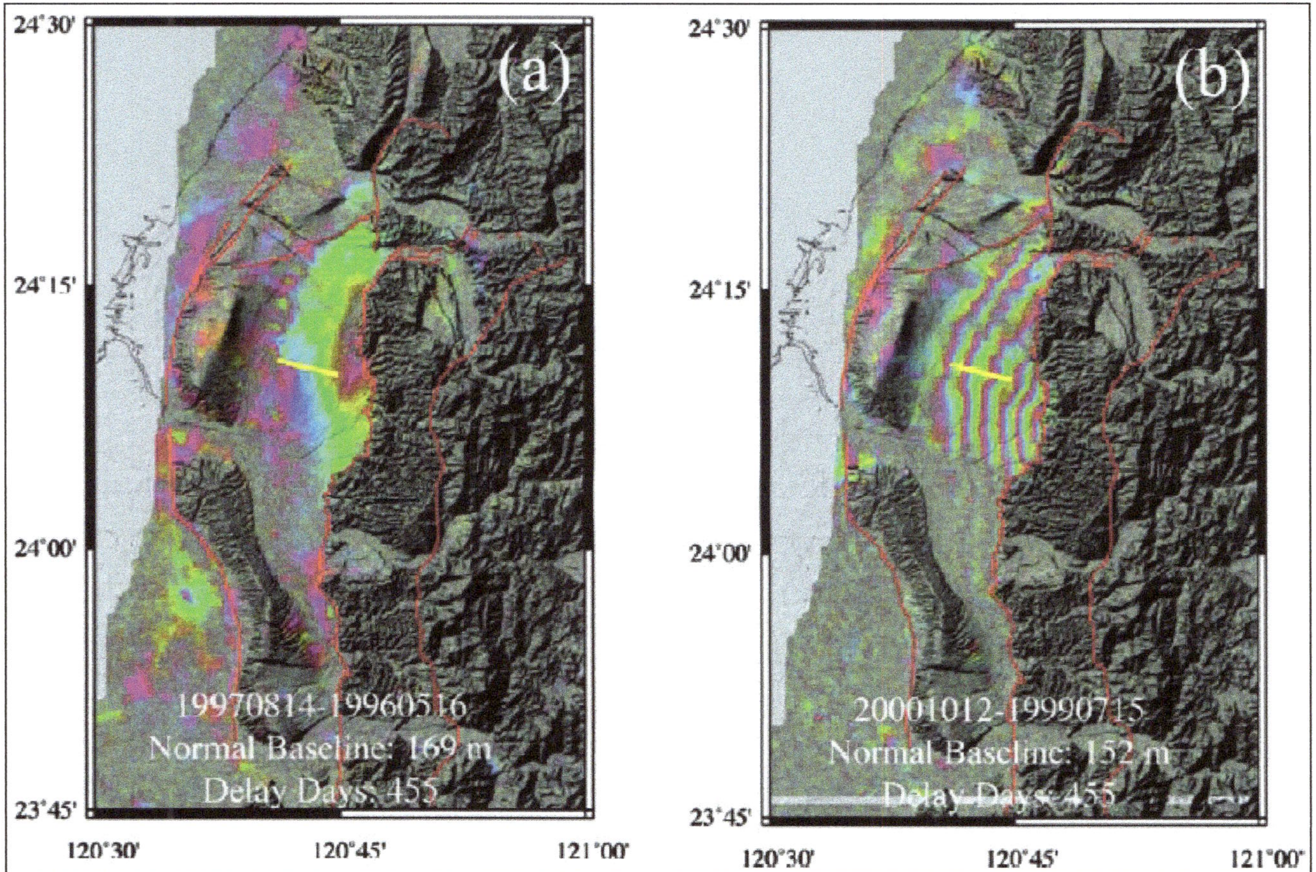

Figure 41.4: Pre-seismic and Co-seismic Ground Deformations for the Chi-chi earthquake (M = 7.6, 20 September, 1999, Taiwan) (a) Pre-seismic deformations between 16 May 1996 and 14 August 1997, (b) Co-seismic ground deformations between 15 July 1999 and 12 October 2000 [11].

possible precursory surface deformation in the areas of the main fault before the Chi-chi earthquake. It was found that surface deformation began at least three years before the Chi-chi earthquake in the areas to the west of the northern segment of the fault where clear co-seismic surface deformation patterns were observed (Figure 41.4).

One of the other applications has described pre-seismic deformations related with the 2004 Parkfield earthquake [13]. Pre-seismic movements are clearly recorded for strike-slips like the San Andreas Fault Zone by InSAR. The relative horizontal speed of the slip is evaluated at about 2 cm/year. The spatial pattern of the pre-seismic InSAR deformation rate (Figure 41.5) indicates that strain distribution is not uniformly partitioned along the fault strike.

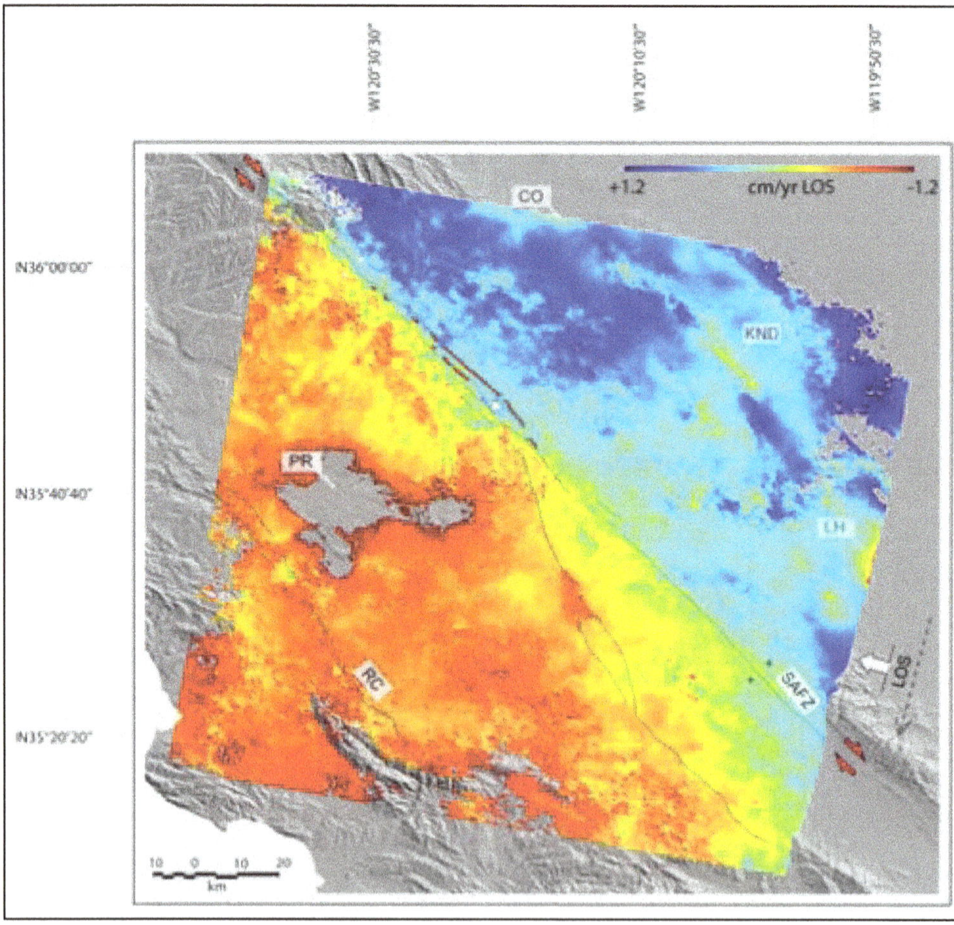

Figure 41.5: Pre-seismic Strain Buildup on the San Andreas Fault Zone at Parkfield. The map is composed by averaging 20 unwrapped ERS-1/2 interferograms in a stack. Each 20 × 20 m pixel is then a measure of the average in the line of sight direction velocity in the period 1992 – 2004 prior to the earthquake. The scale-bar values represent centimeters parallel to the fault [13].

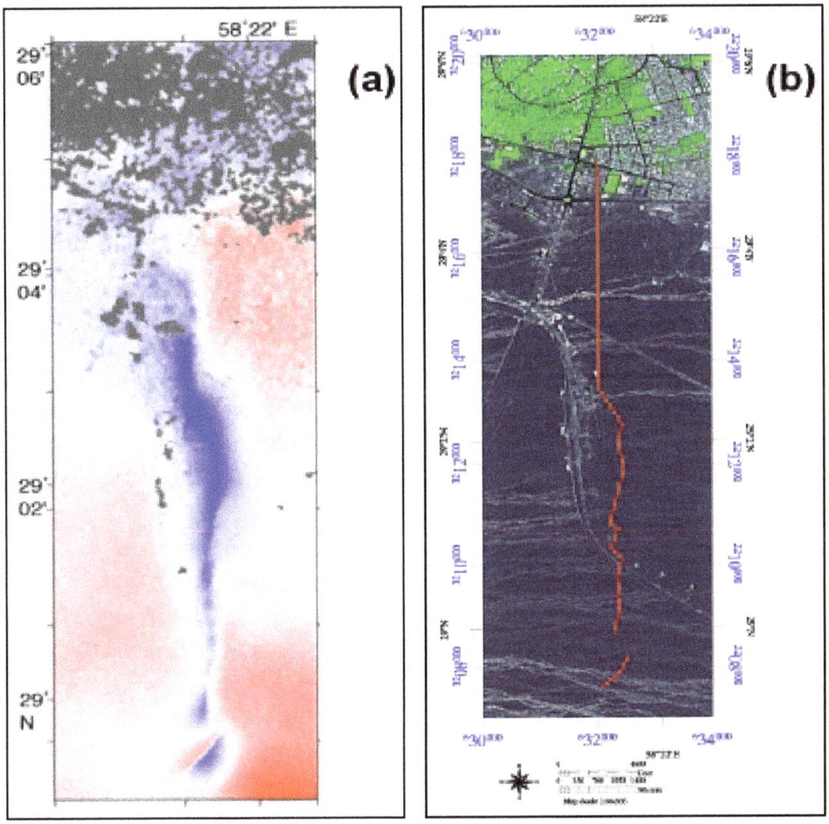

Figure 41.6: Post-seismic Deformations for the Bam Earthquake (M = 6.6, 26 December 2003, Iran) (a) Vertical displacement of the land surface south of Bam, during the three and half years after the earthquake derived from analysis of SAR images. The dark blue area sank a total of more than 3 cm (1.2 inches), revealing a zone of rock that was damaged during the earthquake and then healed afterwards, (b) False color Landsat Thematic Mapper image taken on October 1, 1999 of the area of the earthquake rupture south of Bam. Red line shows location of the fault damage zone above the buried 2003 rupture. Vegetation in the city of Bam is green and stone-covered desert has various tones of gray [14].

Post-seismic deformations also became the object of research. It was possible to separate co-seismic and post-seismic deformations for the Bam earthquake (Iran). A subsidence of about 3 cm was observed (Figure 41.6) in the area of the destructive fault three years after the main shock [14].

The further enhancement of the InSAR technology will allow us to record very fine difference in surface displacement. The currently operated COSMO/SkyMed mission aims to provide daily observations overcoming limited observational frequency, by using a constellation of four satellites. The current constellation consists of three medium-size satellites, each one equipped with a microwave high-resolution

SAR operating in X-band (31 mm), having ~600 km single side access ground area, orbiting in a sun-synchronous orbit at ~620 km height over the earth surface, with the capability to change attitude in order to acquire images at both right and left side of the satellite ground track. TERRASAR-X (wavelength – 31 mm) radar also provides relable information for surface deformation measurements.

Existing satellite InSAR instruments have X-band and C-band (a wavelength of 56.6 mm), offering high resolution, but they only provide reliable interferograms for coherent, non-vegetated surfaces. Data from the JERS-1 satellite demonstrated during its lifetime that L-band (236 mm wavelength) satellites propose reduced resolution but provide interferograms over a far greater range of surface cover types. Modern L-band (236 mm wavelength) SAR PALSAR onboard Japanese satellite ALOS supplies high quality data for interferometry. Unfortunately, this instrument is designed to test applications other than interferometry, so it provides only limited support for deformation analysis. Future missions like MOSART (Monitoring of Surface Deformation in Active Tectonic Zones) and TERRASAR-L are still not realized.

The InSAR technique opens new fields of remote sensing applications in seismology. InSAR methods are the best tool to study earthquake deformations at the moment after the shock. New systems will allow us to shift to pre-seismic deformation measurements and earthquake forecasting. Epicenter area deformations are still a problem for InSAR due to coherence loss and permanent scatter destruction.

4. Application of InSAR in Volcanology

Volcanic processes such as magma accumulation in subsurface reservoirs, magma transport, and emplacement beneath volcanic structures will result in surface deformation, which InSAR can be used to detect. For instance, some offshore mud volcanoes in the south Caspian region possess shallow mud chambers which refill prior to eruption. The inflation of chamber caused by the refill may produce measurable surface deformation that can be detected by interferogram [15]. The measured deformation can then be used as constraints to inverse the depth, size, shape of the magma chamber (and pressure changes) and magma supply dynamics, because point sources [16], dikes and sills [17-18] result in different surface deformation patterns in the modeling of deformation caused by magma intrusion [19]. Therefore, interferogram derived deformation is widely used as constraints for modeling studies of volcano deformation to determine magma sources and dynamics [20], to identify faults caused by the volcanic activiteies [21].

Spatio-temporal evolution of volcanic processes can be derived from analysis of surface deformation derived from a temporal series of InSAR interferograms. For example, the spatio-temporal evolution of a magmatic intrusion associated with the 1999 earthquake swarm is investigated via analysis of produced surface deformation. A series of InSAR images (Figure 41.7), spanning various periods of the intrusion, show that in 1999 surface deformation occurred mainly on the southern flanks of the volcano. The deformation amounted to more than 20 cm of range change [22].

The temporal characteristics of the magma supply system beneath the Okmok caldera is examined using InSAR and statistical seismology. Surface deformation

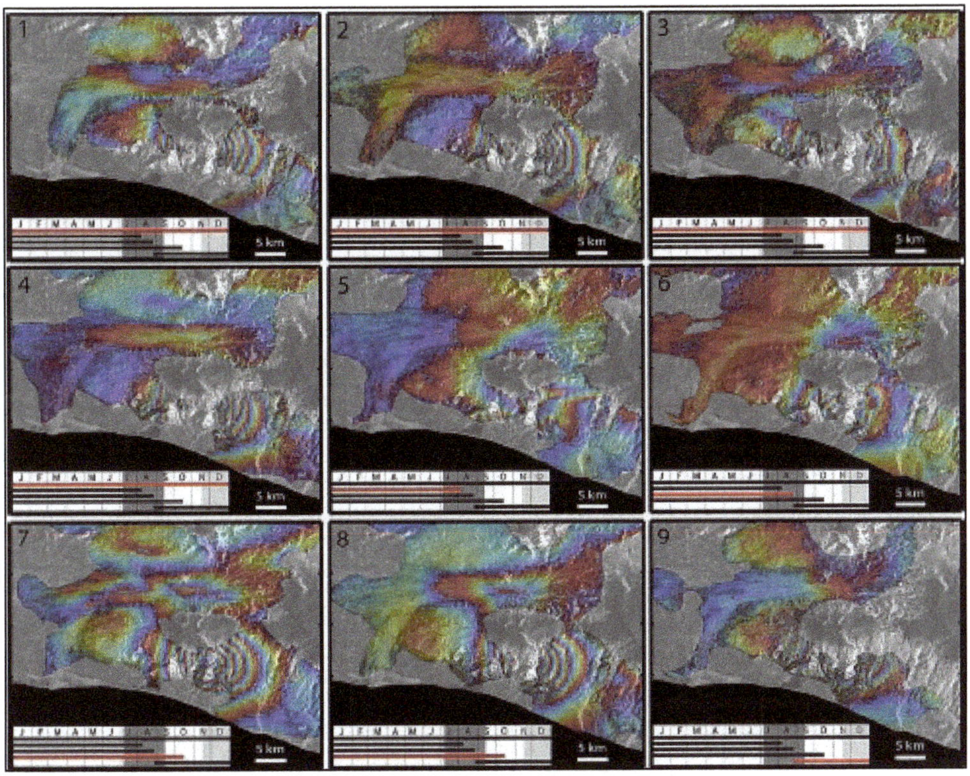

Figure 41.7: The Nine Interferograms Selected for Modeling the 1999 Intrusive Episode. Incoherent areas are masked. The time span insets show the relation to the period (s) of elevated seismic activity [22].

produced by a shallow magma chamber is studied through time series analysis of InSAR imagery acquired between 1995 and 2008 by the ERS-1/2 satellites. Okmok is an active shield volcano with a large caldera that spans approximately 10 km in diameter. It comprises the northeastern portion of Umnak island, which is located off Alaska's west coast near the center of the Aleutian arc. Okmok has a rich eruption history, where 16 documented eruptions have occurred since 1805. Previous deformation studies using InSAR and GPS detected both subsidence and uplift near the center of the caldera from a source approximately 3.0 km below sea level. In Junek *et al.*, [23], a spherical source model (Mogi source) is utilized to simulate the deformation pattern produced by the shallow magma chamber's fluctuating geometry. Statistical seismology provides an independent estimate of the Magma chamber's depth to reduce the non-uniqueness of the Mogi source model solutions. Seismic results show that a shallow magma chamber is located approximately 4.0 km below the caldera floor. InSAR imagery indicates that the chamber has expanded continuously between 1997 and 2008. Modeling results suggest that magma replenishment has occurred at an average rate of approximately 7.0 × 10-3 km^3/year and that the volume within the chamber returned to its pre-1997 eruption state by June 2008. Figure 41.8 shows a

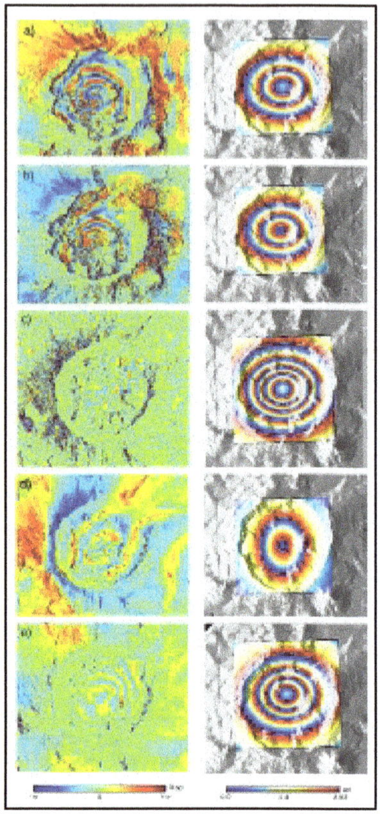

Figure 41.8. Multi-year interferograms highlighting deformation within the Okmok caldera. Color-scale for images shown in column one and two highlight phase wrapping intervals which span ±π and 0 – 2.83 cm for the actual and modeled interferograms, (a) 1997-1998 (b) 1998-1999 (c) 2003-2004 (d) 2004-2005 (e) 2007-2008 23].

collection of selected interferograms, where the first and second columns are the actual and modeled images.

Knowledge of temporal and spatial distribution of magma accumulation, transport, and emplacement can help locate magma reservoirs and the formation of eruptive vents, pave the path of understanding volcanic processes, development of sophisticated and comprehensive models for magma intrusion, recharging, and volcano spreading, and competent forecasting of future volcanic eruptions, flank instabilities and sector collapses. GPS is one of the most suitable techniques to measure ground surface deformations because of high accuracy and provision of three dimensional components of deformation field. But the limitation of GPS station density (GPS stations per unit area) and the continuous coverage of interferogram make the integration of InSAR and GPS a useful approach to map highly accurate deformations (*i.e.*, at sub-centimeter levels) with unprecedented spatial coverage. Interferogram alone, or in combination with other data sets of such as GPS, geodimeter, and micro-

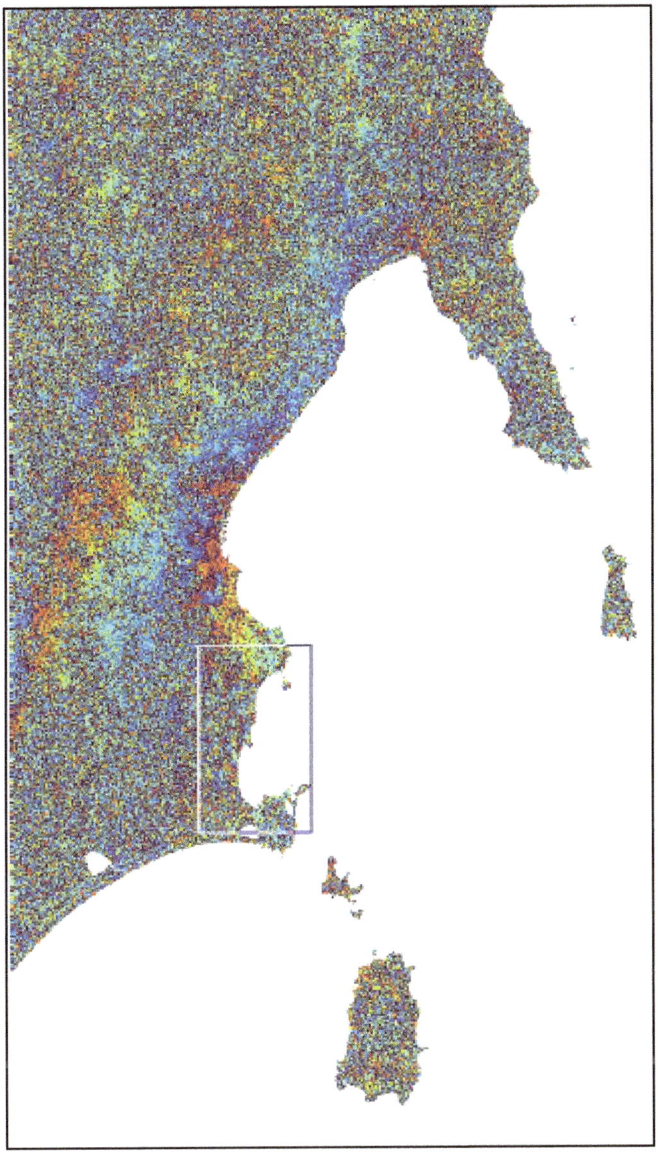

Figure 41.9: Best Interferogram Obtained from the Phase Difference of Two SAR Images (1993-02-14 and 1996-04-03). Three cycles of phase (fringes) are clearly visible, corresponding to about 84 mm of total displacement along the sensor line of sight. Inset selection puts in evidence the deflating area and the modeled one [26].

gravity data are now widely used for pre-, during-, and post-eruption observation of volcanoes (Figure 41.9). Examples include the measurement of deflation induced by the activation of Etna volcano from May 17, 1992 to October 24, 1993 by Massonnet *et al.*, [24], the measurement of the deformation field associated to the 1997 eruption of Okmok volcano in Alaska [25] and the Campi Flegrei caldera activation [26].

Flalko *et al.*, [27] used the interferogram data to constrain the dynamics and morphology of the deformation source responsible for the 1997-1998 inflation crises in the Long valley caldera (California) (Figure 41.10). Joint inversion of the interferogram and the two-color geodimeter data for the period between 1996 and 1998 suggests that the deformation source has the shape of a steeply dipping prolate spheroid with a depth of 7 km, major and minor axes of 4.2 and 1.8 km, respectively, and an excess pressure of several megapascals. Using InSAR, deformation field of Nisyros volcano (Greece) was studied by several authors [28]. The seismic unrest without eruption started at the end of 1995, peaked in August 1997, and ended at the end of 1998. For these studies, InSAR provides the only tool to assess the deformation fields of the volcano before 1997 when the first GPS stations were set up.

5. Applications of InSAR in Land Subsidence and Landslides

5.1 Land Subsidence and Uplift

Land subsidence, lowering of the land surface by a variety of subsurface displacement processes, affects an aggregate area of more than 44,000 km^2 in 45 states in the U.S. [29]. Common causes of land subsidence include sediment consolidation due to its own weight and tectonic movements (geological or natural subsidence), withdrawal of groundwater and geothermal fluid, oil, and gas extraction from underground reservoirs; dissolution of limestone aquifers (sinkholes); collapse of underground mines; drainage of organic soils; and initial wetting of dry soils (hydro compaction). Measuring the spatial and temporal changes in the subsidence pattern using interferogram can provide constraints on the permeability and compressibility properties of the compacting formulations, which are important parameters for predicting future subsidence. Measuring the spatial and temporal changes in the subsidence pattern using interferogram can provide constraints on the permeability and compressibility properties of the compacting formations, which are important parameters for predicting future subsidence.

Land subsidence can be caused by groundwater pumping, which results in compaction of aquitards during the slow process of aquitards drainage. Subsidence detected from interferogram can be used for parameter estimates in simulations of aquifer system compaction [30]. Subsidence from excess groundwater withdrawal has impacted costly on both the operation and design of canals in California [31]. The spatial scale of the land subsidence can be well monitored using InSAR. For instance, during 1950-1970, the city of Venice (Italy) was affected by serious land subsidence with highest rates due to groundwater withdrawals [32]. Since 1970, drastic measures to reduce both industrial and other artesian extractions were taken to stop the subsidence due to groundwater withdrawal. But subsidence due to consolidation processes may still be continuing.

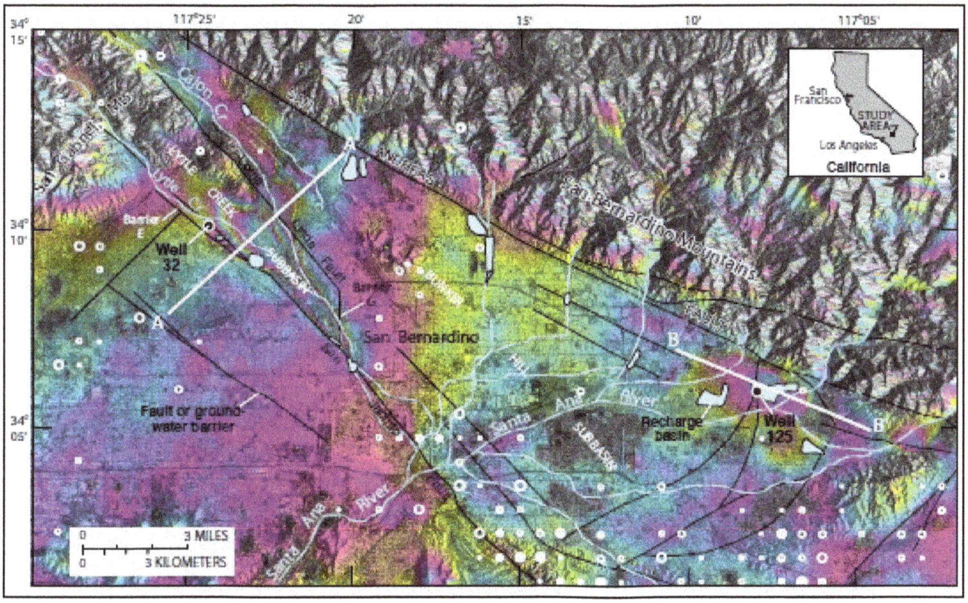

Figure 41.10: SAR Interferogram of Land-Surface Deformation over the San Bernandino Area, California, during the Period from December 25, 1992 to August 27, 1993. Uplifts of about 7 cm and 3 cm are indicated by concentric colour patterns in the middle of sections A-A' and B-B' respectively. Earthquake epicenters (white circles) were obtained from the South California Earthquake Information Center [33].

In semi-arid and arid regions, groundwater is the main water resource for civilian use and agricultural irrigation. Excessive withdrawal of groundwater can cause large scale subsidence. InSAR application will continue to be very useful in monitoring the land subsidence due to excessive withdrawal in the next decades. Conversely, during a wet season, groundwater recharge makes the ground-water levels to rise. Originally, unsaturated sediment shifts from the granular skeleton to the pressurized pore fluid, causing the skeleton to expand and the surface to uplift. From interferogram analysis, Lu and Danskin [33] successfully detected an uplift of several centimeters in the Bernardino basin of southern California due to ground-water recharge using interferogram analysis of ERS-1 and ERS-2 images.

Using InSAR, Bawden *et al.*, [34] showed that seasonal fluctuations of 50 mm of basin uplift due to groundwater recharge during the wet winter season, and 60 mm of subsidence due to groundwater withdrawal during the dry summer season in Santa Ana basin, California, with the greatest fluctuations near the city of Santa Ana and at the northwestern end of the basin (Figure 41.11).

Warm or hot groundwater in the areas of hot springs or volcanic zones is a useful energy resource. Hot water is pumped for its geothermal energy. Like groundwater exhaust, however, excess pumping of the ground hot water at a speed higher than recharging rate will cause ground subsidence both vertically and horizontally, in severe cases can damage property and environment. Land subsidence

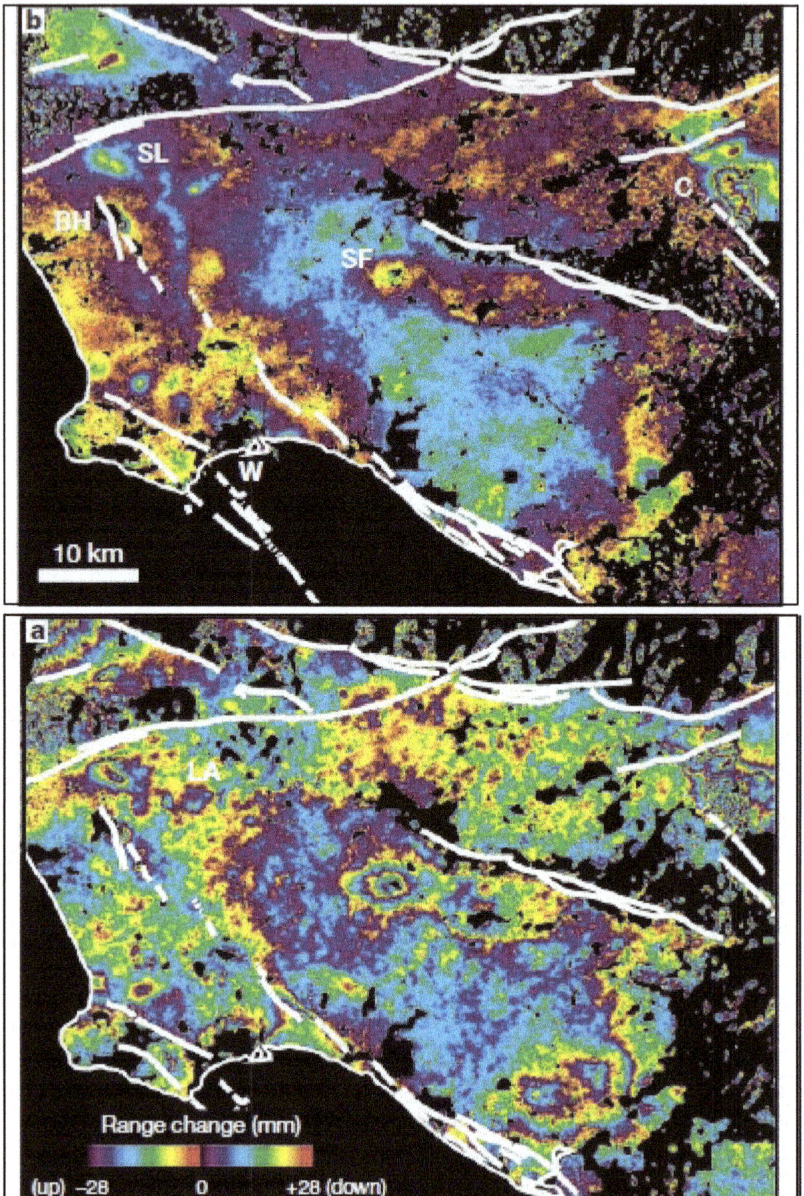

Figure 41.11: Long-term Differential Interferograms for Metropolitan Los Angeles (a) Five year interferogram between October 1993 an October 1998 shows over 60 mm of subsidence at the southeastern end of the Santa Ana basin during this time period (fringes in the upper left-hand corner are from the uplift associated with the 1994 Northridge earthquake), LA, downtown Los Angeles (b) Two-year interferogram between October 1997 and October 1998 showing ~25 mm of subsidence within the basin, BH: Baldwin Hills, C: Chino, LA: Los Angeles, SF: Santa Fe Springs, SL: Salt Lake, and W: Wilmington oil fields. Both (a) and (b) show significant deformation near these labeled sites [34].

is induced by volume changes or reduction of subsurface pore pressure in the geothermal reservoir by depletion of fluid storage as well as thermal contraction. Monitoring the subsidence in the geothermal field using precise level network or interferogram is important so that measure can be taken to prevent damage to infrastructure and environment. The objective of mapping the subsidence field of the geothermal fields using InSAR is four-fold:

1. To assess the environmental and structural damage due to excess extraction of geothermal fluids.
2. To derive important parameters such as permeability and compressibility properties of the compacting formations, the scale and rate of subsidence so that the future subsidence rate and scale can be predicted.
3. To identify areas of infrastructure potentially at risk of structural damage and to reduce the impact of future development of the geothermal fields, and
4. To develop future strategy including control of extraction rate and reinjection of fluids back to the geothermal fields to minimize environmental impact.

5.2 Landslides

InSAR appears attractive for landslide hazard investigations and possibly for preliminary warning. However, the steep and rough topography typical of landslide-prone areas, relative large rates of movement causing phase ambiguity problems and signal decorrelation, together with the fact that local atmospheric variations can be particularly pronounced in regions with strong topographic relief, leading to strong atmospheric phase artifacts, often hamper the interferometric pre-processing, making it difficult to estimate displacements. Thus, in many cases, the spaceborne interferometric monitoring of mass movements was suggested to be used together with ground-based SAR interferometer systems [35]. Successful applications of InSAR technique to landslide monitoring and slope instability studies include [36]. On an average, the slope angles computed via interferometric methods are generally in better agreement with landslide locations than DEM slope angles. This indicates the usefulness of InSAR to help categorize the landslide hazard levels of slope.

6. Applications of InSAR in Glaciology

Glaciers and ice sheets are sensitive to internal instabilities or climate fluctuations. Continuous retreat of glaciers and disintegration of ice shelves cause concerns of rise of sea level and severe weather due to changes of climate patterns [37]. To understand well the dynamics of glaciers and physical mechanism of motion of ice streams and ice sheets, accurate measurements of displacement and subsequent velocity field of glaciers and ice shelves are important because they provide a better knowledge of the rheological parameters that control the flow of glaciers and ice shelf. Through measurement of displacement and velocity field of glaciers, InSAR has been used successfully to monitor glacier surges, ice sheet motion, ground lines of marine ice sheets, ice sheet topography, glacier ice-flow velocity field and motion patterns and flux, patterns of wind-drifted snow and snow accumulation, uplifting

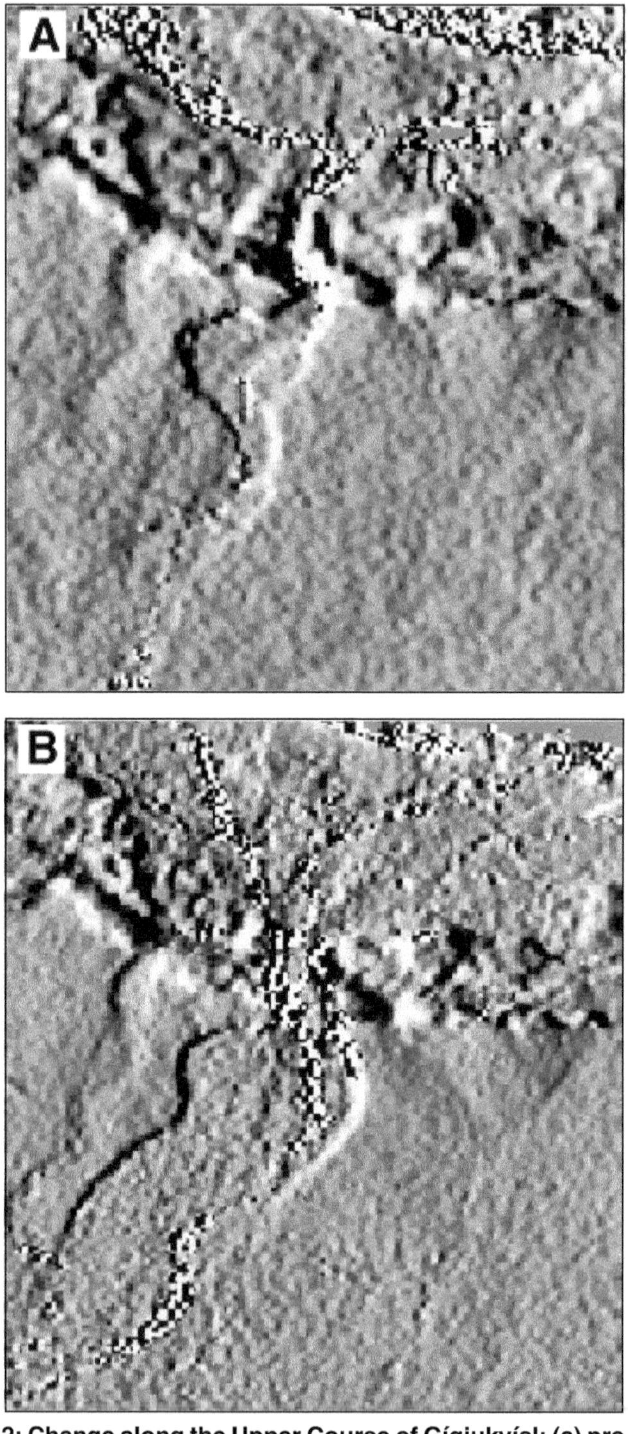

Figure 41.12: Change along the Upper Course of Gígjukvísl: (a) pre-jökulhlaup interferometric DEM, (b) post-jökulhlaup interferometric DEM [39].

and infilling of ice cauldrons, the effects of the subglacial floodwater on glacier flow during draining episodes, and calibration of numerical modeling of ice shelf and dynamics.

InSAR has also been used to infer the water level or volume change of supraglacial lakes through DEM construction and then comparison of height change [38-39]. Using repeat-pass satellite synthetic aperture radar interferometry, a methodology is developed by Smith *et al.*, [39] to measure flood-induced erosion and deposition (Figure 41.12) and applied it to a record 1996 glacier outburst flood (jökulhlaup) on Skeiõarársandur, Iceland.

Non-polar and alpine-style glaciers are recognized as being particularly sensitive to the current trend of climatic warming. Mountain glaciers are usually associated with small sizes but larger deformation rates than large ice caps, ice fields, and the Greenland and Antarctic ice sheets. However, due to the displacement gradient *(i.e.,* strain) threshold of InSAR, only a limited number of studies have successfully measured the ice-flow velocity field with InSAR. For this reason, monitoring icefalls or marginal shear zones of glaciers and mountain glaciers using InSAR requires short periods of SAR images acquisitions so that correlation between the two images are preserved. Offset fields between pairs of JERS-1 satellite SAR data acquired in winter with 44 days time interval were employed for the estimation of Arctic glacier motion over Svalbard, Novaya Zemlya and Franz-Josef Land (Figure 41.13) [40].

7. Application of InSAR in Hydrology

7.1 Soil Moisture Monitoring (Figure 41.14)

Estimates of erosion, deposition, soil moisture, water level change, and net volumetric change of discharges may be achieved by using InSAR. The dependence

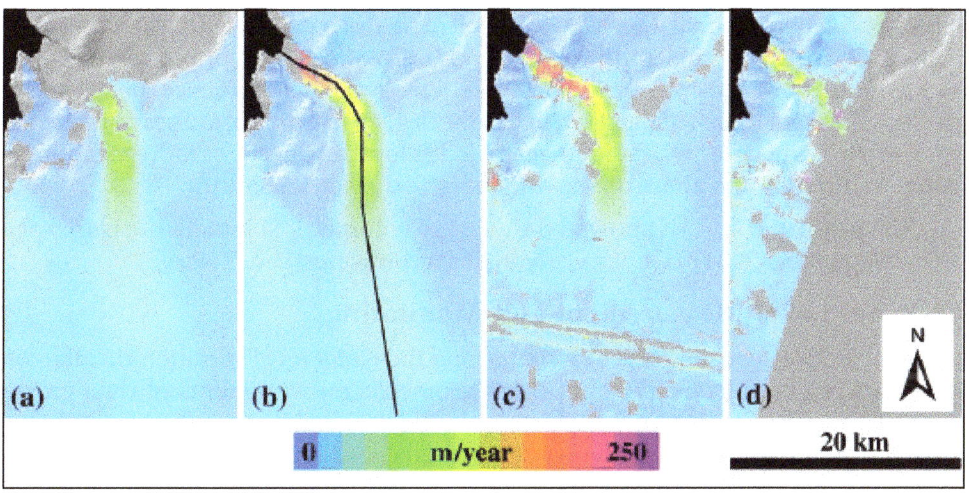

Figure 41.13: Comparison between Ice-Surface Velocities of Duvebreen (Austfonna, Svalbard) from ERS SAR interferometry in (a) Jaunary 1994, (b) January 1996, (c) JERS-1 offset tracking in January 1998 and, (d) April 1994. The approximate main flow of Duvebreen is shown in (b) [40].

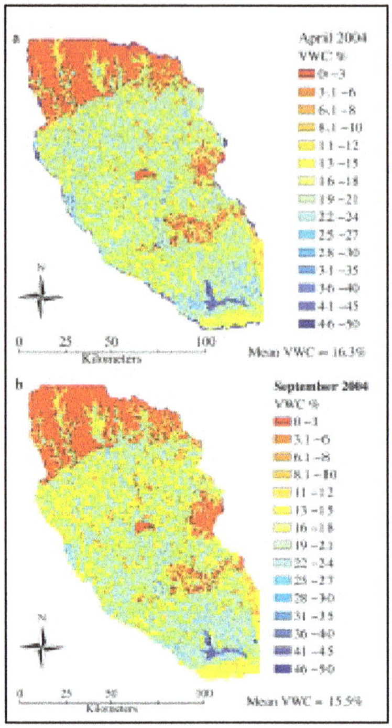

Figure 41.14: (a) Soil moisture map derived from RADARSAT-1 SAR data acquired in April 2004 and the supplemental data. The average volumetric soil moisture is 16.3 per cent, (b) Soil moisture map derived from RADARSAR-1 SAR data acquired in September 2004 and the supplemental data. The average volumetric soil moisture is 15.5 per cent [41].

of the backscattered signal intensity to the soil moisture has been widely documented in the literature. Soil dielectric constant not only affects the amplitude of radar backscattering but also the phase of backscattered radar echoes. Since the dielectric constant of water is much higher than that of the dry soil matrix, soil moisture difference in the ground when the two radar images used to form an interferogram are acquired results in difference in phase due to difference in dielectric constant.

Makkeasorn *et al.*, [41] showed the average volumetric soil moisture is 15.5 per cent in September 2004 in Choke Canyon Reservoir watershed, Texas.

7.2 Water Level Measurement and Monitoring

In principle, temporal river discharge and the water level variations in lakes or reservoirs can be calculated from purely geometric measurements of river cross-sections, water levels, and surface slopes for both rivers and lakes or reservoirs. Thus, water level (stage) for lakes/reservoirs and major rivers can be retrieved from spaceborne remote sensing that have the capabilities of geometric measurements [42]. These instruments include imaging radars. Spaceborne water level measurement is greatly improved with current measurement [42], which was in turn demonstrated by using along-track InSAR (ATI) techniques [43]. For example, if there are persistently existing reflectors (*e.g.*, leafless trees) sticking out of a water surface, interferograms can be formed even over the water that can result in cm-scale estimates in water-level [43].

8. Applications of InSAR in Forestry (Figures 41.15 and 41.16)

In forestry, canopy height is often used to estimate forest biomass and above-ground forest carbon stock as the quantities are allometrically related. The phase of a specific pixel containing vegetation depends on vegetation structure, scattering mechanisms, and sensor characteristics (wavelength, polarization, looking angle, etc). Because the phase of a pixel is a sum of the returns from a collection of scatterers including stems, branches, twigs, leaves or needles, trunks, and surface soil, the type of scatterers interact most strongly with the radar wave depending on the wavelength, polarization, incidence angle, and vegetation subpixel fraction. Forest canopy height can be estimated using InSAR, including polarimetric interferometry (PollnSAR), because the interferometric phase relates to terrain height and vegetation canopy height [45].

The use of InSAR for estimating canopy height was demonstrated using C-band interferometric height discontinuities at forest edge by Hagberg *et al.*, [46] and using

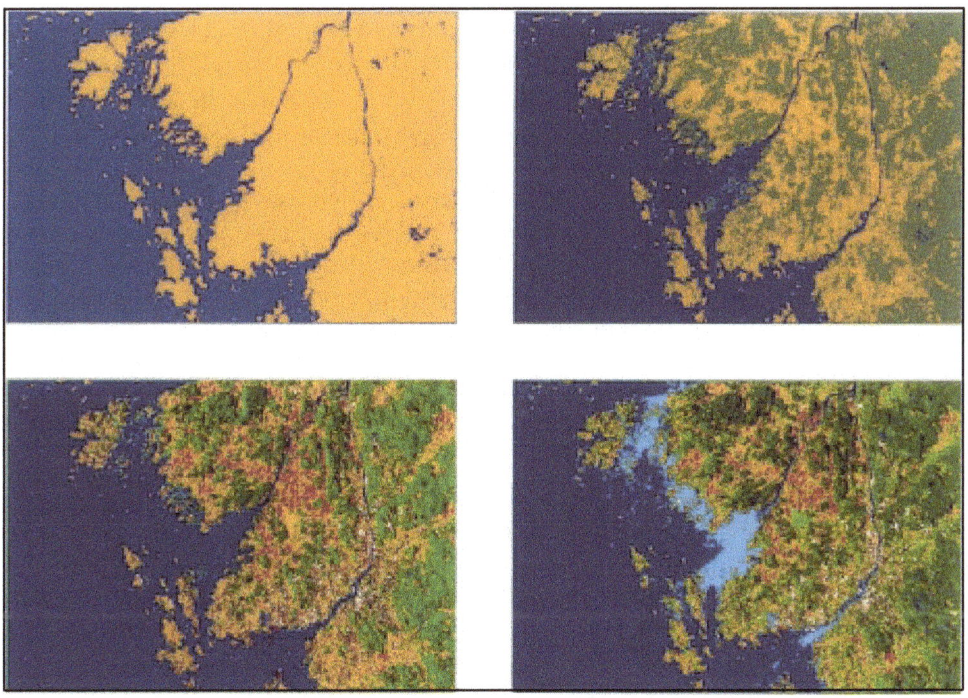

Figure 41.15: Four Clustering Results for Gothenburg, Number of Clusters are Two (top left), Nine (bottom left), and Ten (bottom right). Shades of blue: water, shades of green: forests, white: man-made structures (urban), orange: open areas (other), red: farmlands, and black: system effects. The light blue is the sea ice from the winter images [49].

effective interferometric C-band height from ERS-1 by Askne *et al.*, [47]. Besides the capability of estimating canopy height and subsequent carbon content, InSAR can also be used to improve biomass estimation from radar backscattering coefficient because the fringe frequencies of the interferogram can be used to correct radar backscattering coefficient for terrain effects and thus improve radiometric calibration [48]. Other applications of InSAR in forestry include classification of forest types and land cover [49].

Remote Sensing in Support of the Kyoto Protocol

The Kyoto protocol to the United Nations Framework Convention on Climate Change (UNFCCC) contains quantified and legally binding commitments to limit or reduce greenhouse gas emissions to 1990 levels. The protocol allows sinks associated with vegetation growth and expansion to be included to offset carbon emissions, which in turn raises debate about the inadequacy of existing methods for establishing reliable estimates of 1990 carbon stocks/sinks levels and for measuring and monitoring current and future carbon stocks/sinks. The role of remote sensing is therefore being considered given its potential capacity for systematic observations at scales ranging from local to global and for the provision of data archives extending back over several decades. These issues also bring forth the need for the exchange of information between remote sensing scientists and organizations *(e.g.,* national and international policy makers, government agencies and both legal and scientific bodies involved in the development and implementation of the protocol). [d].

Article 3.1 of the Kyoto protocol states that, "The Parties included in Annex-I shall, individually or jointly, ensure that their aggregate anthropogenic carbon dioxide equivalent emissions of the greenhouse gases listed (in Annex A) do not exceed their assigned amounts..." "... with a view to reducing their overall emissions of such gases by at least 5 per cent below 1990 levels in the five-year commitment period 2008 to 2012." Emissions of six greenhouse gases (measured in terms of carbon dioxide (CO_2) equivalents as a function of the gases global warming potential) are covered by the Kyoto protocol; CO_2, methane (CH_4), nitrous oxide (N_2O), hydrofluorocarbons (HFCs), perfluorocarbons (PFCs) and Sulphur Hexafluoride (SF_6). Of these gases, remote sensing is (at present) best equipped to quantify changes in CO_2 and, to a certain extent, CH_4, particularly in relation to land use, land use change and forestry (LULUCF) activities.

A major benefit of using remote sensing data to support calculations of emissions is that systematic observation systems are provided and historical archives of data exist, which can be augmented through current and future data acquisitions. As such remote sensing can (to some extent) meet the requirements of Article 10, which highlights the importance of developing such systems and archives for monitoring and assessing status and trends in the global terrestrial carbon budget.

Remote sensing can also play a role in the establishment of national carbon stock baseline datasets and assessment of change in stocks. Such datasets are implicit requirements under Article 3, which introduces the need to quantify afforestation, reforestation, and deforestation (ARD; Article 3.3), revegetation and land use management (under specified categories; Article 3.4) and to establish a baseline of carbon stocks for 1990 (Article 3.1; where 1990 is regarded as the starting point for accounting) against which to assess change.

Over the past half century, a range of airborne and space-borne sensors has acquired remote sensing data, with the number of sensors and their diversity of capability increasing over time. Today a large number of satellite sensors observe the earth at wavelengths ranging from visible to microwave, at spatial resolutions ranging from sub-meter to kilometers and temporal frequencies ranging from 30 min to weeks or months. In addition, archives of remotely sensed data are increasing and provide a unique, but not complete, chronology of the earth during this time period. New sensors are continually being launched and existing sensors are often replaced to ensure continuity in the data record. Given this enormous resource, remote sensing provides a huge potential to assist countries to meet their obligations under Kyoto protocol. Five major areas can be identified where remote sensing technology is applied to support implementation of the treaty:

1. Provision of systematic observations of relevant land cover (Articles 5 and 10)
2. Support to the establishment of a 1990 carbon stock baseline (Article 3)
3. Detection and spatial quantification of change in land cover (Articles 3 and 12)
4. Quantification of above-ground vegetation biomass stocks and associated changes therein (Articles 3 and 12)
5. Mapping and monitoring of certain sources of anthropogenic CH_4 (Articles 3, 5, and 10)

Supporting the requirements of the Kyoto protocol, remote sensing can play an important role in providing systematic observations of land cover, supporting establishment of a 1990 carbon stock baseline, detecting and quantifying rates of land cover change, quantifying above ground biomass stocks and changes in these, and mapping and monitoring certain sources of anthropogenic CH_4. The greenhouse gases considered relevant in the context of remote sensing are essentially CO_2 and CH_4.

Systematic observations of the world's forested regions are presently provided at a global level by coarse and medium resolution sensors. For detecting hotspots of change, data from such sensors are most useful due to their frequent temporal coverage, although for quantifying change at the local to regional level, data provided by finer resolution instruments, such as Landsat and SPOT sensors are required. Space-borne SAR is suitable for detection and quantification of land cover change, particularly in areas with persistent cloud cover.

For establishing a 1990 carbon stock baseline, historical Landsat and sensors as well as JERS-1 SAR are most useful. Within their range of biomass sensitivity (< 100 Mg ha^{-1}), SAR have the greatest potential for quantifying biomass (carbon stocks) at regional scales. At more local levels and for a restricted range of environments, airborne LIDAR, polarimetric SAR and VHF band radar have shown significant potential for

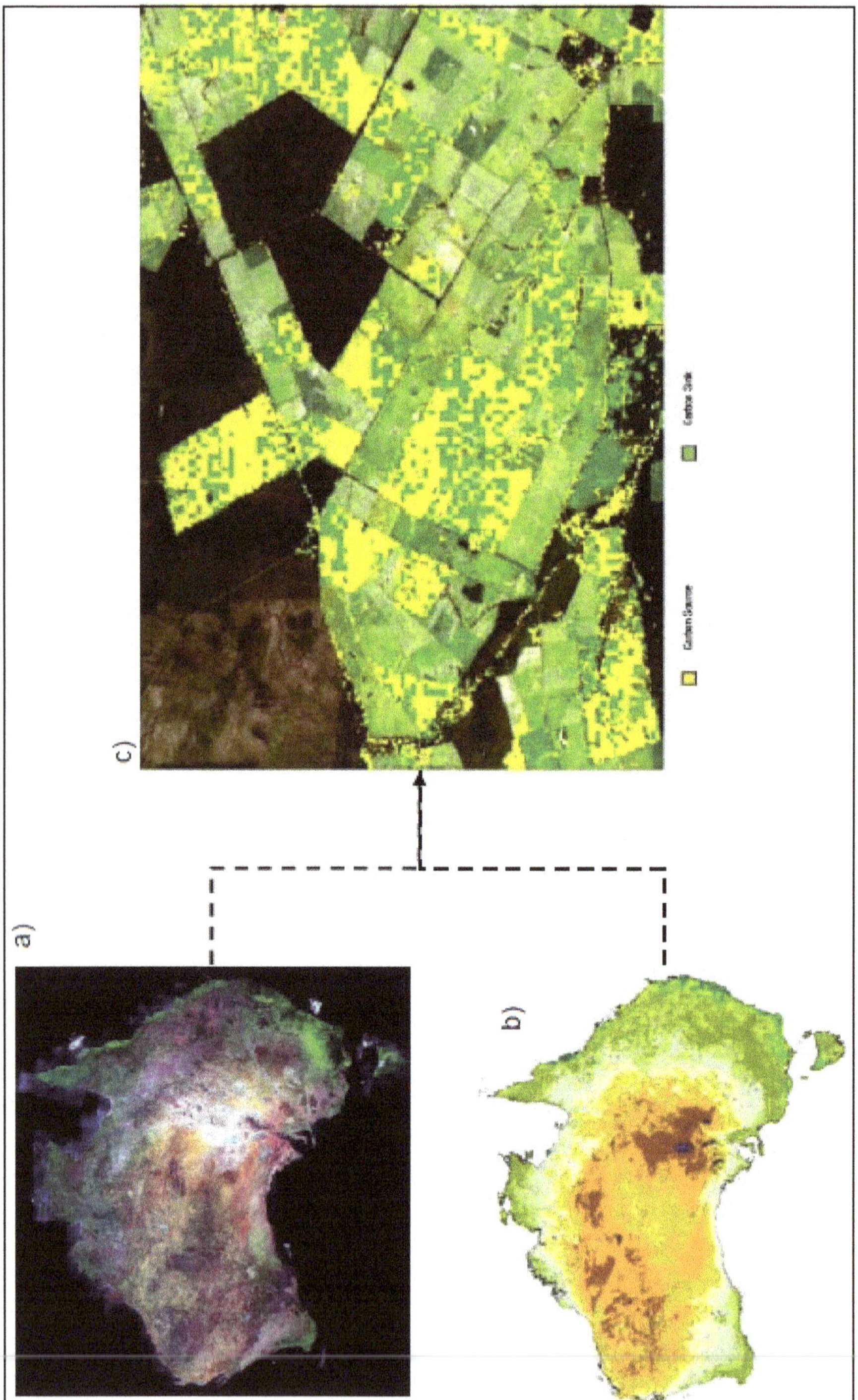

Figure 41.16: The Australia National Carbon Accounting System (NCAS) methods for estimating above ground biomass where (a) shows landsat imagery over Australia (b) shows an Net Primary Production (NPP) index used to quantify biomass, and (c) illustrates the partitioning of NPP at 30 m spatial resolution.

biomass estimation. For detecting anthropogenic sources of CH_4, both space-borne SAR and optical data are deemed important.

Summary

The above varied applications of remote sensing particularly InSAR show the breadth and depth of this technology. InSAR has proven itself to be a valuable tool for measuring earth surface deformation due to earthquakes, volcanic unrest, groundwater migration, and antropogenic activity. New techniques and applications of InSAR are being developing all the time, including stacking, time-series analysis of surface deformation, ScanSAR, polarimetric InSAR, and along-track interferometry. InSAR also provides valuable assistance to support the requirements of the Kyoto protocol.

References

1. C. Elachi and J. van Zyl, "*Introduction to the physics and techniques of remote sensing*," Wiley Interscience, Second Edition, 2006.
2. J. G. Liu and P. J. Mason, "*Essential image processing for GIS and remote sensing*," Wiley-Blackwell, 2009.
3. J. A. Richards, "*Remote sensing with imaging radar*," Springer, 2009.
4. D. Massonnet and K.L. Feigl, "Radar interferometry and its application to changes in the Earth's surface," *Rev. Geophys.*, vol. 36, pp. 441 – 500, 1998.
5. D. Massonnet, M. Rossi, C. Carmona, F. Adragna, G. Peltzer, K. Feigl, T. Rabaute, "The displacement field of the Landers earthquake mapped by radar interferometry," *Nature*, vol. 364, pp. 138 – 142, 1993.
6. G. Peltzer, K. Hudnut, and K. Feigl, "Analysis of coseismic surface displacement gradients using radar interferometry: New insights into the Landers earthquake," *J. Geophys. Res.*, vol. 99, pp. 21971 – 21981, 1994.
7. M. Tobita, S. Fujiwara, S. Ozawa, P. A. Rosen, E. J. Fielding, C. L. Werner, M. Murakami, H. Nakagawa, K. Nitta, and M. Murakami, "Deformation of the 1995 North Sakhalin earthquake detected by JERS-1 / SAR interferometry," *Earth Planets Space*, vol. 50, pp. 313 – 325, 1998.
8. D. A. Schmidt and R. Bûrgmann, "InSAR constraints on the source parameters of the 2001 Bhuj earthquake," *Geophys. Res. Lett.*, vol. 33, LO2315, 2006.
9. S. Stramondo, M. Moro, C. Tolomei, F. R. Cinti, and F. Doumaz, "InSAR surface displacement field and fault modeling for the 2003 Bam earthquake (Southeastern Iran)," *J. Geodynamics*, vol. 40, pp. 347 – 353, 2005.
10. M. Chini, C. Bignami, S. Stramondo, and N. Pierdicca, "Uplift and subsidence due to the 26 December 2004 Indonesian earthquake detected by SAR data," Int. *J. Remote Sens.*, vol. 29, pp. 3891 – 3910, 2008.
11. Y. B. Tsai, J. Y. Liu, K. F. Ma, H. Y. Yen, K. S. Chen, Y. I. Chen, and C. P. Lee, "Precursory phenomena associated with the 1999 Chi-Chi earthquake in Taiwan as identified under iSTEP Program," *Phys. Chem. Earth*, vol. 31, pp. 365 – 377, 2006.

12. S. Kuzuoka and T. Mizuno, "Land deformation monitoring using PSInSAR Technique," In *Proceedings of International Symposium on Monitoring, Prediction, and Mitigation of Disasters by Satellite Remote Sensing*, Awaji, Hyogo, Japan, pp. 176 – 181, January 2004.

13. M. de Michele, D. Raucoulesa, J. Salichan, A. Lemoine, and H. Aochia, "Using InSAR for seismotectonic observations over the Mw 6.3 Parkfield earthquake (28/09/2004), California," In *Proceedings of Commission IV ISPRS Congress*, Beijing, China, July 2008.

14. E. J. Fielding, P. R. Lundgren, R. Bûrgmann, and G. J. Funning, "Shallow fault-zone dilatancy recovery after the 2003 Bam, Iran earthquake," *Nature*, vol. 458, pp. 64-68, 2009.

15. R. Mellors, T. Bunyapanasarn, and B. Panahi, "InSAR analysis of the Absheron Peninsula and nearby areas, Azerbaijan," in Mud Volcanoes, Geodynamics, and Seismicity, G. Martinelli and B. Panahi, Eds., NATO Science Series, Springer, pp. 201 – 209, 2005.

16. K. Mogi, "Relations between the eruptions of various volcanoes and the deformation of the ground surfaces around them," *Bull. Earthquake Res. Institute*, Univ. Tokyo, vol. 36, pp. 99 – 134, 1958.

17. Y. Okada, "Surface deformation due to shear and tensile faults in a half-space," *Bull. Seismol. Soc. Am.*, vol. 75, pp. 1135 – 1154, 1985.

18. Y. Okada and E. Yamamoto, "Dyke intrusion model for the 1989 seismovolcanic activity off Ito, Central Japan," *J. Geophys. Res.*, vol. 96, pp. 10361 – 10376, 1991.

19. T. Árnadóttir, F. Sigmundsson, and P. T. Delaney, "Sources of crustal deformation associated with the Krafta, Iceland, eruption of September 1984," *Geophys. Res. Lett.*, vol. 25, pp. 1043 – 1046, 1998.

20. Z. Lu, T. Masterlark, D. Dzurisin, R. Rykhus, and C. Wicks Jr., "Magma system dynamics at Westdahl volcano, Alaska modeled from satellite radar interferometry," *J. Geophys. Res.*, vol. 108, 2003.

21. F. Amelung, S. Jonsson, H. Zebker, and P. Segall, "Widespread uplift and 'trapdoor' faulting on Galapagos volcanoes observed with radar interferometry," *Nature*, vol. 407, pp. 993 – 996, 2000.

22. R. Pederson and F. Sigmundsson, "Temporal development of the 1999 intrusive episode in the Eyja Fjallajökull volcano, Iceland, derived from InSAR images," *Bull. Volcanol*, vol. 68, pp. 377 – 393, 2006.

23. W. N. Junek, "Temporal analysis of the magma supply system beneath the Okmok caldera by Interferometric Synthetic Aperture Radar and Statistical Modeling," *IEEE International Symposium on Geoscience and Remote Sensing*, 2010.

24. D. Massonnet, P. Briole, and P. Arnaud, "Deflation of Mount Etna monitored by spaceborne radar interferometry," *Nature*, vol. 375, pp. 567 – 570, 1995.

25. Z. Lu, D. Mann, and J. Freymueller, "Satellite radar interferometry measures deformation at Okmok volcano," *EOS*, vol. 79, pp. 461 – 468, 1998.

26. A. Avallone, A. Zollo, P. Briole, C. Delacourt, and F. Beauducel, "Subsidence of Campi Flegrei (Italy) detected by SAR interferometry," *Geophys. Res. Lett.*, vol. 26, pp. 2303 – 2306, 1999.

27. Y. Fialko, M. Simons and Y. Khazan, "Finite source modeling of magmatic unrest in Socorro, New Mexico, and Long Valley, California," *Geophys. J. Int.*, vol. 146, pp. 191 – 200, 2001.

28. I. Parcharidis and E. Lagios, "Deformation in Nisyros volcano (Greece) using differential radar interferometry," *Bull. Geol. Soc. Greece*, vol. 34, pp. 1587 – 1594, 2001.

29. National Research Council, *"Mitigating losses from land subsidence in the United States,"* Natl. Acad. Press., Washington, D.C., 1991.

30. D. L. Galloway, K. W. Hudnut, S. E. Ingebritsen, S. P. Phillips, G. Peltzer, F. Rogez, and P. A. Rosen, "Detection of aquifer system compaction and land subsidence using interferometric synthetic aperture radar, Antelope Valley, Mojave Desert, California," *Water Resour. Res.*, vol. 34, pp. 2573 – 2585, 1998.

31. N. P. Prokopovich and M. J. Marriott, "Cost of subsidence to the Central Valley Project, California," *Assoc. Eng. Geol. Bull.*, vol. 20, pp. 3325 – 3332, 2004.

32. L. Carbognina, P. Teatini, and L. Tosi, "Eustacy and land subsidence in the Venice lagoon at the beginning of the new millennium," *J. Marine Syst.*, vol. 51, pp. 345 – 353, 2004.

33. Z. Lu and W. R. Danskin, "InSAR analysis of natural recharge to define structure of a ground-water basin, San Bernardino, California," *Geophys. Res. Lett.*, vol. 28, pp. 2661 – 2664, 2001.

34. G. Bawden, W. Thatcher, R. S. Stein, K. W. Hudnut, and G. Peltzer, "Tectonic contraction across Los Angeles after removal of groundwater pumping effects," *Nature*, vol. 412, pp. 812 – 8151, 2001.

35. D. Leva, G. Nico, D. Tarchi, J. Fortuny-Guasch, and A. J. Sieber, "Temporal analysis of a landslide by means of a ground-based SAR interferometer," *IEEE Geosci. Remote Sens.*, vol. 41, pp. 745 – 752, 2003.

36. P. Bernardino, M. Costantini, G. Franceschetti, A. Iodice, L. Pietranera, and V. Rizzo, "Use of differential SAR interferometry in monitoring and modeling large slope instability at Maratea (Basilicata, Italy), *Eng. Geol.*, vol. 68, pp. 31 – 51, 2003.

37. R. S. Nerem, É. Leuliette, and A. Cazenave, "Present-day sea-level change: a review," *C. R. Geoscience*, vol. 338, pp. 1077 – 1083, 2006.

38. M. Furuya and J. M. Wahr, "Water level changes at an ice-dammed lake in west Greenland inferred from InSAR data," *Geophys. Res. Lett.*, vol. 32, L14501, 2005.

39. L. C. Smith, D. E. Alsdorf, F. J. Magilligan, B. Gomez, L. A. K. Mertes, N. D. Smith, and J. B. Garvin, "Estimates of erosion, deposition, and net volumetric change caused by the 1996 Skeidarársandur jökulhlaup, Iceland, from synthetic aperture radar interferometry," *Water Resour. Res.*, vol. 36, pp. 1583 – 1594, 2000.

40. T. Strozzi, A. Kouraev, A. Wiesmann, U. Wegmûller, A. Sharov, and C. Werner, "Estimation of Arctic glacier motion with satellite L-band SAR data," *Remote Sensing of Environment*, vol. 112, pp. 636 – 645, 2008.

41. A. Makkeasorn, N. B. Chang, M. Beaman, C. Wyatt, and C. Slater, "Soil moisture prediction in a semi-arid reservoir watershed using RADARSAT satellite image and genetic programming," *Water Resour. Res.*, vol. 42, pp. 1 – 15, 2006.

42. D. M. Bjerklie, S. L. Dingman, C. J. Vorosmarty, C. H. Bolster, and R. G. Congalton, "Evaluating the potential for measuring river discharge from space," *J. Hydrol.*, vol. 278, pp. 17 – 38, 2003.

43. D. M. Bjerklie, D. Moller, L. C. Smith, and S. L. Dingman, "Estimating discharge in rivers using remotely sensed hydraulic information," *J. Hydrol.*, vol. 309, pp. 191 – 209, 2005.

44. R. Romeiser, H. Runge, S. Suchandt, J. Sprenger, H. Weilbeer, A. Sohrmann, and D. Stammer, "Current measurements in rivers by spaceborne along-track InSAR," IEEE Trans. Geosci. *Remote Sens.*, vol. 45, pp. 4019 – 4031, 2007.

45. J. S. Lee, S. R. Cloude, K. P. Papathanassiou, M. R. Grunes, and I. H. Woodhouse, "Speckle filtering and coherence estimation of polarimetric SAR interferometry data for forest applications," *IEEE Trans. Geosci. Remote Sens.*, vol. 41, pp. 2254 – 2263, 2003.

46. J. O. Hagberg, L. M. H. Ulander, and J. Askne, "Repeat-pass SAR interferometry over forested terrain," IEEE Trans. Geosci. *Remote Sens.*, vol. 33, pp. 331 – 340, 1995.

47. J. I. H. Askne, P. B. G. Dammert, L. M. H. Ulander, and G. C. Smith, "C band repeat-pass interferometric SAR observations of the forest," *IEEE Trans. Geosci. Remote Sens.*, vol. 35, pp. 25 – 35, 1997.

48. L. M. H. Ulander, "Radiometric slope correction of synthetic aperture radar images," IEEE Trans. Geosci. *Remote Sens.*, vol. 34, pp. 1115 – 1122, 1996.

49. T. Castel, J. M. Martinez, A. Beaudoin, U. Wegmûller, and T. Strozzi, " ERS INSAR Data for remote sensing hilly forested areas," *Remote Sens. Environ.*, vol. 73, pp. 73 – 86, 2000.

Chapter 42

Maximizing of Parking Space in Multistoried Complex with Reference to Environmental and Seismic Consideration

Priyanka H. Jagadale

SET College of Engineering, Satra

ABSTRACT

Open first storey is a typical feature in the modern multistory constructions in urban India. Such features are highly undesirable in buildings built in seismically active areas; this has been verified in numerous experiences of strong shaking during the past earthquakes. Though multistoried buildings with open (soft) ground floor are inherently vulnerable to collapse due to earthquake load, their construction is still widespread in the developing nations like India. Social and functional need to provide car parking space at ground level far out-weighs the warning against such buildings from engineering community. This paper highlights the importance of explicitly recognizing the presence of the open storey in the analysis of the building. The error involved in modeling such buildings as complete bare frames, neglecting the presence of infill's in the storey's, is brought out through the study of an example building with different analytical models. This paper argues for immediate measures to prevent the indiscriminate use of soft storeys in buildings, which are designed without regard to the increased displacement, ductility and force demands in the storey columns.

Keywards: *Concrete, Earthquake, Multistory, Soft story, Space.*

Introduction

Many urban multistory buildings in India today have open first storey as an unavoidable feature. This is primarily being adopted to accommodate parking or reception lobbies in the first storeys. Due to parking in the ground floor all space occupied by vehicles so there is no space for playing ground for children.

Some national codes like the Indian Seismic Code (IS 1893) requires members of the soft story (story stiffness less than 70 per cent of that in the story above or less than 80 per cent of the average lateral stiffness of the three stories above) to be designed for 2.5 times the seismic story shears and moments, obtained without considering the effects of MI in any story. The factor of 2.5 is specified for all the buildings with soft stories irrespective of the extent of irregularities; and the method is quite empirical and may be too conservative and thus have further scope of improvement. The upper storey's have brick infilled wall panels. As per ASCE7-05 [5], a stiffness-related soft story irregularity is defined to exist when there is a story in which the lateral stiffness is less than 70 per cent of that in the story above or less than 80 per cent of the average stiffness of the three stories above, whereas a stiffness-extreme soft story irregularity is defined when target reference values of the conditions mentioned above are respectively 60 per cent and 70 per cent. For design purposes, there are some restrictions on the permitted analytical procedures to use, as well as for some seismic design categories.

To safeguard the soft first storey from damage and collapse, clause 7.10 of IS: 1893-2002 (Part 1) provides two alternative design approaches:

(*i*) The dynamic analysis of the building is to be carried out which should include the strength and stiffness effects of infill's as well as the *inelastic deformations* under the design earthquake force disregarding the Reduction Factor R.

(*ii*) The building is analyzed as a bare frame neglecting the effect of infill's and, the dynamic forces so determined in columns and beams of the soft (stilt) storey are to be designed for 2.5 times the storey shears and moments: OR the shear walls are introduced in the stilt storey in both directions of the building which should be designed for 1.5 times the calculated storey shear forces.

These past researches demonstrate the poor performance of buildings with soft ground story under seismic loading and advocates against construction of such buildings. Despite such poor performance, construction of multistoried buildings with soft ground story is being continued. It appears that the practical need of an open space to provide car parking facility far overweighs the warnings issued by the engineering community and provision of such an open space seems to be unavoidable. Under such circumstances, attention should be directed to device some guideline or methodology readily adoptable by practicing designers which shall minimize the danger to some extent. In this paper seismic performance of soft storey in multistoried building analyses and maximization of parking space in multistoried complex with reference to environmental and seismic consideration are studied.

Building Under Study (Table 42.1)

Table 42.1: Analysis Data for a Residential Building

Zone	IV
Response reduction factor	5
Importance factor	1
Soil condition	Medium
Plan size	21m x 15m
Height of building	32.5m
Depth of foundation	1.5m
Height of parapet	1m
External wall	230mm
Internal wall	230mm
Thickness of shear wall	150mm
Brick masonry	20KN/m^3
R.C. material	25KN/m^3
Depth of slab	120mm
Floor to floor height	3.1m
Size of column	C1-500 x 500
	C2-300 x 600
Size of beam	300mm x 450mm
Grade of steel	Fe 280
Grade of concrete	M 20
Floor finish	1.875 KN/m^2
Live load	3KN/m^2

Soft-first Storey

Open ground storey (stilt floor) used in most severely damaged or, collapsed R.C. buildings, introduced 'severe irregularity of sudden change of stiffness' between the ground storey and upper storey's since they had infilled brick walls which increase the lateral stiffness of the frame by a factor of three to four times. Such a building is called a building with *'soft' ground storey*, in which the dynamic ductility demand during the probable earthquake gets concentrated in the soft storey and The Building Sank evenly about 1 m due to soil liquefaction. The displaced soil caused a bulge in the road. The upper storeys tend to remain elastic. Hence whereas the 'soft' storey is severely strained causing its total collapse, much smaller damages occurs in the upper storey's, if at all. Behavior of soft first storey buildings (buildings on *stilts* or with *open plinth*) during earthquakes may be seen in Figures 42.1–42.7.

Recommendation

In view of the functional requirements of parking space under the buildings, more and more tall buildings are being constructed with stilts. To safeguard the soft

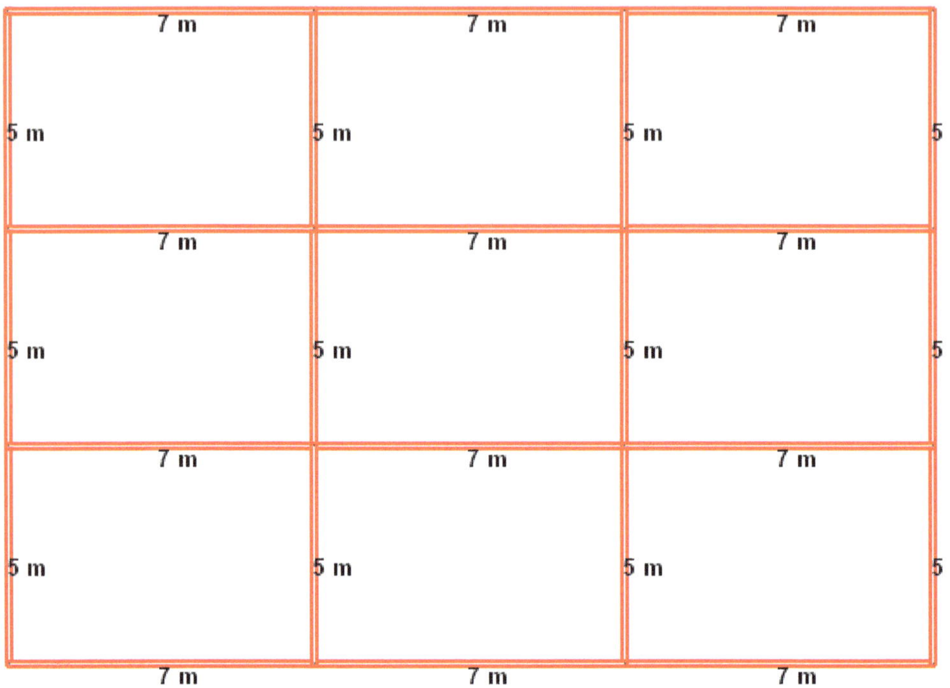

Figure 42.1: Plant at a Ground Floor of the Example Building Considered in the Study.

first storey from damage and collapse, clause 7.10 of IS: 1893-2002 (Part 1) provides two alternative design approaches

1. The dynamic analysis of the building is to be carried out which should include the strength and stiffness effects of infill's as well as the *inelastic deformations* under the design earthquake force disregarding the Reduction Factor R.
2. The building is analyzed as a bare frame neglecting the effect of infill's and, the dynamic forces so determined in columns and beams of the soft (stilt) storey are to be designed for 2.5 times the storey shears and moments: OR the shear walls are introduced in the stilt storey in both directions of the building which should be designed for 1.5 times the calculated storey shear forces.

Sometimes a soft storey is created somewhere at mid-height of the multi-storey building, for using the space as restaurant or gathering purposes, Such soft storey in building also collapsed in Kutch and Kobe earthquakes. For such a case also, the storey columns should be designed for the higher forces OR a few shear walls introduced to make up for the reduced stiffness of the storey.

Equivalent Static Analysis

The natural period of the building is calculated by the expression, T=0.09H/"D given in IS: 1893-2002, wherein *H* is the height and *D* is the base dimension of the

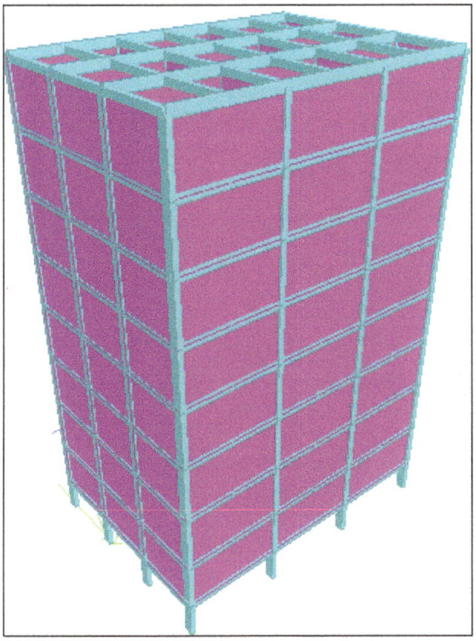

Figure 42.2: Elevation 1: Building with No Soft Storey.

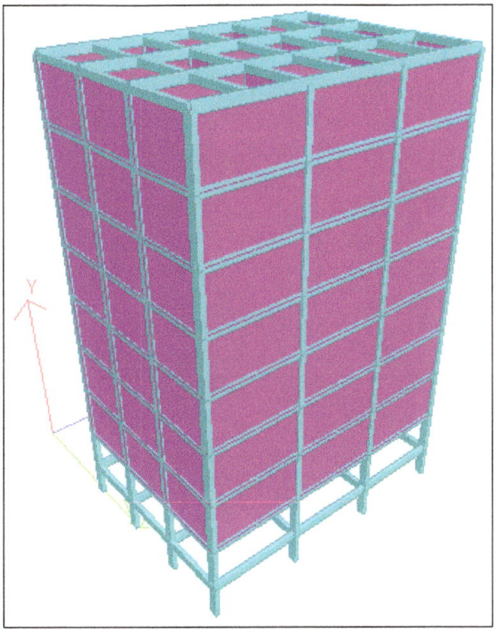

Figure 42.3: Elevation 2: Building with Ground Floor as Soft Storey.

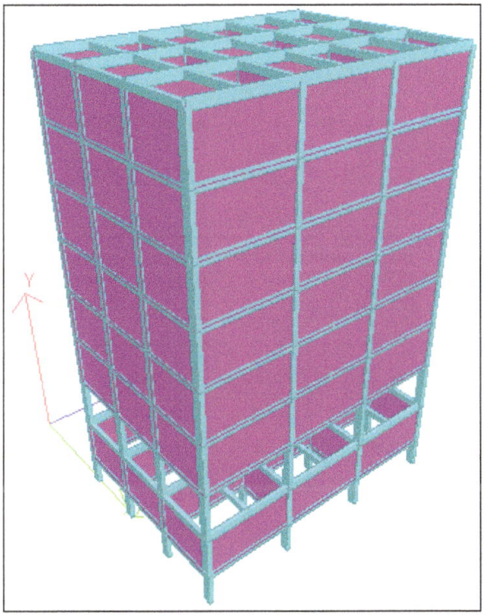

Figure 42.4: Elevation 3: Building with First Floor as Soft Storey.

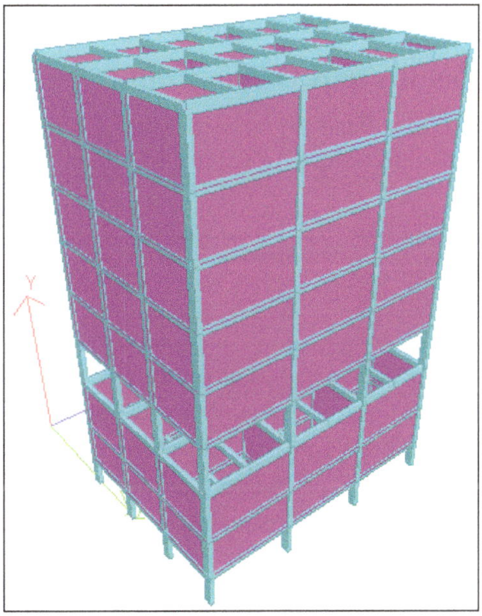

Figure 42.5: Elevation 4: Building with Second Floor as Soft Storey.

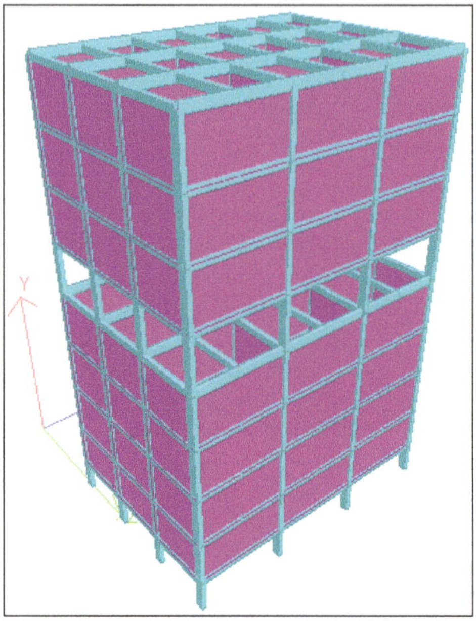

Figure 42.6: Elevation 5: Building with Fourth Floor as a Soft Storey.

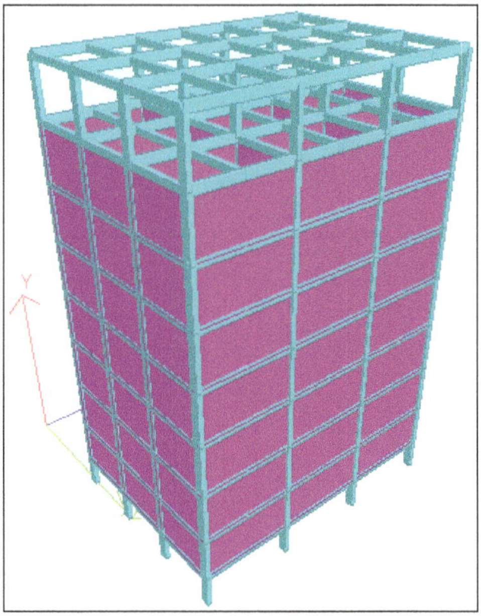

Figure 42.7: Elevation 6: Building with Top Floor as Soft Storey.

building in the considered direction of vibration. Thus, the natural periods for all the models in this method is the same. The lateral load calculation and its distribution along the height is done as per IS: 1893-1984.

Results and Discussion

The Base Shear and Mass Participation Factors for different elevations are summarized in the Table 42.2

Table 42.2

Sl.No.	*Elevation*	*Base Sear (in KN)*		*Mass Participaton Factors (In Per cent)*	
		X	*Z*	*X*	*Z*
1.	Building with No Soft Storey	1225.25	1706.98	99.998	99.960
2.	Building with Ground Floor as Soft Storey	555.38	857.34	100.0	98.632
3.	Building with First Floor as Soft Storey	570.86	861.17	99.998	97.602
4.	Building with Second Floor as Soft Storey	679.71	940.22	99.997	99.791
5.	Building with Fourth Floor as Soft Storey	782.63	1113.06	99.984	99.628
6.	Building with Top (7th) Floor as Soft Storey	1177.96	1668.79	99.843	96.950

Conclusion

RC frame buildings with open first storey's are known to perform poorly during in strong earthquake shaking. In this paper, the seismic vulnerability of buildings with soft first storey is shown through an example building. The drift and the strength demands in the first storey columns are very large for buildings with soft ground storey's. It is not very easy to provide such capacities in the columns of the first storey. Thus, it is clear that such buildings will exhibit poor performance during a strong shaking. This hazardous feature of Indian RC frame buildings needs to be recognized immediately and necessary measures taken to improve the performance of the buildings. The open first storey is an important functional requirement of almost all the urban multi-storey buildings, and hence, cannot be eliminated. Alternative measures need to be adopted for this specific situation. The under-lying principle of any solution to this problem is in (a) increasing the stiffness's of the first storey such that the first storey is at least *50 per cent* as stiff as the second storey, *i.e.*, soft first storey's are to be avoided, and (b) providing adequate lateral strength in the first storey. Due to parking in the ground floor all space occupied by vehicles so there is no space for playing ground for children due to this there is some maximum space gained due to soft storey at upper floors. Another important factor is pollution, due to soft storey on upper side then air pollution, soil pollution also avoided.

References

1. AIJ, 1995, "Preliminary Reconnaissance Report of the 1995 Hyogoken-Nanbu Earthquake, Architectural Institute of Japan, Tokyo, Japan.
2. EQEI, 1994, "The January 17, 1994 Northridge, California Earthquake–An EQE Report," *EQE International*, San Francisco, USA.
3. Hall,J.F. (Editor), 1994, "Northridge Earthquake January 17, 1994 – Preliminary Reconnaissance Report," Report No.94-01, Earthquake Engineering Research Institute, Oakland, USA.
4. IS: 13935-1993, *Repair and Strengthening of Buildings–Indian Standard Guidelines*, Bureau of Indian Standards, New Delhi, 1993.
5. Murty,C.V.R., and Jain,S.K. (1996), "Draft IS: 1893 Provisions on Seismic Design of Buildings," Bureau of Indian Standards, New Delhi.
6. Smith, B. S. and Coull. (1991). "*Infilled-Frame Structures, Tall Building Structures–Analysis and Design*", John Wiley and Sons, inc. 168-174.
7. IS 1893: Part 1 (2002); "Criteria for earthquake resistant design of structures", Bureau of Indian Standards, New Delhi.
8. IS 456: 2000; "Plane and Reinforced Concrete-Code of Practice", Bureau of Indian Standards, New Delhi.
9. Hirde, S.K. and Prasad, R.Vaidya, 2006 "Effectiveness of Reinforced Concrete Shear wall in seismic performance of multi storied buildings", *Earthquake analysis and design of structures*, A179-A184.
10. ASCE 7-05, "*Minimum design loads for buildings and other structures*", ASCE Standard ASCE/SEI 7-05, American Society of Civil Engineers, ISBN 0-7844-0809-2, 2005.

11. R. Kanitkar and V. Kanitkar (2004), "Seismic performance of conventional multi-story building with open ground floors for vehicular parking". *The Indian Concrete Journal*, pp. 99-104.

12. Uniform Building Code (UBC) (1997). International Council of Building.

Chapter 43

Laboratory and Field Assesment of Arsenic Testing Field Kits in Bangladesh and West Bengal

S.P. Pande, L.S. Deshpande and S.N. Kaul

CSIR-NEERI, Nagpur

ABSTRACT

High concentrations of arsenic in groundwaters in West Bengal and Bangladesh have become a major cause for concern in recent years. Given the enormity and the severity of the problem of arsenic poisoning, a task of evaluating the commercially available arsenic detection field kits for their capabilities was undertaken. In the light of the findings, generic specifications were recommended which could form the basis for indigenous manufacture of these kits in the arsenic affected countries. This article presents the results of the laboratory and field evaluation conducted in Bangladesh and West Bengal of five arsenic testing field kits. The salient features of the kits, their merits and limitations have been brought out. Based on the criteria of kit design, quality of chemicals used, colour comparator charts, detection range, time required for analysis, cost etc., a comparative ranking of the kits has been made to facilitate the choice of the kit to meet specific requirements.

***Keywords**: Arsenic, Field kits, Groundwater, Interferences.*

1. Introduction

Arsenic contamination in groundwater has been widely reported in Bangladesh (Dhar *et al.*, 1997) and West Bengal, India (Chatterjee *et al.*, 1995; Das *et al.*, 1995). It is well documented that prolonged intake of arsenic contaminated water has detrimental health effects. Because about 95 per cent of the drinking water in the study areas comes from aquifers, people who get their water from cased wells may have serious concerns about arsenic contamination. In a study (1996–1997) conducted by the School of Environmental Studies (SOES) of Jadavpur University, Calcutta, the Rajshahi University, the National Institute of Preventive and Social Medicine (NIPSOM), and the Dhaka Community Hospital in Bangladesh, well water in 27 of the 34 districts surveyed had arsenic levels in excess of 0.05 mg L^{-1}. Groundwater samples from six districts of West Bengal (India), which has a population of approximately 30 million people, had arsenic concentration levels over the maximum permissible level (0.05 mg L^{-1}). Major health concerns due to arsenic toxicity are skin diseases such as hyperpigmentation or depigmentation and/or keratosis which ultimately lead to cancer in more advanced stages. There

is no cure for arsenic toxicity except providing arsenic free water to the affected people. The two most likely sources of arsenic pollution of groundwater are: (a) from anthropogenic activities, and (b) from release of arsenic in arsenic-bearing sub- strate material. From the reports of distribution and extent of arsenic pollution of the groundwater, it is considered highly unlikely that the primary cause of the present sets of problems in the border region of West Bengal and Bangladesh is due to antropogenic activities. The initial factors considered in respect to sources and processes leading to release of arsenic from arsenic bearing geological material appear to be: (1) presence of arsenic in the alluvium, and (2) the use of cased wells as a means for abstracting water. Reported literature indicate that the high concentrations of arsenic are found primarily in the upper 150 m of the alluvial sediments that form the alluvial aquifer underlying the border region of West Bengal and most of Bangladesh. Two models exists in respect to possible mechanisms for release of arsenic from the arsenic bearing materials: (a) reduction mechanisms and (b) oxidation processes. In the former process, it is reasoned that reductive dissolution of arseniferrous iron oxyhydroxides releases the arsenic responsible for pollution of the groundwater. The other model for arsenic release from the alluvium relies on oxidation of arsenopyrites as the principal mechanism.

A major problem related to the availability of potable water for millions of people in rural regions ofWest Bengal and Bangladesh concerns the lack of civil infrastructure for selecting, screening, testing, and monitoring groundwater samples for arsenic concentration levels. Such testing and monitoring will contribute significantly to the public health and safety of people who inhabit regions where aquifers are contaminated by arsenic.

2. Materials and Methods

2.1. Procurement of Arsenic Field Testing Kits

Five arsenic detection field kits, one each from M/s E.Merck, Germany; M/s Asia Arsenic Network (AAN), Japan, M/s Aqua Consortium, Calcutta, the National Institute of Preventive and Social Medicine (NIPSOM), Bangladesh and the All India Institute of Hygiene and Public Health (AIIH and PH), Calcutta, were procured for assessment.

2.2. Salient Features of Field Kits

The chemistry of arsenic estimation used in all the field kits, evaluated in the present study, is the standard mercuric bromide stain method. This method, suitable only for qualitative and semi-quantitative determination of arsenic, requires care and experience in its use. Arsenic in water normally occurs as arsenite (trivalent state) or arsenate (pentavalent state). The basic principle of the mercuric bromide stain method is that the arsenic present in a water sample is converted to the trivalent form by the addition of potassium iodide and stannous chloride. The subsequent addition of zinc and hydrochloric acid to the water sample liberates nascent hydrogen which reacts with the trivalent arsenic to release arsine gas. The arsine thus released reacts with the mercuric bromide paper to form complex salts of arsenic and mercury, producing yellow to brown to black stain depending on the arsenic concentration in the sample. This colouration is due to the formation of the compounds $H(HgBr)_2As$ (yellow), $(HgBr)_3As$ (brown) and Hg_3As_2 (black). At low arsenic concentration a yellow stain is produced while high levels yield a black stain.

3. Results and Discussions

3.1. Laboratory Evaluation of the Field Kits

The arsenic field test kits were evaluated in the laboratory for their performance using blanks and standard arsenic solutions including US-EPA Quality Control (QC) samples. Some of the inorganic constituents, when present in water along with arsenic, interfere with the estimation of arsenic. Accordingly, the effects of acidity, sulfides, chlorides and iron on the estimation of arsenic using the various kits was evaluated.

3.1.1. Analysis of Blank and Arsenic Standards

No colour stain is developed in any of the five kits when tested with the blank.With 0.02 mg L^{-1} standard, no stain was developed when tested with the E.Merck, Aqua and AIIH and PH kits, indicating that they can not detect arsenic at this concentration. With the 0.05 mg L^{-1} standard, the E.Merck kit was the only one that did not detect the presence of arsenic. With standard solutions of 0.1, 0.5, 1.0, 2.0 and 3.0 mg L^{-1} all of the tested kits produced a stain comparable to that of the colour charts provided.

3.1.2. Analysis of EPA-QC Samples

When tested with the EPA-QC sample having 0.04 mg L^{-1} arsenic concentration, none of the kits produced a stain on the mercuric bromide paper. With 0.2 mg L^{-1} QC sample, all the kits produced a stain. Both the QC samples showed arsenic values less than the true certified values when estimated using the field kits. The lower values obtained for the EPA samples are probably due to the acidic nature of the samples, the EPA-QC samples are preserved by adding concentrated HNO_3.

3.1.3. Interference of Nitric Acid

When water samples are analysed on-the-spot using a field kit, there is no need to preserve the samples by acidification. However, if the samples are to be brought to a centralised location for analysis, and there is a significant time lag between sampling and analysis, then the samples are generally acidified by adding HNO_3. This brings the samples to a pH value of 2 or less, as recommended in Standard Methods. The addition of HNO_3 to the water samples results in lower measured values of arsenic than the true/actual values for all the five field kits.

3.1.4. Interference of Sulfide

Even when no arsenic is present in the sample, the AAN (Japan) and NIPSOM kits show a positive test for arsenic at 0.1 mg L^{-1} sulphide, while all the kits are sensitive to sulphide interference at 1.0 and 10.0 mg L^{-1} concentration. Thus there is a positive interference of sulfide when tested with these field kits (*i.e.*, the values indicated by the kits are higher than the true values of arsenic present). The Aqua and AIIH and PH kits are designed to eliminate interference due to the presence of sulfide. A small cotton plug soaked in lead acetate acts as an efficient sulfide trap in these two kits.

3.1.5. Interference of Chlorides and Iron

1000 mg L^{-1} of chloride and 2 mg L^{-1} of iron had no interference on arsenic results at the 0.5 mg L^{-1} level when tested with the field kits.

3.2. Evaluation of Field Test Kits in Bangladesh

Water samples were collected from cased wells (which were fitted with hand pumps) in the Chandpur district of Bangladesh. Some of the wells sampled were known to have water that contained arsenic and other wells were known to have arsenic-free water. The selection of cased wells was made in consultation with the officials of Department of Public Health Engineering (DPHE), Bangladesh. Due care was taken to work the hand-pump for about 5 min before collecting the water samples in 100 mL polythene bottles. One aliquot of the sample was not acidified while another aliquot was acidified with five drops of concentrated nitric acid for preservation.

The preserved samples were subsequently analysed using an ICP spectrometer (Jobin-Yvon, France, Model JY-24). All the unacidified samples were tested for arsenic using the E.Merck, AAN and the NIPSOMkits. The same samples were also tested concurrently using the AQUA kit. The details of locations from which the samples were collected.

The presence of sulfide was tested qualitatively using the lead acetate scrubber provided in the Aqua kit. None of the water samples tested in Bangladesh showed the presence of sulfide. The arsenic results in the acidified samples analysed using ICP are always higher than the arsenic results obtained in the unacidified samples tested using the field kits. This can be explained by the fact that all the water samples collected in Bangladesh contained high concentrations of iron (4.1 to 8.9 mg L^{-1}). Also, because of the time difference between when the samples were collected and when they were analysed at the Chandpur analytical facility, some of the arsenic present in the unacidifed samples may have co-precipitated with iron.

3.3. Evaluation of Field Test Kits in West Bengal

Field visits were undertaken in the villages of 24 Parganas (North) and Nadia Districts of West Bengal. Water samples were collected from cased wells (which were fitted with hand pumps) in these villages. Some of the wells sampled were known to have water that contained arsenic and other wells were known to have arsenic-free water. The selection of cased wells was made in consultation with the officials of the Public Health Engineering Department (PHED), West Bengal. Due care was taken to work the hand-pump for about 5 min before collecting the water samples in 100 mL polythene bottles. One aliquot of the samples was not acidified while another aliquot was acidified with five drops of concentrated nitric acid for preservation, for analysis by the ICP spectrometer (Jobin-Yvon, France, Model JY-24). All the unacidified samples were tested for arsenic using the E.Merck, AAN, NIPSOM, and AIIH and PH kits. The details of locations from where the water samples from West Bengal were collected. The presence of sulfide was tested qualitatively using the lead acetate scrubber in the Aqua and AIIH and PH kits. Sulfide was detected in four of the water samples tested in West Bengal. The presence of sulfide was confirmed by the black stain developed on the cotton plug of lead acetate placed in the sulfide trap (scrubber) of the kits. The arsenic results in the acidified samples analysed using ICP are always higher than arsenic results obtained in the unacidifed samples tested using the field kits. This is due to the fact that all the water samples, collected in West Bengal, also contained high iron (0.09 to 7.4 mg L^{-1}). Also, because of the time difference between the collecting of the samples and the analyzing at the Barasat and Kalyani analytical facility, some of the arsenic present in the unacidified samples may have co-precipitated with iron.

3.4. Ranking of the Field Kits

For want of well defined specifications a qualitative ranking of the kits has been attempted on the following broad criteria.

1. Design of the kit
2. Presence of essential chemicals
3. Colour comparator chart
4. Provision to remove interference
5. Time required for analysis
6. Detection limit of the kit
7. Occupational hazards
8. Cost of the kit
9. Capability of the kit

The available technical information on the various field kits, the results of the laboratory and field evaluations, and users' experience, have been used to make a comparative assessment of the kits.

4. Salient Findings

- ☆ The performances of these arsenic detection field kits were critically evaluated both in the laboratory and in the arsenic affected areas of West Bengal and Bangladesh.
- ☆ All the five field kits evaluated in the present study are based on the principle of the standard mercuric bromide stain method of arsenic estimation, and are capable of detecting arsenic in water in a qualitative/semi-quantitative manner.The detection limits of these kits are different and accordingly, their utility will depend on the specific purpose which they are intended to serve.
- ☆ If water supply sources with arsenic contamination >0.05 mg L^{-1} are to be screened, then the Aqua and the AIIH and pH kits can serve the purpose in a qualitative manner. The AAN and NIPSOM kits can screen sources in a semiquantitative manner to detect a minimum concentration of 0.02 mg L^{-1}.
- ☆ If the sources with arsenic concentration >0.10 mg L^{-1} are to be screened, then the E.Merck kit, in its present form, will also serve the purpose, and with a modified procedure as suggested by the firm, the detection limit could be improved to 0.05 mg L^{-1}.
- ☆ All the field kits evaluated are subject to inference (negative) if samples are acidified with nitric acid for preservation. Use of 1 mL of 5 per cent ascorbic acid per 100 mL of sample in place of nitric acid can eliminate the interference of acidity. Laboratory studies with the field kits have shown that ascorbic acid has no interference in arsenic analysis.
- ☆ None of the kits, in its present form, is capable of detecting arsenic concentration of 0.01 mg L^{-1}, the recommended WHO guideline value for drinking water.

5. Desired Characteristics of Field Kit for Arsenic Testing in Water

- ☆ The portable kit should be compact, with the kit box and contents made of light weight durable materials, convenient to carry in the field and user friendly.
- ☆ The kit should contain all the reagents required for estimation of arsenic in durable, leak proof plastic/glass bottles or tubes.
- ☆ The minimum detectable concentration of arsenic using the kit should be equal to the prescribed national standard or in its absence the WHO Guideline value of 0.01 mg L^{-1}.
- ☆ The kit should be capable of detecting total arsenic in the test sample irrespective of the form (arsenite, arsenate) in which it is present.
- ☆ The inner architecture/configuration of the kit should be such as to retain the contents in place without being disarranged during transit.
- ☆ The kit should include in it a stand/rack to hold the reagent bottles and reaction tube in place while performing the test in the field.

☆ The consumables (reagents/$HgBr_2$ paper) in the kit should last at least for 100 tests before they must be replenished.

☆ The zinc used in the test should preferably be in powder form to enable rapid reaction. If used in granular form, it should be free from arsenic contamination.

☆ The reagents, their concentration and sequence of addition should be such as to complete the reactions as fast as possible.

☆ The shelf life (validity) of the $HgBr_2$ paper should be as long as possible. The date of manufacture and the date of expiration should be indicated.

☆ The colour comparator chart should be of a permanent type, unaffected by exposure to light during normal use; the useful lifetime, if any, should be indicated, and the concentration values should be legible, easy to read. The colour intensity and hue in the chart should match those produced by the sample for the same concentration of arsenic.

☆ The kit should have provision to store the used mercuric bromide paper for subsequent safe disposal.

☆ The test should not involve direct contact of $HgBr_2$ paper with the fingers so as to eliminate any health hazard to the user.

☆ The kit should have in-built provision (lead acetate trap) for elimination of interference due to sulphide, if present, in the sample.

☆ The volume of the test sample, and the concentrations and volumes/weights of the reagents should be such as to enable detection of the desired minimum concentration of arsenic.

☆ The components of the arsine generator assembly should not be susceptible to damage by the reagents used in the test or the reaction products leading to inaccurate results.

☆ The test kit should contain illustrated instructions in simple language (English and local language) to enable the user perform the test scientifically and accurately.

In the backdrop of the evaluation of arsenic detection field kits, a prototype kit has been developed at NEERI which has most of the desirable features as described under Section 5 above. This field kit is presently being subjected to field investigation.

References

Chatterjee, A., Das, D., Mandal, B. K., Choudhury, T. R., Samanta, G. and Chakraborti, D.: 1995,'Arsenic groundwater in six districts of West Bengal, India: The biggest arsenic calamity in the world, Part 1 – Arsenic species in drinking water and urine of the affected people', *Analyst* **120** (3), 643–650.

Das, D., Chatterjee, A.,Mandal, B. K., Samanta, G., Chakraborti, D. and Chanda, B.: 1995, 'Arsenic in groundwater in six districts of West Bengal, India: The biggest arsenic calamity in the world, Part 2 – Arsenic concentration in drinking water, hair, nails, urine, skin-scale and liver tissue (biopsy) of the affected people', *Analyst* **120** (3), 917–924.

Dhar, Ratan Kr., Biswas, Bhajan Kr., Samanta Gautam, Mandal Badal Kr., Chakraborti, D., Roy Shibotosh, Jafar Abu, Islam Ariful, Ara Gulshan, Kabir Saiful, Khan Wadud, A., Ahmad Akhtar Sk. and Hadi Abdul Sk.: 1997, 'GroundWater Arsenic Calamity in Bangladesh', *Current Science* **73** (1), 48–59.

Chapter 44

Evaluation of Direct Contact Membrane Distillation for Concentration of Orange Juice

S.K. Deshmukh[1], V.S. Sapkal[2] and R.S. Sapkal[1]

[1]*Sant Gadge Baba Amravati University, Amravati, Maharashtra*
[2]*Rashtrasant Tukadoji Maharaj Nagpur University, Nagpur, Maharashtra*

ABSTRACT

In this work Membrane Distillation is applied to concentrate orange Juice. Clarified orange juice (11.5°Brix) obtained from fresh fruits was subjected to membrane distillation. The experiments were performed on a flat sheet module using orange juice as feeds. The concentration was carried out in a direct contact membrane distillation using hydrophobic PTFE membrane of pore size 0.2 µm and porosity 70 per cent. The influences of the feed temperature, feed concentration, flow rate, operating time on the transmembrane flux were studied.

Keywords: *Membrane distillation, Transmembrane flux, Orange juice, Polytetrafluoroethylene.*

1. Introduction

Membrane distillation (MD) is a membrane technique that involves transport of water vapor through the pores of hydrophobic membranes due to a vapor pressure driving force provided by temperature and/or solute concentration differences across the membrane. A variety of methods may be employed to impose this vapor pressure difference [1-7]. In the present work, the direct contact membrane distillation method is considered. In this configuration the surfaces of the membrane are in direct contact with two liquid phases, the feed (warm solution) and the permeate (cold solution), kept at different temperatures. A liquid vapor interface exists at the pore entrances where liquid-vapor equilibrium is established. Inside the pores only a gaseous phase is present through which vapor is transported as long as a partial pressure difference is maintained. The vaporization takes place at the feed membrane interface. The vapor diffuses through the membrane pores and condenses at the permeate membrane interface. Thus, MD relies on vapor-liquid equilibrium as a basis for separation and requires that the latent heat of vaporization be supplied to achieve the characteristic phase change. Membrane distillation offers advantages like techniques suitable for heat-sensitive products, modularity, easy scale-up, possibility to treat solutions with high level of suspended solids, Possibility of using modules in series, low temperatures, low operating pressures, no fouling problems, constant permeate flux in time, new technologies based on the use of conventional well-tested materials and low investment cost. Drawbacks of the process can be compensating by enhancing the flux rate. This technology work with certain disadvantages like low evaporative capacity with a long time of treatment, necessity of an inactivation enzyme pre-treatment and low flux rate. The goal of the present article is to enhance flux rate by surface modification of membrane surface to make process more efficient and commercially viable.

DCMD is not a simple process of mass transfer through the membrane, but a complex process combination of several interrelated heat and mass transfer steps. In fact, as vaporization takes place at the feed membrane interface and condensation at the permeate membrane interface, membrane distillation requires the heat of vaporization to be supplied to the feed vapor–liquid interface, and the heat of condensation to be removed from the vapor–liquid interface in the permeate side. Conductive heat transport through the thin membrane also takes place.

As a consequence, thermal boundary layers develop at both sides of the membrane, that is, temperature polarization arises. On the other hand, concentration boundary layers develop in the liquid phases (that is, concentration polarization arises) if there is solute rejection by the membrane [7].

Orange juice is probably the best known and most widespread fruit juice all over the world, particularly appreciated for its fresh flavour and considered of high beneficial value for its high content in vitamin C and natural antioxidants, such as flavonoids and phenylpropanoids. The advantages of the concentration of the liquid foodstuffs include the reduction in packaging, storage, transport cost and prevention of deterioration by microorganisms. For these reasons, many concentration techniques have been developed and used for the food industries. They include evaporative concentration, freeze concentration, and membrane processes such as reverse osmosis (RO) and ultrafiltration (UF) [9-10].

Nevertheless, when concentration is carried out by traditional multi step vacuum evaporation, a severe loss of the volatile organic flavour/fragrance components occurs as well as a partial degradation of ascorbic acid and natural antioxidants, accompanied by a certain discolouration and a consequent qualitative decline. These effects are mainly attributable to heat transfer to the juice during evaporation. In order to overcome some of these problems and to better preserve the properties of the fresh fruits, several new mild technological processes have been proposed in the last years for juice production [8]. MD has many significant advantages, such as high system compactness, possibility to operate at low temperatures (30–90°C) which makes it amenable for use with low temperature heat sources, including waste or solar heat, and, when compared with say reverse osmosis or electrodialysis, the simplicity of the membrane which allows it to be manufactured from a wide choice of chemically and thermally resistant materials, and much larger pores than of reverse osmosis membranes (and typically larger than in ultra-filtration membranes, that aren't nearly as sensitive to fouling [1-12].

Table 44.1: Module Design

Module	
Configuration	Flat Plate
Membrane area, m^2	0.0115
Material	PTFE
Membrane thickness	160 μm
Nominal pore diameter	0.05–0.2 μm
Porosity (%)	70

2. Material and Methodology

2.1 Module Development (Table 44.1)

Cross flow module of hydrophobic PTFE 0.2μm has been developed with the help of viton gasket, polyester mesh and adhesive. Module has length 11.5cm, breadth 10 cm and hydraulic diameter 2.28 mm is supported with stainless steel holding device. Module has effective membrane area 0.0115 m^2.

2.2 Orange Juice Samples

Marmalade orange were purchased from a local market. Fruits (47) were manually washed in water and then peeled by hand, with a knife, obtaining 4.06 kg of peeled fruits (yield 47.1 per cent). Seeds and mesocarp fibres were removed with a squeezer and then washed with water. Orange juice is obtained by squeezing; TSS concentration of the raw juice was about 11.5°Bx with a pH = 3.5 and vitamin C is 285.64 PPM. Orange juice of 11.5°Brix is used as feed. The suspended matter was removed from juice by using masline cloth, filter paper and through wattman filter paper no. 40 for primary filtration and Sodium azide (Aldrich, Germany) was added to both sucrose and orange juice solutions (0.2 per cent w/w), in order to prevent their fermentation during the experiments.

3. Experiment Setup (Table 44.2)

Concentration of clarified orange juice by DCMD was carried out using a flat-sheet membrane cell with an effective membrane area 0.0115 m^2. The membrane cell was made of stainless steel and was placed in a vertical configuration. The system to be studied consists of a porous hydrophobic membrane, which is held between two

Table 44.2: Experimental Setup

No. of oranges	47	36
Wt. of pulp	0.976 kg	0.571 kg
Wt. of juice	2.883 kg	1.582 kg
Total wt.	4.06 kg	2.486 kg
Brix	11.5	11.2
pH	3.35	3.5
Carotene mg/100 g	0.7	0.7
Acidity, per cent anhydrous citric acid	28.56	28.56
Viscosity cp	29.50	29.50
Reducing sugar per cent	3.87	3.87
Total suagar	8.86	8.86
Alcoholic insoluble solids per cent	2.80	2.80
Clarity per cent transmittance at 660 nm	–	–

symmetric channels. Hot feed is circulated through one of the channels and cold permeate through the other one. The hot and cold fluids counter-flow tangentially to the membrane surface in a flat membrane module. In our experiments, the membrane is sandwiched between two equal stainless steel manifolds. Microporous hydrophobic PTFE membrane of 0.2 µm pore size and thickness 160 µm was placed between polyester mesh (0.28mm), polyviton gasket (3 mm) on both side which create the two identical flow channels, the membrane and the manifolds create spacer-filled flow channels for hot feed and cold permeate liquids.

Feed tank with thermostat, peristaltic pump and temperature and flow indicator is arrange in feed side, where as peristaltic pump, and temperature and flow indicator is arrange in permeate side. Module is supported with stainless steel holding device. The schematic arrangement is shown in Figure 43.1. Clarified juice as feed solution and distilled water as receiving phase were contained in two jacketed reservoirs and were circulated through the membrane cell by one two-channel peristaltic pump. The feed and distillate streams flow counter currently from the bottom to the upper part of the membrane cell. Different experiments were carried out for fixed temperatures in the membrane module. The average feed temperature Tf varied for the different experiments from 40 to 70°C and permeates temperature Tp varied for the different experiments from 20 to 30°C.

The linear velocity feed and permeate was also varied. Different experiments were carried out applying different recirculation rates. A drainage tube in the upper part of the receiving reservoir confined the total volume of receiving phase to about 100 ml. Excessive liquid due to permeate transferred across the membrane escaped from receiving reservoir and was collected in a graduated cylinder. The permeate volume was measured continuously as a function of time and these data were used for calculation of the permeate flux.

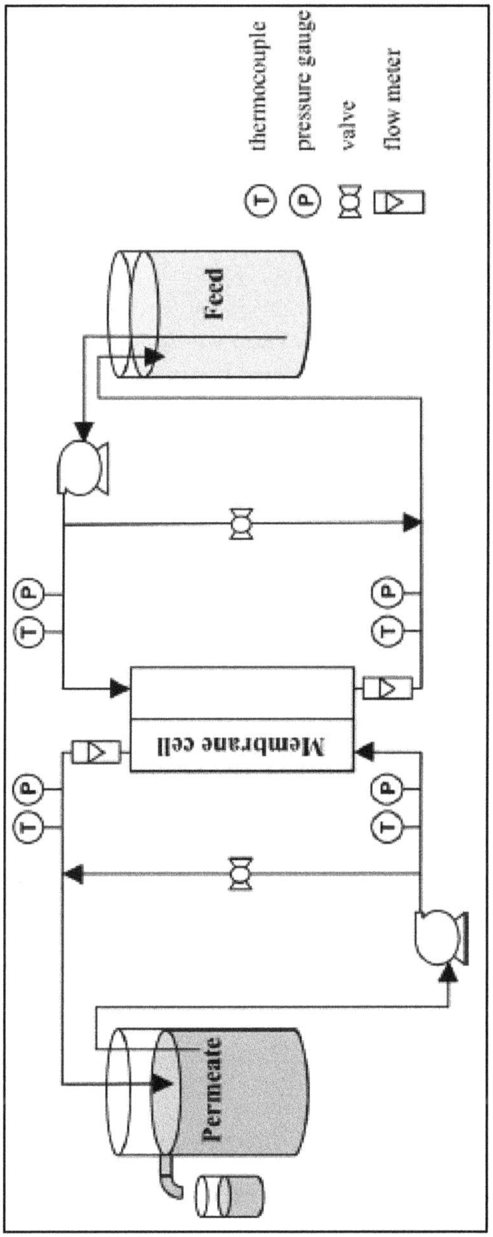

Figure 43.1: Flow Diagram of Membrane Distillation.

4. Results and Discussion

The effect of the various parameters on permeate flux has been investigated in the DCMD configuration. The temperature has been varied from 30 to 70°C (below the boiling point of the feed solution) and cold side temperature *Tp* has been varied from from 20°C to 30°C maintaining all other MD parameters constant (Table 44.3).

4.1 Effect of Feed Velocity

Effect of feed flow rate on transmembrane flux for Orange juice is estimated and presented in Figure 44.2. During experiments, the feed side flow rate is varied between 48 to 84 L/hr and permeate side flow rate (30 L/hr), Feed temperature (50°C), Permeate temperature (30°C), temperature difference (ΔT = 30°C), and concentration was maintained constant (11.5°Bx). The transmembrane flux increases with increase in flow rate. The increase is mainly due to the reduction in temperature polarization and because of reducing concentration polarization and fouling phenomenon. Figure 2 shows as the permeate flux increase when the recirculation rate increases. The effect of high recirculation rate is to increase the heat transfer coefficient and thus reduce the effect of temperature polarization. This means that the temperature at the membrane surface more closely approximated to the bulk temperature, and thus the transmembrane temperature difference is greater. This produces greater driving force and consequently enhanced the flux.

4.2 Effect of Permeate Velocity

The effect of permeate velocity resulted into minimal effect (not noticeable) on the permeate flux. The increase in velocity from 30 to 120 L/h resulted in a 16 per cent increase of the permeate flux. Whereas in feed velocity increases from 48 to 84 L/h resulted in a 48 per cent increase of the permeate flux. This implies that the effect of permeate velocity is less significant than that of feed velocity. This implies that at permeate side the concentration polarization does not exist (Figure 44.3).

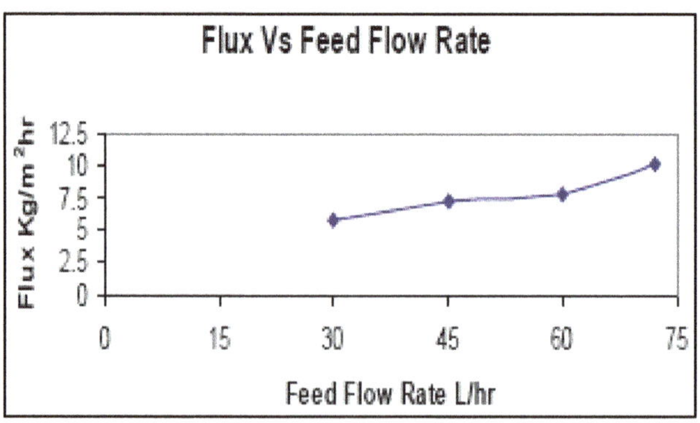

Figure 44.2: Effect of Feed Velocity on Transmembrane Flux Using PTFE Membrane (Permeate flow rate 30L/hr, feed temperature 50ºC, permeate temperature 30ºC and 11.5ºBrix).

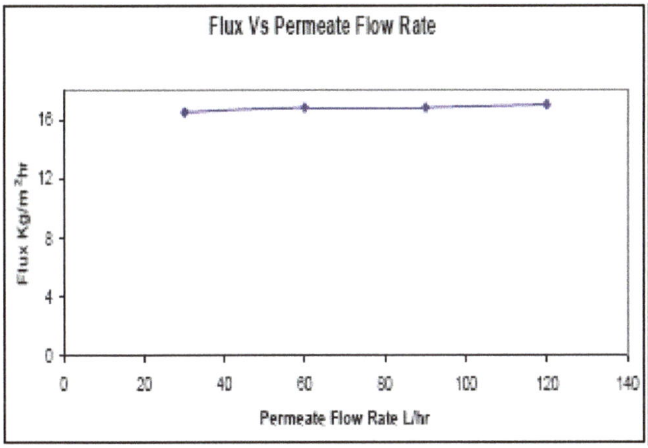

Figure 44.3: Effect of Permeate Velocity on Transmembrane Flux using PTFE Membrane (Feed flow rate 60L/hr, feed temperature 50ºC, permeate temperature 30ºC and 11.5 ºBrix).

4.3 Effect of Feed Temperature

The effect of the feed temperature on permeate flux has been investigated in the DCMD configuration. The feed temperature has been varied from 40 to 70°C (below the boiling point of the feed solution) maintaining all other MD parameters constant. Figure 44.4 shows that in DCMD configuration there is an exponential increase of the MD flux with the increase of the feed temperature. This is due to the exponential increase of the vapor pressure of the feed solution with temperature, which increases the transmembrane vapor pressure (*i.e.* the driving force) as all the other involved MD parameters are maintained invariables. It was stated that it is better to work under high feed temperature as the internal evaporation efficiency, defined as the ratio of

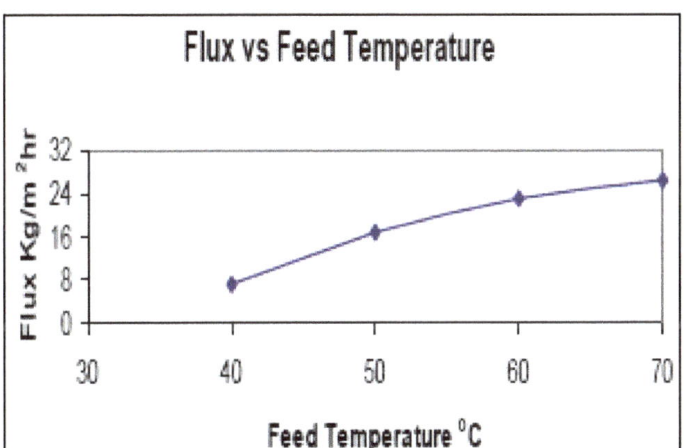

Figure 44.4: Effect of Feed Temperature on Transmembrane Flux using PTFE Membrane (Feed flow rate 72L/hr, permeate flow rate 30L/hr, permeate temperature 30ºC and 11.5 ºBrix).

the heat that contributes to evaporation and the total heat exchanged from the feed to the permeate side is high although the temperature polarization effect increases with the feed temperature. Primarily while deciding feed temperature, the quality of feed is foremost important.

4.4 Effect of Feed Concentration

The concentration of orange juice was varied over 7 – 22 oBrix. During the experiments the feed flow rate (72 l/hr), permeate flow rate (30L/hr), feed temperature (50°C), permeate temperature (30°C) and temperature difference (20°C) are maintained constant. The values of transmembrane flux observed at different concentration of feed solution and are shown in Figure 44.5. The transmembrane flux decreases with increase in concentration. The flux decay observed at high concentration is related mainly to the significant increase in juice viscosity. This is also due to increase in concentration and temperature polarization and thus decrease the driving force. Accordingly, concentration polarization may be significant at high concentration, high temperature and low feed velocity. In other words, the evaporation rate of water at the membrane surface decreases with increase in concentration Flux decline with time was observed, but it was more significant at high concentration.

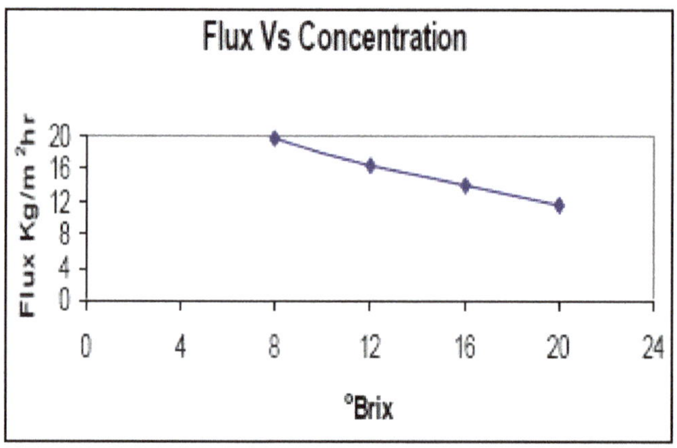

Figure 44.5: Effect of Concentraton of Orange Juice on Transmembrane Flux using PTFE Membrane (Feed flow rate 72L/hr, permeate flow rate 30L/hr, feed temperature 60ºC, and permeate temperature 30ºC).

4.5 Effect of Temperature Difference

Figure 44.6 shows the results obtained at four constant temperatures of juice in the hot cell (40°C, 50°C, 60°C and 70°C) with constant cold cell temperatures (30°C). During experiments the feed (72l/hr) and permeate velocity (30L/hr) were maintained constant. The flux was calculated based on experimental data using the following equation:

$$J = \frac{\text{Measured permeate flow rate}}{\text{Membrane area}}$$

The fluxes exhibit an exponential dependence on temperature–as would be expected when considering the Antoine equation for vapor pressure of water:

$$pmi = \exp\left(23.238 - \frac{3841}{Tmi - 45}\right), i = 1,2$$

where, p is the vapor pressure of water in Pa and *T* is the temperature in K.

The flux increased exponentially with temperature. The temperature difference creates vapor pressure difference and thus the membrane distillation flux rises. This leads to water vapor diffusion through the membrane. At lower temperature difference across the membrane, the transmembrane flux decreases as expected due to lower vapor pressure difference between feed and permeate.

Table 44.3: Bulk Temperature, Surface Tempeature, Vapour Pressure difference and Flux for hm = 503w/mK, hf=6642.09 w/mK and hp=4337.07 w/mK.

$T_f(K)$	$T_p(K)$	$T_1(K)$	$T_2(K)$	$P^o_1(Pa)$	$P^o_2(Pa)$	$\Delta P(Pa)$	$J\ kg/m^2h$
313	303	312.1	303.81	7030.4	4438.18	2592.22	7.2
323	303	321.1	305.07	11475.84	4807.82	6668.02	16.5
333	303	330.67	306.04	17997.71	5054.32	12943.39	23.0
343	303	339.96	307.00	27391.82	5313.46	22078.36	26.1
353	303	349.24	308.14	40863.8	5642.03	35221.77	31.13

4.6 Effect of Vapour Pressure Difference

Figures 44.6 and 44.7 display flux measurements of the PTFE 0.015 m^2 flat sheet membranes tested with a feed solution at a series of different input temperatures under the same flow conditions. The same data are expressed as flux as a function of average temperature difference (Figure 44.6), and as a function of average vapour

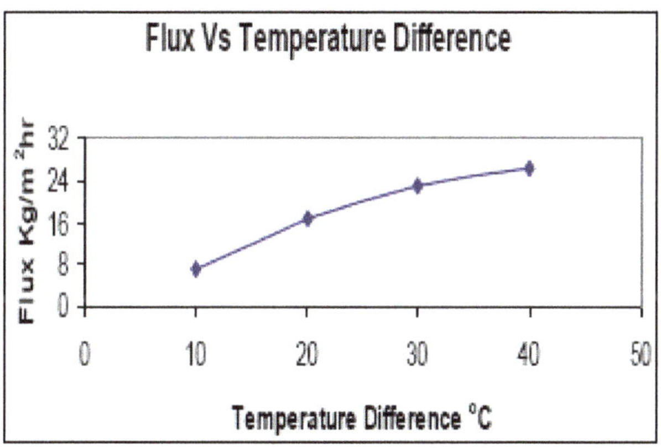

Figure 44.6: Effect of Temperature difference on Transmembrane Flux using PTFE Membrane (Feed flow rate 72L/hr, permeate flow rate 30L/hr, permeate temperature 30ºC and 11.05 ºBrix).

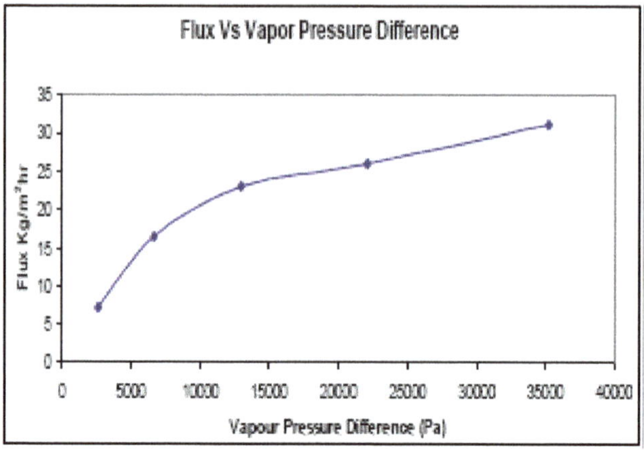

Figure 44.7: Effect of Vapour Pressure difference on Transmembrane Flux using PTFE Membrane (Feed flow rate 72L/hr, permeate flow rate 30L/hr, permeate temperature 30ºC and 11.05 ºBrix).

pressure difference (Figure 44.7). It is widely understood that application of a temperature difference across a MD membrane will induce water vapour to pass and some amount of permeate to be generated. Furthermore, developing significant temperature differences should lead to greater production rates. However, the actual driving force for MD is the vapour pressure difference across the membrane, which is induced by this temperature difference. The relationship between flux and DINESHDDDT shows the expected trend but with significant deviations, whereas the relationship with DINESHDDDP shows a correlation closer to Schofield's model.

4.7 Flux Declining Rate

In addition to above experiments, another experiment was performed to study flux decline rate with respect to time. With feed at flow rate 72 L/hr, permeate flow rate 30L/hr, Temperature difference 20°C, feed temperature 50°C and permeate temperature 20°C. The aim of these experiments was to study the flux decay in

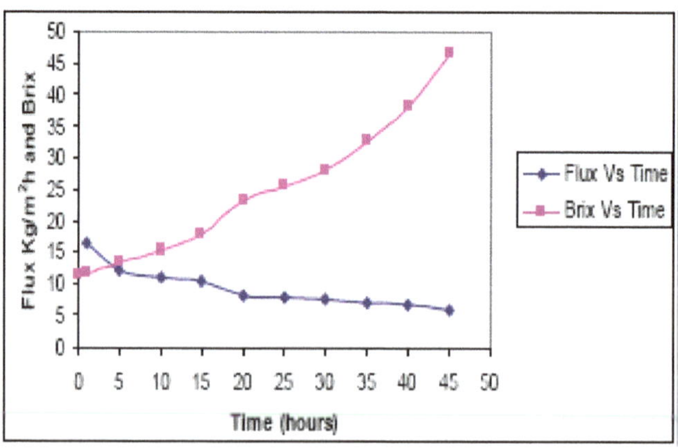

Figure 44.8: Flux Declining Rate.

membrane distillation. The result indicates that it was possible to consistently remove water at steady value of approximately 10- 12 kg/m^2 hr. Flux decline over time was observed, but it was more significant at high concentration. This suggests a possible effect of both concentration and temperature polarization. Accordingly, concentration polarization may be significant at high concentration, high temperature and low feed velocity. The flux was found around 12 kg/m^2h and has not decreased significantly for first 15 hours. It value started to drop when the juice concentration has reached 23°Brix levels during the measurement due to the significant viscosity increase (Figure 44.8).

5. Conclusion

The concentration of orange juice was carried out by direct contact membrane distillation using PTFE membrane. The influence of various parameters such as feed flow rate, temperature difference, Feed Temperature and concentration of orange juice with respect to transmembrane flux were studied for real system. Membrane distillation experiments shows that transmembrane flux gradually increases with increase of feed juice temperature at constant flow rate and constant permeate temperature. At lower temperature difference across the membrane, the transmembrane flux decreases as expected due to lower vapor pressure difference between feed and permeate. The increase in feed flow rate could increase the transmembrane flux in the membrane distillation process, reducing concentration polarization and fouling phenomenon. The transmembrane flux decrease with an increase in juice concentration. The flux decay observed at high concentration is related mainly to the significant increase in juice viscosity.

References

1. K.W. Lawson, D.R. Lloyd, Review on Membrane distillation, *Journal of Membrane Science* 124 (1997) 1–25.
2. J.I. Mengual, L. Pe˜na, Membrane distillation, *Colloid Interf. Sci.* 1 (1997) 17–29.
3. R.W. Schofield, A.G. Fane, C.J.D. Fell, Heat and mass transfer in membrane distillation, Journal of Membrane Science 33 (1987) 299–313.
4. M. Khayet, M.P. Godino, J.I. Mengual, Theory and experiments on sweeping gas membrane distillation, *Journal of Membrane Science* 165 (2000) 261–272.
5. F.A. Banat, J. Simandl, Theoretical and experimental study in membrane distillation, *Desalination* 95 (1994) 39–52.
6. S. Bandini, C. Gostoli, G.C. Sarti, Separation efficiency in vacuum membrane distillation, *Journal of membrane Science* 73 (1992) 217–229.
7. L. Martýnez, J.M. Rodrýguez-Maroto, On transport resistances in direct contact membrane distillation, *Journal of Membrane Science* 295 (2007) 28–39.
8. Gianni Galaverna, Gianluca Di Silvestro, Alfredo Cassano, Stefano Sforza,Arnaldo Dossena, Enrico Drioli, Rosangela Marchelli, A new integrated membrane process for the production of concentrated blood orange juice: Effect on bioactive compounds and antioxidant activity, *Food Chemistry* 2007.

9. Thanedgunbaworn, R. Jiraratananon, M.H. Nguyen, Mass and heat transfer analysis in fructose concentration by osmotic distillation process using hollow fibre module, *Journal of Food Engineering* 78 (2007) 126–135.

10. Rico Bagger-Jorgensen, Anne S. Meyer, Camilla Varming, Gunnar Jonsson, Recovery of volatile aroma compounds from black currant juice by vacuum membrane distillation, *Journal of Food Engineering* 64 (2004) 23–31.

11. A.M. Alklaibi, Noam Lior, Transport analysis of air-gap membrane distillation, *Journal of Membrane Science* 255 (2005) 239–253.

12. Fadi Ali, Manuel Dornier, Albert Duquenoy, Max Reynes, Evaluating transfers of aroma compounds during the concentration of sucrose solutions by osmotic distillation in a batch-type pilot plant, *Journal of Food Engineering* 60 (2003) 1–8.

13. B. Ravindra Babu, N.K. Rastogi, K.S.M.S. Raghavarao, Mass transfer in osmotic membrane distillation of phycocyanin colorant and sweet-lime juice, *Journal of Membrane Science* (2005).

14. Sanjay Nene, Suhkvender Kaur, K. Sumod, Bhagyashree Joshi, K.S.M.S. Raghavarao, Membrane distillation for the concentration of raw cane-sugar syrup and membrane clarified sugarcane juice, *Desalination* 147 (2002) 157-160.

15. Ganapathi Patil, K.S.M.S. Raghavarao, Integrated membrane process for the concentration of anthocyanin, *Journal of Food Engineering* 78 (2007) 1233-1239.

16. A.F.G. Bailey, A.M. Barbe, P.A. Hogan, R.A. Johnson, J. Sheng, The effect of ultrafiltration on the subsequent concentration of grape juice by osmotic distillation, *Journal of Membrane Science* 164 (2000) 195–204.

17. Oleksandr S. Lukanin, Sergiy M. Gunko, Mikhaylo T. Bryk, Rinat R. Nigmatullin, The effect of content of apple juice biopolymers on the concentration by membrane distillation, *Journal of Food Engineering* 60 (2003) 275–280.

18. A.M. Barbe, P.A. Hogan, R.A. Johnson, Surface morphology changes during initial usage of hydrophobic, microporous polypropylene membranes, *Journal of Membrane Science* 172 (2000) 149–156.

19. L.Martinez, Comparison of membrane distillation performance using different feed, *Desalination* 168 (2004) 359-365.

20. W.T. Hanbury and T. Hodgkiess, Membrane Distillation–An Assessment *Desalination*, 56, (1985) 287-297.

21. A.S. Jonsson, R.Wimmerstedt and A.C. Harrysson, Membrane Distillation: A theoretical study of evaporation through microporous membrane, *Desalination*, 56, (1985) 237-249.

22. E. Drioli and Yonglie Wu, Membrane Distillation ë Që An Experimental Study, *Desalination*, 56, (1985) 339-346.

23. M. Tomaszewska, Membrane Distillation–Examples of Applications in Technology and Environmental Protection, *Polish Journal of Environmental Studies* Vol. 9, No. 1 (2000), 27-36.

24. M.S. El-Bourawi, Z. Ding, R. Ma, M. Khayet, A framework for better understanding membrane distillation separation process, *Journal of Membrane Science* 285 (2006) 4–29.

25. B.L. Jiao, V. Calbro and E. Drioli, Concentration of Orange and Kiwi Juice by integrated Ultrafiltration and Membrane Distillation, Paper presented at IMSTEC' 92 held at Sydney Australia, 10-12 November 1992, 7-8.

26. Sophie Cerneaux, Izabela Struzynsk, Wojciech M. Kujawski, Michel Persin, Andre Larbot, Comparison of various membrane distillation methods for desalination using hydrophobic ceramic membranes, *Journal of Membrane Science* 337 (2009) 55–60.

27. V.D. Alves, I.M. Coelhoso, Orange juice concentration by osmotic evaporation and membrane distillation: A comparative study, *Journal of Food Engineering* 74 (2006) 125–133.

28. A. Cassano, B. Jiao, E. Drioli, Production of concentrated kiwifruit juice by integrated membrane process, *Food Research International* 37 (2004) 139–148.

Chapter 45

Evaluation of Kinetic Constants for Anaerobic Fixed Bed Reactor System

B. Chakradhar, Alka Mishra, Prabha Padmakaran and S.N. Kaul

CSIR-NEERI-,Nagpur

ABSTRACT

Anaerobic fixed bed reactor system is used for the treatment of cotton digestion wastewater which is generated during the cotton digestion with alkali. The wastewater is highly alkaline, with high concentrations of organic matter with pale yellow colour. Substrate removal kinetics was developed using Dewalle and Chain model and pseudomechanistic models. The values of kinetic rate constant K_2 (A/V), loading rate constant K_L, and the inhibition coefficient (K_s) were determined for the system. The maximum loading rate obtained from Lineweaver-Burk relationship is 20 kg $COD_{r/m3}$.d with optimum conditions of 35°C temperature and at a pH of 7.0.

Keywords: *Anaerobic reactor, Fixed bed, Substrate kinetics, Loading rate, Kinetic rate constants.*

Introduction

During last 40 years there has been tremendous growth and there were about 305 paper and paper board mills in India with an installed capacity of 3.0 million tonnes per annum[1]. The highly diversified paper industry includes the primary production of wood pulp and paper and production of pulp. A high quality pulp is required for the manufacture of security paper used for printing of bank notes, debentures, bonds and other documents. The paper industry use cotton linters and raw cotton in the manufacture of cotton fibre paper. Raw stable cotton, direct from the cotton fields is used in the production of highest quality rug content bond, ledgers and bank note papers.

Cleaned cotton is cooked in static digesters with caustic and hydrogen peroxide. The wastewater generated during the digestion process is pale yellow in colour, with high pH, alkalinity, BOD, COD and would pose serious water and land pollution problems if discharged without treatment. Adequate treament to meet the regulatory standards would thus become imperative.

Wastewater Sources and Characterisation

During digestion process cotton at a temperature of 100°C and 1.5 kg/m^2 pressure, vegetable fats, waxes, resins, germs etc. get removed from the fibres and find way into the wastewater. Processing of 20 tonnes of cotton yields about same quantity of pulp and generates around 200 m^3/d wastewater. The wastewater from digester along with washes are collected and analysed. The characteristics are presented in Table 45.1.

Table 45.1: Characteristics of Cotton Digestion Wastewater

Parameter	*Concentration (Range)*
pH	11.4-12.6
Total Solids (per cent)	1.33-1.66
Total Volatile Solids (per cent)	0.61-1.03
Alkalinity, mg/L	
pH 8.3	2200–2970
pH 4.5	4200-4800
COD, mg/L	2080-9630
BOD_5, mg/L	1240-5000
Sodium, mg/L	1435-2070
T.Phosphorus, mg/L	8-20
T.NH3-N, mg/L	39-53
Total Kjeldhal nitrogen, mg/L	126-154
BOD: N: P	100: 5.2: 0.54
BOD_5 rate constant k, d^{-1}	0.12

The analysis of the digestion wastewater reveals that the wastewater is highly alkaline with pH between 12.3 and 12.4 COD ranged from 4360 to 6280 mg/l with an average of 5500 mg/l. On an average BOD was 2700 mg/l. BOD to COD ratio of 0.49 indicates the ready biodegradability of the organic matter present. Nitrogen present in the wastewater is adequate as nutrient for anaerobic treatment, but phosphorous needs to be supplemented for aerobic treatment. In view of the high concentrations of organic matter in the digester spent liquour an attempt was made in the study to treat the wastewater by using anaerobic fixed bed system.

System Analysis

With the development of anaerobic systems in the field of treatment technology renewed interest is being developed in their modelling because these models help to understand the mechanism of the particular system as well as provide the designer with necessary design information. A review of some substrate kinetic models for anaerobic fixed bed reactor is presented with an attempt to apply the same to the present investigated data.

The substrate removal kinetics in fixed bed biological system can either be described in terms of substrate concentration existing in bulk liquid phase or simulated substrate concentration in each layer of the biofilm, the sum of which predicts the response of the entire film. In biological film reactor at study state, a mass balance within the film can be made by equating.

Input–output = uptake

According to Fick's Law of molecular diffusion[2,3], the mass transfer rate dF/dt through a surface area, A is proportional to the concentration gradient of the substrate S at the interface:

$$dF/dt = -ADs\ ds/dz \tag{1}$$

where, dF/dt = rate of mass transfer at interface, M/T

D_s = diffusion coefficient, L^2/T

ds/dz = substrate concentration gradient, M/L

z = depth of biofilm L

A = surface area of biofilm, L^2

The biological uptake of any substrate material within the biological layer can be expressed as:

$$ds/dt = US \times K_S + S \tag{2}$$

where,

U: Maximum substrate removal rate,

M: Substrate/M-biomass T

S: Substrate concentration at depth Z, M/L^3

X: Biomass concentration within the biolayer, and

Ks: Half velocity concentration, M/L^3

for a given waste. U and X are not expected to vary greatly. The rate is directly proportional to the thickness of the biolayer. In this case the rate equation can be modified to indicate that the rate of substrate removal/reactor volume is related to the specific surface area, A/V and the square root of the bulk substrate concentration:

$$1/V\ dF/dt = K_1\ (A/V) \qquad Se \tag{3}$$

where, K is the coefficient based on zero order kinetics and V is the reactor volume. Dewalle and Chain describe the rate of biological regeneration of activated carbon particles due to the presence of activated sludge solids. When S is much small than Ks, equation 2, reduces to a first order kinetic equation.

$$d^2z/dz^2 = -UXS/Dk_S \tag{4}$$

Which after intergration yields the mass transfer rate equation at the interface *i.e.*

$$dF/dt = Se \quad DUX/Ks. \tag{5}$$

Since D, U, X and Ks are not expected to vary greatly with a given suhstrate, equation 5 reduces to:

$$1/V = dF/dt = K_2 (A/V). Se \tag{6}$$

where,

K_2 = coefficient based on first order kinetics, L/T

The equation 6 indicates that the specific surface area is the most important factor determining the rate of substrate removal.

Pseudo Mechanistic Models

Literature is still being written regarding the role played by major trickling filter parameters, *viz.* depth, volume, hydraulic loading, recycle rate, and volumetric organic loading. The plug flow nature of the three phase flow regime prohibits adequate mechanistic explanation of packed bed reactor removal mechanisms and for us the designed to look for empirical mechanistic solutions such as widely used *viz.*

$$Se/Si = \exp(-KD/Qn) \tag{7}$$

K = constant

Se/Si = Effluent and Influent substrate concentration

n = exponent

The other model used by Oleszkiewicz has been found to fit experimental data quite well.

$$Se/Si = \exp(-K/L) \tag{8}$$

Equation (8) relates directly food/microorganisms, flow rate *i.e.*

$$Se/Si = \exp(-K/F/M)$$

To obtain the maximum rate of substrate removal it has to be assumed that (1/v) (dF/dt) and Se follow a mood hyperbolic relationship, where after the constants can be determined from a Lineweaver Burk Plot.

$$1/v. dF/dt = L_r \; LmSe/Ks + Se \tag{9}$$

where, Lm = maximum loading rate

Experimental Set Up

In the present work, a laboratory model anaerobic fixed bed bio reactor was used to assess the treatability of cotton digester spent liquor with a feasibility of energy recovery in the form of methane. The reactor is constructed using PVC column of 110 mm (OD) dia with 2000 mm height. The empty volume of the reactor is 17.5 litres. The type of media used in the reactor was plastic scrubbers which is made of plastic thread. The void volume was determined by water displacement technique and was found that the reactor had a void volume of 12.5 litres with a porosity of 0.715. Wastewater was fed to the unit using a variable speed peristatic pump. The salient features of the entire experimental reactor system is described in Table 45.2.

Table 45.2: Salient Features of the Experimental Setup

Description	*Values*
Reactor Details	
Diameter of PVC packed column, mm	110
Height of the packed column, mm	2000
Volume of packed column, L	17.5
Type of media	Synthetic Nylon Pads
No. of pads	144
Height of packed media, mm	1840
Volume of void space, L	12.5
Porosity, per cent	71.5
Inlet and Outlet pipes, mm	25
Gas outlet, mm	15
Diameter of sampling ports, mm	6
No. of sampling ports	6
Distance between the ports, mm	305
Media Details	
Length of each nylon pad thread, m	60
Breadth of each nylon pad thread, mm	0.75
Thickness of the thread, mm	0.33
Area of each pad, m^2	0.136
Area of media within the reactor, m^2 (144 pads)	19.6
Specific surface area, m^2/m^3	1120

Biologically active growth was obtained with sewage, 2 per cent cowdung and 1 per cent night soil. It took about 4 weeks for the anaerobic micro-organisms to grow and attach to the media. After acclimation different proportions of sewage and spent liquor was applied to the reactor and COD removals were observed. Finally only cotton digester liquor was fed to the system at required flow rate. The wastewater used in the present investigation was obtained from M/s. Security Paper Mills, Hoshangabad. After some days of acclimation run, the unit was initially operated at an organic loading of about 0.5 kg COD/m^3/d and subsequently increased. Data were collected at different temperatures and pH. The influent COD concentrations used in the reactor were 2000 to 9000 mg/l at various detention periods ranging from 0.25 days to 7 days based on void volume of the reactor.

Performance data for each loading were collected during steady state periods. Influent and effluent samples collected were analysed according to Standard Methods (1971) for parameters of prime importance. Daily total gas production was also recorded and methane content determined using Orsat Gas Analysis Apparatus.

Results and Discussion

The performance data of the reactor system operated at different organic loading rates and deduction times at a constant temperature of 35°C is given in Table 45.3. The studies were carried out at different pH with a influent concentration of 9000 mg/l. Plot of organic loading rate (1/v dF/dt) versus effluent substrate concentration (Se). The values of kinetic rate constant (K_2 A/V) were determined. A linear relationship exists between organic loading rate and effluent substrate concentration as defined by equation 6. At a pH of 7.0 and 35°C the reaction rate constant is maximum with a value of 2.6 day^{-1}. The expression developed based on the above model is as follows:

1/V dF/dt = 2.6 Se (Temperature = 35°C and pH = 7.0)

Table 45.3: Performance Data of Anaerobic Fixed Bed Reactor COD = 9 kg/m^3, Temperature = 35ºC

pH = 6.6								
Flow, L/d	50.00	25.00	16.50	12.30	8.35	6.20	4.60	3.10
Detention Time, d	0.25	0.50	0.75	1.00	1.50	2.00	2.70	4.00
Efficiencly, per cent	27.00	35.00	46.00	65.00	70.00	76.00	80.00	83.00
Effluent COD, kg/M^3	6.57	5.85	4.86	3.15	2.70	2.16	1.80	1.53
Loading Rate, kg.COD$_r$/m^3.d	9.72	6.30	5.52	5.85	4.20	3.42	2.67	1.87
COD Removal, kg.COD$_r$/d	0.12	0.08	0.07	0.07	0.05	0.04	0.03	0.02
Methane Gas, m^3/d	0.04	0.02	0.03	0.03	0.02	0.01	0.01	0.01
Biogas Yield, m^{3}CH$_4$/m^3.d	2.80	1.92	2.00	2.16	1.28	1.04	0.80	0.56
Influent SS, kg/m^3	0.45	0.44	0.43	0.43	0.42	0.41	0.40	0.39
Effluent SS, kg/m^3	0.81	0.77	0.75	0.71	0.63	0.59	0.56	0.55
Biomass Volume, L	1.80	1.95	1.95	2.30	2.45	2.53	2.53	2.53
Biomass Conc., kg/m^3	11.09	12.01	12.01	14.17	15.09	15.55	15.55	15.55
SRT, d	7.66	18.15	28.15	50.60	105.79	167.25	257.64	393.77
pH = 7.0								
Flow, L/d	50.00	25.00	16.50	12.30	8.35	6.20	4.60	3.10
Detention Time, d	0.25	0.50	0.75	1.00	1.50	2.00	2.70	4.00
Efficiencly, per cent	30.00	38.00	53.00	73.00	76.00	81.00	86.00	88.00
Effluent COD, kg/M^3	6.30	5.58	4.23	2.43	2.16	1.71	1.26	1.08
Loading Rate, kg.COD$_r$/m^3.d	10.80	6.84	6.36	6.57	4.56	3.65	2.87	1.98
COD Removal, kg.COD$_r$/d	0.13	0.09	0.08	0.08	0.06	0.05	0.04	0.02
Methane Gas, m^3/d	0.04	0.03	0.03	0.03	0.02	0.02	0.01	0.01
Biogas Yield, m^{3}CH$_4$/m^3.d	3.28	2.16	2.24	2.40	1.68	1.28	1.12	0.72

Contd...

Table 45.3–*Contd...*

Influent SS, kg/m^3	0.42	0.41	0.40	0.40	0.40	0.39	0.38	0.36
Effluent SS, kg/m^3	0.77	0.73	0.71	0.66	0.60	0.56	0.53	0.51
Biomass Volume, L	1.80	1.95	1.95	2.30	2.45	2.53	2.53	2.53
Biomass Conc., kg/m^3	11.09	12.01	12.01	14.17	15.09	15.55	15.55	15.55
SRT, d	7.92	18.77	29.54	54.49	113.19	182.99	270.94	432.06
pH = 7.5								
Flow, L/d	50.00	25.00	16.50	12.30	8.35	6.20	4.60	3.10
Detention Time, d	0.25	0.50	0.75	1.00	1.50	2.00	2.70	4.00
Efficiencly, per cent	28.00	36.00	50.00	68.00	73.00	78.00	82.00	85.00
Effluent COD, kg/M^3	6.48	5.76	4.50	2.88	2.43	1.98	1.62	1.35
Loading Rate, kg.COD_r/m^3.d	10.08	6.48	6.00	6.12	4.38	3.51	2.73	1.91
COD Removal, kg.COD_r/d	0.13	0.08	0.07	0.08	0.05	0.04	0.03	0.02
Methane Gas, m^3/d	0.04	0.03	0.03	0.03	0.02	0.02	0.01	0.01
Biogas Yield, m$^3CH_4$/m^3.d	2.96	2.00	2.16	2.24	1.44	1.20	0.88	0.64
Influent SS, kg/m^3	0.43	0.42	0.41	0.40	0.40	0.39	0.39	0.38
Effluent SS, kg/m^3	0.78	0.75	0.72	0.67	0.61	0.57	0.55	0.53
Biomass Volume, L	1.80	1.95	1.95	2.30	2.45	2.53	2.53	2.53
Biomass Conc., kg/m^3	11.09	12.01	12.01	14.17	15.09	15.55	15.55	15.55
SRT, d	7.79	18.48	28.88	52.87	110.43	176.75	262.47	409.32

Pseudomechanistic model is used to determine the organic loading rate constant. The effluent to influent COD ratio versus reciprocal of organic loading rates under various pH and at a constant temperature of 35°C. From this the loading rate constant (K_2) is determined. The equation developed for a pH of 7.0 and a temperature of 35°C is:

$$Se/Si = \exp(-4.61/Lr)$$

Anaerobic fixed bed reactor can work even at very high flow rates because of the biomass attachment to the media. In order to determine the maximum applicable loading rate using Monod's hyperbolic expression, reciprocal of organic loading rate versus reciprocal of effluent concentration was plotted and shown in Figure 45.3. The reciprocal of the intercept of the plot gives the maximum loading rate that can be applied to the reactor. It is found that the maximum loading rate (Lm) that can be applied to fixed bed reactor is 20 kg COD_r/m^3.d. The value of inhibition coefficient (Ks) found to be 275 mg/l depending upon the experimental conditions. The influent COD concentration applied to the bioreactor was in the range of 2000 to 9000 mg/l.

The expression developed for maximum loading rate is as follows:

$$L_r = 20\,Se/275 + Se \text{ (Temperature} = 35°\text{C and pH 7.0)}$$

Conclusion

Studies on treatment of cotton digestion wastewater using anaerobic fixed bed reactor led to the following conclusions.

The substrate removal efficiency decreases linearly as the loading rate increases having maximum removal efficiency of 88 per cent at a lowest loading of 1.98 kg COD_r/m^3.d to a minimum of 30 per cent at 10.8 kg $COD/m^3/d$ at optimum conditions. The rate of decrease in COD reduction was higher upto a loading rate of 6 kg $COD_{r/}$ $_{m3}$.d and thereafter there was a decrease in the rate of per cent COD reduction. The detention time at this loading rate is one day.

Kinetic constants for substrate removal were determined using Dewalle and Chain model. The results reveal that substrate removal rate (K_2 A/V) increases with increase in specific surface area of the media.

The maximum loading rate (Lm) obtained from Lineweaver Burk relationship suggest that fixed bed reactor can be loaded upto a maximum of 20 kg COD_r/m^3.d.

References

1. B.Chakradhar, S.B. Deshukh and J.S. Gadgil, *Journal of Indian Association for Environmental Management*, Conference Issue, Volume 18, No. 1 and 2, February–June, 1991.
2. J.A. Oleszkiewics, Low Temperature Anaerobic Bio-filterations, *Journal of Water Pollution Control Federation (JWPCF)*, 1465, 1982.
3. F.B. Dewalle and E.S.K. Chain, Kinetics of Substrate Removal in a Completely Mixed Anaerobic Filter, *Biotech and Bioengg*, Vol XVIII, 1275-1295, 1975.

Chapter 46

Poly Aluminium Chloride (PAC): An Emerging Promising Coagulant in Water Treatment Practice

S.P. Pande, L.S. Deshpande, P.S. Kelkar and S.N. Kaul

CSIR-NEERI-Nagpur

ABSTRACT

Performance of commercial Poly Aluminium Chloride manufactured in India vis-à-vis ferric alum was studied using Jar test experiments. Research was conducted on removal of turbidity, heavy metals, cyanide, silica etc. The study showed that consumption (optimum dose) of PAC was 40 to 70 per cent that of ferric alum and it acted as a good coagulant over a turbidity range of 5 NTU to 1000 NTU. Floc formation with PAC was larger and heavy compared to the alum flocs with low generation of sludge volume. During coagulation process, PAC and alum, both showed marginal removal of heavy metals, cyanide, silica and arsenic contaminants. Literature review on the use of PAC for coagulation in water treatment has also been presented in the paper.

Keywords: *Poly aluminium chloride, Coagulation, Heavy metals removal, Turbidity removal.*

Introduction

The present trend in India under water sector is to construct more number of piped water supply schemes, even for rural installations. Regional Piped Water Supply Schemes (RPWS) cover a number of villages with a central water resource having treatment facility to provide safe potable water. In Ninth Five Year Plan, the Government of India, has allocated more than 206,000 million rupees for the water sector. The operation and maintenance cost of these water supply schemes range from Rs. 8 per capita per year for a handpump to Rs. 50 per capita per year for a regional PWS[1]. The comprehensive action plan of the State Governments indicate that the total requirement of funds for providing safe water within five years period would be around Rs. 440 billion [2].

Invariably in India, commercial alum is in use for coagulation purpose in all the technologies for water treatment. An in-depth survey of water treatment plants was carried out by NEERI [3] to bring out the 'state-of-art' of operation and maintenance of water treatment plants. The evaluation study had revealed that the addition of the chemicals like alum, lime and chlorine is indiscriminate and is not supported by proper laboratory testing of water quality. The stirring mechanism was also defunct at many places leading to improper doses of alum. Maintenance of alum solutionizing facilities and dosing equipment invariably leaves much to be desired and is the weakest link in proper O and M of the water treatment facility. In view of the above, replacement of commercial alum by another chemical was sought which is comparatively easier to handle, properly dosed and have a good efficiency for coagulation. Poly Aluminium Chloride (PAC) in one of such chemicals, now manufactured in India on commercial scale. Performance of such commercial PAC products was studied in-depth in the laboratory and plant scale for coagulation and removal of turbidity, heavy metals, cyanide and coliform organisms and was compared with that of alum.

PAC: Manufacture

Poly Aluminum Chloride (PAC) is the generic term for a group of coagulants produced by partial neutralization of aluminium chloride solution by some form of basic solution. This neutralization leads to the formulation of hydroxy aluminium species which may be monomeric, dimeric or polymeric depending on the degree of neutralization [4]. The process for preparation of polyaluminium hydroxy chloride complexes having general formula $Al(OH)_xCl_y$ in solution and intended to be used as flocculants and coagulants in water treatment have been patented [5] in Europe. Another process patented for presumed formula of $Al_a(OH)_b(SO_4)_cCl_{3a-b-2c}$ where a, b and c may have the values 13, 34 and 1 respectively has been used in water purification technology[6]. There are a number of patented formula for Poly Aluminium Chloride formulation particularly from Japan, Europe and U.S. The polymeric species $Al_{13}O_4(OH)_{24}^{7+}$ is thought to be most effective species [7]. At the international workshop [8] on 'Aluminium in drinking water' held at Hong Kong, January 15-17, 1992 important findings were presented on use of PAC in water treatment. PAC is now manufactured on commercial basis in India and products from two industries were studied in detail for their efficacy in coagulation and flocculation.

PAC in Water Treatment

Using aluminium based compounds, water of widely different chemical characteristics and biological quality can be successfully treated. Upon addition of Al coagulant in water treatment, multiple reaction pathways are possible. Aluminium monomers, Al-polymers and $Al(OH)_3$ may take place depending on pH [9]. Alum reacts with alkalinity in the water in form of calcium and magnesium bicarbonate producing Al $(OH)_3$

$Al_2 (SO_4)_3 18 H_2O + 3 Ca (HCO_3)_2 = 3 CaSO_4 + 2 Al (OH)_3 + 6 CO_2 + 18 H_2O$ $Al_2 (SO_4)_3 18 H_2O + 3 Mg (HCO_3)_2 = 3 MgSO_4 + 2 Al (OH)_3 + 6 CO_2 + 18 H_2O$

Coagulation occurs by interaction of aluminium hydrolysis products with contaminants such as dissolved organic matter and colloidal inorganic or organic polymers. Coagulation of colloidal particles in the presence of Al $(OH)_3$ is termed as enmeshment or sweep floc. Dissolved organics can be removed by adsorption on Al-precipitate. Chemical aspects of hydrolysis reactions during coagulation using alum and PAC studied by Benschoten *et al.*, [10] showed that alum did not form polymers in the water. On the other hand PAC consists of stable polymers below pH 6 and over the time period present in water treatment. Polyaluminium chloride consists of cationic hydrolysis polymers of aluminium. The polymeric structure remains intact within the PAC precipitate and particles are mere positively charged and produce low turbidity than alum floc. PAC is much less affected by temperature than alum.

Studies on coagulation kinetics of PAC and alum using kaolin clay suspensions indicated that increased alum doses reduced the time required for particle destabilization. PAC worked faster than alum. Increasing the Al doses accelerated the flocculation rate because it increased collision–attachment efficiency. PAC was less sensitive to change in pH and effectively coagulated fine particles than alum. Modelling studies on coagulation of organic contaminants like Fulvic Acid (FA) by alum and PAC showed dose requirements are nearly identical because Al binding by FA is similar for both Alum and PAC [11]. Dempsey *et al.*, [12] demonstrated that PAC is more efficient than alum or $AlCl_3$ for removal of humic substances (FA) at low pH. When a lower concentration of PAC than alum is effective in forming a precipitate with FA, the resulting sludge volume is also low compared to the use of alum. A literature review [13] on use of PAC for removal of organic contaminants show that coagulation is generally found to preferentially remove colour, THM and TOC precursors. At low to moderate concentration of contaminants PAC is likely to be a better coagulant than alum especially at pH values [< 5.5 or 7.0], may be due to greater charge neutralizing capacity of PAC for the stated conditions [14]. The laboratory experiments showed that PAC produced fastest settling flocs when rapid mix conditions were correctly controlled, however, these flocs were very sensitive to shear. Sludge volume per mass of contaminant removed was also low compared to alum sludge. Polymeric aluminium hydroxide solutions remain stable for few months without undergoing modifications in the properties. Thus, it is possible to store polymers for at least few months without significant loss in reactivity.

In another full scale plant study [15] demonstrated that at the Buffalo Pound Water Treatment Plant PAC coagulation could replace alum using approximately one third the dose normally required with alum. PAC consistently removed 20 per cent more turbidity than alum over a wide range of water temperatures at one third the dose. The financial considerations were not attractive at the prevailing high cost of the PAC.

Experimental Method

The coagulation studies using alum and PAC were confined to the laboratory experiments using Jar tests. The performance of PAC may vary with different brand name and even batch to batch manufacturing process.

PAC manufactured by two different firms using the latest generation state-of-art technology imported from Italy were tested. These are described as PAC (I) and PAC (II). As mentioned earlier the coagulation is influenced by a large number of interrelated factors such as pH, alkalinity, turbidity, sequence of addition of coagulant, the degree of agitation during flocculation etc. Since alum and commercial PAC were to be compared, the approach in the present study was to determine the most effective doses of these chemicals for specific test conditions.

A series of Jar tests were conducted using different turbid waters for evaluating the performance of PAC vis-à-vis ferric alum. Natural turbid water was collected from the river Kanhan which flows 15 km away from Nagpur city. Dosing of these natural and synthetic coagulants was done by (i) preparing 1 per cent solution in distilled water and using graduated pipette and (ii) neat with the help of microsyringe. It was noted that when the dosing was done with 1 per cent solution of PAC then the treated water showed slight haziness.

Measured volumes (1000 ± 10 mL) of samples were flocculated using the Jar test apparatus. After addition of coagulant, a rapid mix was maintained at 120 (10 rpm for 30 sec followed by slow mix at 25 ± 3 rpm for 10 minutes and the contents were allowed to settle for 20 minutes. The floc size and its settleability were observed visually with illuminating device at the base of Jar test apparatus. The test waters used for turbidity removal were either natural or synthetic waters prepared with dried powdered fine (below 100 μm sieve) black cotton soil. All the analytical estimations were done as per the procedures described in the Standard Methods [17].

Results and Discussion

Studies were conducted on natural and synthetic waters upto 1000 NTU turbidity and the results are presented in Table 46.1. The optimum dose of alum and PAC to achieve comparable settled water treatability was found to vary with different test waters. In general the consumption of PAC was 40 per cent–70 per cent that of ferric alum (Figure 46.1) and reduction in pH was less as compared to alum.

Sludge volume generated by alum and PAC was compared by studying their performance on naturally available turbid waters from Kanhan river as well as synthetic turbid waters prepared in the laboratory. Optimal Jar test doses were applied so that the settled water turbidity is around 5 NTU. The results of the settling test is given in Table 46.2. It was observed that with 3 hrs settling period, the reduction in sludge volume with PAC (I) as compared to that obtained with alum was 16 per cent in case of the river water and 11 per cent in case of synthetic turbid water. The PAC coagulated water produced longer and readily settleable flocs than alum.

Table 46.1: Performance of Coagulant for Different Turbid Waters

Ferric Alum			PAC (I)			PAC (II)		
Dose mg/L	Settled Water		Dose mg/L	Settled Water		Dose mg/L	Settled Water	
	Turbidity	pH		Turbidity	pH		Turbidity	pH
Raw Water Turbidity: 5 NTU, pH: 8.0								
6	4.0	7.9	4.0	3.0	8.0	2	2.8	8.0
7	3.8	7.9	5.0	3.0	8.0	3	2.6	7.9
8	3.2	7.8	6.0	2.5	7.9	4	2.5	7.9
10	2.7	7.8	7.0	2.3	7.9	5	2.2	7.9
Raw Water Turbidity: 50 NTU, pH 7.9; (Artificial Colloidal Turbidity)								
10	25	7.6	8	25	7.8	6	22	7.8
15	15	7.5	12	15	7.7	8	15	7.7
20	7	7.4	14	7	7.6	10	6.5	7.7
25	7.1	7.4	16	6.5	7.6	12	6.2	7.6
Natural Water Turbidity: 150 NTU, pH: 8.0								
20	25	7.0	15	20	7.6	10	15	7.7
30	12	6.8	20	10	7.6	15	10	7.6
40	5.5	6.5	28	5	7.5	20	4	7.6
50	6.0	6.3	35	4	7.5	25	3	7.5
Natural Water Turbidity 300 NTU, pH 8.0								
40	22	7.5	25	24	7.9	20	18	7.8
45	15	7.3	30	20	7.8	22	15	7.8
50	10	7.0	35	12	7.7	25	12	7.7
55	6	6.7	39	5.5	7.6	28	5	7.7
Artificial Raw Water Turbidity: 500 NTU, pH: 7.8								
25	15	7.4	16	20	7.8	12	15	7.7
30	12	7.1	20	16	7.8	15	10	7.7
35	6.5	7.0	22	12	7.7	17.5	7.0	7.6
40	8.0	6.8	25	7.5	7.6	20	6.0	7.6
Artificial Raw Water Turbidity: 800 NTU, pH: 7.8								
30	18	7.4	22	16	7.7	18	15	7.7
40	9.0	7.0	25	12	7.7	20	10	7.6
45	6.5	6.8	32	5.5	7.6	23	5	7.5
50	7.5	6.7	40	6.0	7.6	30	4	7.4
Natural (Kanhan River) Water Turbidity: 950 NTU, pH 8.0.								
60	58	7.5	40	38	7.9	35	25	7.8
70	40	7.3	50	30	7.8	40	20	7.7
80	25	6.5	55	25	7.8	45	5.0	7.6
90	7.5	6.3	65	6.5	7.7	50	5.0	7.5

Table 46.2: Sludge Volume Produced by the Coagulant

Sl.No.	*Time Hour*	*River Water Turbidity: 2010 NTU*		*Synthetic Water Turbidity: 1250 NTU*	
		Sludge Volume (mL/Litre of suspension)			
		Alum Dose 90 mg/L	*PAC (I) Dose 36 mg/L*	*Alum Dose 50 mg/L*	*PAC (I) Dose 20 mg/L*
1	0	0	0	0	0
2	0.5	19	15	35	31
3	1.0	14	12	31	27
4	1.5	14	12	28	24
5	2.0	13	11	26	21
6	2.5	13	12.5	25	20
7	3.0	12.5	10	23	20
8	3.5	11	9.5	20	18
9	4.0	11	9.5	20	18
Sl.No.	*Time Hour*	*River Water Turbidity: 950 NTU*		*Synthetic Water Turbidity: 1000 NTU*	
		Sludge Volume (mL/Litre of suspension)			
		Alum Dose 90 mg/L	*PAC (I) Dose 36 mg/L*	*Alum Dose 50 mg/L*	*PAC (I) Dose 20 mg/L*
1	0	0	0	0	0
2	0.5	24	20	46	36
3	1.0	18	16	36	32
4	1.5	18	16	33	29
5	2.0	17	15	30	25
6	2.5	17	14.5	29	24
7	3.0	15.5	13	27	24
8	3.5	15	13	24	22
9	4.0	14	12.5	23	22

Jar test experiments were conducted on synthetic waters of different alkalinities prepared in distilled water. Synthetic water were spiked with 1 mg/L of Fe and Mn and 0.55 mg/L of Cd, Cr, Pb and Hg. Settled water was filtered through Whatman 42 filter paper and filtrate was analysed for heavy metals by Jobin Yvon ICP Spectrometer and Coleman Mercury Analyser. There was considerable reduction in the heavy metal levels in the raw water itself due to settling. The actual removal of heavy metals exclusively due to the coagulants was not significant as shown in the Table 46.3.

Alum and PAC doses of 50 to 300 mg/L showed a marginal removal of arsenic from the test water containing 0.5 mg/L of arsenic. Similarly cyanide removal from

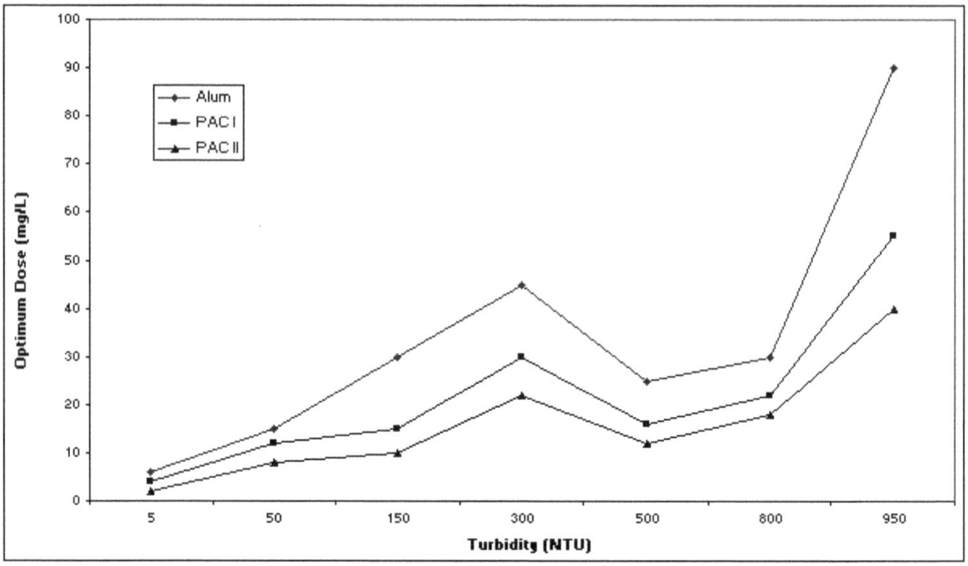

Figure 46.1: Performance of PAC vis-à-vis Ferric Alum as a Coagulant.

Table 46.3: Heavy Metal Removal Studies Synthetic Water Alkalinity (as $CaCO_3$): 600 mg/L

Sl.No.	Alum Dose mg/L	pH	Alkalinity mg/L as $CaCO_3$	Fe mg/L	Mn mg/L	Cd mg/L	Cr mg/L	Pb mg/L	Hg mg/L
1.	0	8.3	625	0.30	0.55	0.25	0.35	0.25	0.50
2.	50	7.7	590	0.25	0.50	0.20	0.32	0.20	0.48
3.	100	7.4	585	0.20	0.45	0.20	0.30	0.18	0.48
4.	300	6.8	510	0.15	0.42	0.20	0.28	0.18	0.48
Sl.No.	PAC (I) Dose mg/L	pH	Alkalinity mg/L as $CaCO_3$	Fe mg/L	Mn mg/L	Cd mg/L	Cr mg/L	Pb mg/L	Hg mg/L
1.	0	8.3	625	0.30	0.55	0.25	0.35	0.25	0.50
2.	50	7.9	600	0.26	0.52	0.22	0.32	0.22	0.50
3.	100	7.7	570	0.22	0.48	0.22	0.32	0.20	0.48
4.	300	7.4	562	0.18	0.42	0.19	0.30	0.18	0.48
Sl.No.	PAC (II) Dose mg/L	pH	Alkalinity mg/L as $CaCO_3$	Fe mg/L	Mn mg/L	Cd mg/L	Cr mg/L	Pb mg/L	Hg mg/L
1.	0	8.3	625	0.30	0.55	0.25	0.35	0.25	0.50
2.	50	7.9	610	0.28	0.52	0.22	0.32	0.22	0.50
3.	100	7.8	580	0.24	0.50	0.20	0.32	0.18	0.48
4.	300	7.3	570	0.20	0.46	0.18	0.30	0.18	0.47

the test water contaning 0.7 mg/L cyanide was also marginal during coagulation with alum and PAC (Doses ranged from 20–120 mg/L).

Synthetic water (Alkalinity: 250 mg/L) spiked with 15 mg/L of silica was used in Jar test experiments. Alum and PAC dose were varied between 0 and 80 mg/L. In case of alum maximum removal of silica was 1.3 per cent at 80 mg/L dose whereas for PAC I and PAC II, the removal of silica was to the extent of 2.7 per cent (at 80 mg/L dose) and 1.3 per cent for 80 mg/L respectively.

Conclusion

- ☆ Studies were carried out to evaluate effectiveness of PAC manufactured in India vis-à-vis use of alum as a coagulant in water treatment Both natural and synthetic turbid water were used in the Jar test experiments. Comparative performance of these coagulants was studied for removal of turbidity, heavy metals and cyanide. The studies concluded that PAC acted as a good coagulant over a wide range of turbidities (5 to 1000 NTU) both in synthetically prepared and in natural waters
- ☆ The average dose required for PAC is 50–70 per cent that of alum when it is prepared in 1 per cent aqueous solution and consumed less alkalinity than alum for equitable performance
- ☆ Flocs produced by PAC were larger and readily settleable compared to the alum flocs
- ☆ Sludge volume generated by PAC was less than that generated by alum for obtaining the same quality of settled water
- ☆ PAC and alum both showed marginal remove of heavy metals for waters having different turbidities
- ☆ Similarly cyanide, silica and arsenic elimination exclusively due to coagulation was equally comparable for both coagulants and had marginal removal efficiency for these contaminants.

References

[1. Mazumdar K. Rural drinking water supply in India–A critical review. *Journal of Indian Water Works Association.* 2001; Vol. XXXIII No. 3: 211.

2. Annual reports published by Rajiv Gandhi National Drinking Water Mission, Ministry of Rural Development.

3. Report on "Evaluation of Water Treatment Plants in India', NEERI, Nehru Marg, Nagpur–440 020 (India); 1989: 32.

4. Bottero JY., Poirier JE. and Fiessinger F; Prog. Study of partially neutralized aqueous aluminium chloride solutions: Identification of aluminium species and relations between the composition of solutions and their efficiency as a coagulant. *Prog. Wat. Tech,* 1980; 12: 601-612.

5. Kemira Kemi Aktiebolag Heisingborg (SE), European Patent, application number 90850050.7 date of filing: 06.02.1990.

6. Carlesson C., Sweden United States Patent, application number 797, 887 date of filing 14.11.1985.

7. O'Melia, CR., Gray KA, and Yao, C. Polymeric inorganic coagulants *AWWA Research Foundation* Denver, 1989.

8. Kaeding UW, Drikas M., Dellaverde P.J., Martin D., A direct comparison between aluminium sulphate and polyaluminium chloride as coagulant in a water treatment plant, *Proceedings of International Workshop on Aluminium in Drinking Water* held at Hong Kong; 1992: 117 – 131.

9. Hundt TR, and O'Melia CR, Aluminium–Fulvic Acid interaction: mechanism and application, *J.A.W.W.A*, 1988; 80 (4): 176-186.

10. Benschoten JE, and Edzwald JK, Chemical aspects of coagulant using aluminium salt–I Hydrolytic reaction of alum and polyaluminium chloride. *Water Research;* 1990; 24 (12): 1519-1526.

11. Benschoten JE, Edzwald JK, Chemical aspects of coagulation using aluminum salt–II: Coagulation of Fulvic acid using alum and polyaluminium chloride, *Water Research*, 1990; 24 (12): 1527-1535.

12. Dempsey BA, Ganho RM, O'Melia CR, The coagulation of humic substances by means of aluminium salts, *J.A.W.W.A.*, 1984; 76 (4): 141-150.

13. Randtke SJ, Organic contaminant removal by coagulation and related process combinations, *J.A.W.W.A.*, 1988; 80 (5): 40-56.

14. Dempsey BA, Shea H, Ahmed TM and Mertink, J Polyaluminium chloride and alum coagulation of clay-fulvic acid suspensions, *J.A.W.W.A*, 1985; (3): 24-80.

15] Parthasarathy N. and Buffle J., Study of polymeric aluminium III hydroxide solutions for applications in wastewater treatment. Properties of the polymer and optimal conditions of preparation. *Water Research*, 1985; 19 (1): 25-36.

16. *Standard Methods for the Examination of Water and Wastewater*, 1998 20th Edn. APHA, AWWA and WEF.

Chapter 47

Electrochemical Oxidation of Tannery Wastewater

N.N. Rao, K.M. Somasekhar, S.N. Kaul[1]*,*
and L. Szpyrkowicz[2]

[1]*CSIR-NEERI, Nagpur*
[2] *University of Venice, Venice, Italy*

Introduction

Tannery wastewater was treated by an electrochemical oxidation method using Ti/Pt, Ti/PbO_2 and Ti/MnO_2 anodes and a Ti cathode in a two-electrode stirred batch reactor. The changes in colour concentration, chemical oxygen demand (COD), ammonia (NH_4^+), sulfide and total chromium have been determined as a function of treatment time and applied current density. Gas chromatography-mass spectrometry (GC-MS) analysis, performed on the wastewater samples before and after treatment, as well as on foam samples, is reported. Anode efficiency, rate constants, and energy consumption were estimated and discussed. The efficiency of Ti/Pt was 0.802 kgCOD h^{-1} A^{-1} m^{-2} and 0.270 kgNH_4^+ h^{-1} A^{-1} m^{-2}, and the energy consumption was 5.77 kWhkg^{-1} COD and 16.63 kWhkg^{-1} of NH_4^+. The order of efficiency of anodes was found to be Ti/Pt » Ti/PbO_2 > Ti/MnO_2. The results indicate that the electro-oxidation method could be used for effective oxidation of tannery wastewater and a final effluent with substantially reduced pollution load can be obtained.

The electrochemical treatment of organic compounds in wastewater is potentially a powerful method of pollution control. Oxidizing agents such as hypochlorite, oxygen-based radicals, ozone and nitrogen oxides can be generated *in-situ* in an electrochemical reactor. These oxidizing agents are utilized for oxidation of organic compounds. Accordingly, the electro-odixation of phenol,[1–4] 1,4-benzoquinone,[5] aliphatic and aromatic acids and alcohols,[6–7] nitration waste produced during the manufacture of TNT,[8] dyes, dye house effluents and textile wastewater,[9–11] tannery wastewater,[12–13] distillery wastewater[14] and landfill leachate,[15] have been reported.

Different anode materials, *viz.* platinum, graphite, carbon, ruthenium oxide and iridium oxide coated Ti substrates, Ti/Pt, Ti/Pt/Ir, Ti/SnO_2, Ti/SnO_2-Sb_2O_5 and Ti/PbO_2 have been used for electrochemical wastewater treatment. Due to the high cost and limited operational range of oxidation potentials for Pi-based anodes, anode materials comprising oxide-based electrocatalysts have received considerable attention[16]. There is an obvious need to examine the oxidation efficiencies of these oxide/based anode materials in the treatment of real wastewater samples and to compare the efficiencies with those of Pt-based anode materials. In view of this, we investigated the electrochemical oxidation of tannery wastewater using various anodes, *viz.* Ti/Pt, Ti/MnO_2 and Ti/PbO_2. The studies revealed that almost all pollution parameters were substantially reduced during this treatment. Furthermore, the Ti/Pt anode was found to be more effective than the oxidee-based anodes for this treatment.

Experimental

The hexachloroplatinic acid, manganese sulfate, sulfuric acid and hydrochloric acid used for this study were of high purity analytical grade. The treated tannery wastewater (secondary clarifier effluent) was obtained from an effluent treatment plant (ETP) engaged in treating about 1250-1350 m^3 of wastewater per day, collected from more than 85 tanneries.

The tanneries linked to the ETP carried out various operations such as the processing of semi-finished to finished leather, based on chrome, as well as vegetable tanning. Some of the important physico-chemical characteristics of the tannery wastewater are given in Table 47.1.

Table 47.1: Some Physico-chemical Characteristics of Tannery Waste-water (Secondary clarifier effluent) Used in the Present Study

Sample No.	*Parameter*	*Magnitude*
1	Colour	Brown 1100-1150 PCU
2	pH	8.00-8.30
3	Chemical Oxygen Demand	480-550
4	Ammonia (NH_4^+)	100-110
5	Sulfide (S^{2-})	1.8-2.0
6	Chloride (Cl^-)	4500-4600
7	Chromium (Cr^{3+} and Cr^{6+})[a]	4.0-4.5

PCU: Pt-Co units.

All other parameters except pH are reported in mg dm^{-3} [a] Trace amounts of Cr^{6+}.

Titanium foils having a thickness of about 1 mm were used for preparing metal or oxide-coated anodes. The Ti/Pt anode was prepared by dabbing a suspension of platonic acid in iso-propanol and heating it in air in a muffle furnace at 500°C for a period of 4 h. This procedure was repeated 10 times so as to attain satisfactory dispersion of Pt on the Ti support. A grey-black deposit of Pt was obtained. The surface area of the Ti/Pt anode was 9.0 cm^2 (3x3 cm). Similarly, manganese from manganese sulfate (0.25 mol dm^{-3}) in 0.2 mol dm^{-3} H_2SO_4 was electro-deposited over a titanium cathode, followed by slow thermal oxidation of Ti/Mn at 400°C for 8 h in a furnace to obtain Ti/MnO_2. A deposit of black MnO_2 was obtained. The surface area of the Ti/MnO_2 anode was 98 cm^2 (10x9.8 cm). Deposition of PbO_2 on titanium plates from an acid medium was achieved at constant potential (1.8 V, 5-20 min) from a solution of 0.1 mol dm^{-3} HNO_3 containing 0.5 mol dm^{-3} Pb $(NO_3)_2$ and 0.04 mol dm^{-3} NaF, according to a reported method [17]. The chocolate-brown PbO_2 deposit was found to be uniform and adherent. The surface area of the Ti/PbO_2 anode was 30 cm^2 (6x5 cm). The cathode was a titanium sheet of size appropriate to that of anode.

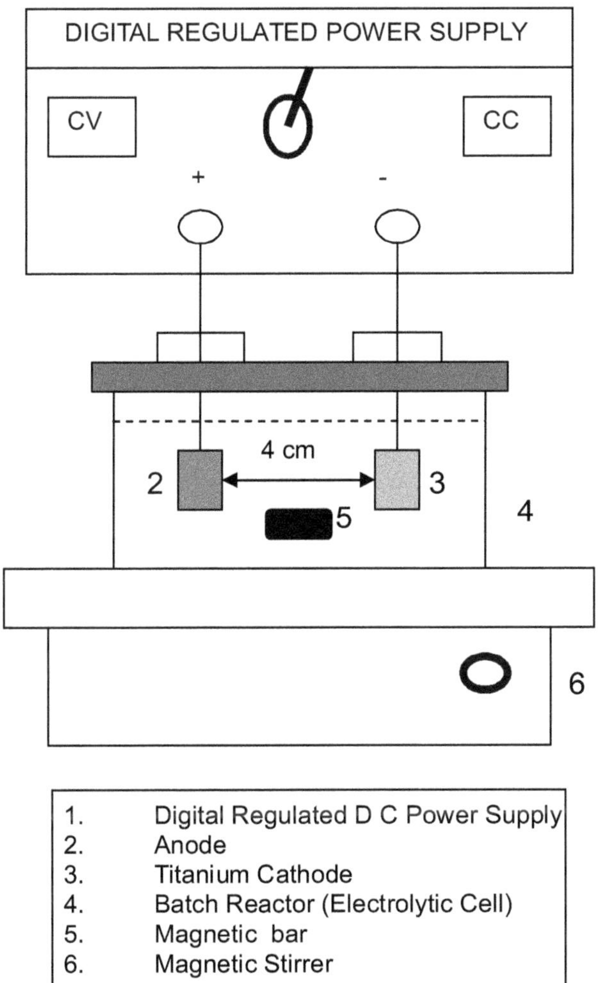

Figure 47.1: Experimental Electro-Oxidation Set-up.

All electro-oxidation experiments were conducted in a Plexiglass batch reactor with a capacity of about 200 cm^3 (5x6x5.5 cm) or 650 cm^3 (11x10z5.9 cm) capacity. The anode and cathode were placed at a fixed distance of 4 cm apart by inserting contact rods (2 mm Ti wire) into holes drilled in the head-plate of the reactor. An experimental set-up comprising a two-electrode stirred batch reactor is shown in Figure 47.1. The working volume of wastewater was 60 cm^3 for a Ti/Pt anode-containing cell 200 cm^3 for a Ti/PbO_2–containing cell and 600 cm^3 for a Ti/MnO_2 –containing cell. A working volume-to-surface area ratio of about six was maintained in the present studies. The electrolyte was magnetically stirred during experiments. The current dependence was evaluated for each case by varying the current in the range 0.1-0.6 A. All experiments were conducted under galvanostatic conditions using a DC Power Supply (Testronix 34D, 0-15V and 0-10A). During electrolysis, 5 or 10 cm^3 aliquots were withdrawn at regular intervals over a period of 4h and the parameters, chemical oxygen demand (COD), ammonium, sulfide and total chromium (Cr^{3+} and Cr^{6+}) were determined according to APHA Standard Procedures. [18] Colour measurements were made according to the Pt-Co standard method. [18] Several runs were performed under each experimental condition to ensure reproducibility. The replication error was found to be normally <5 per cent. The first-order rate constants (k, min^{-1}) were estimated from log[$(COD_t)/(COD_o)$] or log[$(NH_4^+)_t/(NH_4^+)_o$] *versus* time plots according to a first-order kinetic relationship, *i.e.* $\log [C_t/C_o] = kt$. The plots were linear with an R^2 value greater than 0.9. The anode efficiency (kgCOD or NH_4^+ h^{-1} A^{-1} m^{-2}) as well as energy consumption (kWhkg^{-1} COD or NH_4^+) were calculated.

Compounds present in tannery wastewater before and after electrochemical oxidation treatment were identified by GC-MS. The foam that accumulated in the headspace was also collected. The liquid that resulted after settling of foam was processed for extraction of compounds. The compounds were extracted into dichloromethane (25 cm^3 x 3) using a 200 cm^3 wastewater sample. A varian model Gas Chromatograph (Star 3400CX series) coupled with a Mass Spectrometer (Saturn-3) with a DB-5 column (30 m length, 0.25 mm id, 0.25 µm film thickness) was used. Chromatographic separation involved a total run time of 26 min; the column was held initially at 150°C for 1 min and the temperature was allowed to increase gradually at a rate of 5°C min^{-1} up to 250°C.

Results and Discussion

A typical electrochemical treatment led to visible colour removal from the wastewater and formation of foam laden with fine suspended solids. Initially more form was formed at higher current densities due to more rapid gas evolution. Furthermore, the foam was found to be suppressed substantially towards the end of treatment time. There was evidence of some chlorine gas evolution at the anode. The results of the electrochemical experiments using Ti/Pt, Ti/PbO_2 and Ti/MnO_2 anodes are presented in Tables 47.2–47.4. Apart from reductions in colour, COD, Cr, NH_4^+ and S^{2-}, the instantaneous COD removal normalized to initial COD, *i.e.* [$(COD)_{t=0}$ – $(COD)_t$]/ $(COD)_{t=0}$, is also included. The results indicate that, in general, the concentration of all pollutants present in tannery wastewater are simultaneously reduced at all current densities with all types of anode used. The colour reduction

Table 47.2: Electrochemical Treatment of Tannery Wastewater using Ti/Pt Anode

Electrolysis Time (h)	*Ti/Pt*					
	I = 0.09A V = 4.16V		*1.00A dm^{-2a}*		*0.375 Wh^b*	
	Colour (Pt–Co)	*COD (mg dm^{-3})*	*Cr (mg dm^{-3})*	*NH_4^+ (mg dm^{-3})*	*S^{2-} (mg dm^{-3})*	*ICR^c*
0	1115	512	4.3	103	2	—
1	534	454	3.1	45	1.6	0.113
2	377	386	2.3	28	1.1	0.246
3	270	329	1.2	20	0.7	0.357
4	265	252	1	14	0.4	0.507
%R	76	51	76	86	80	
Electrolysis Time (h)	*Ti/Pt*					
	I = 0.2A V = 10V		*2.22A dm^{-2a}*		*2.0 Wh*	
	Colour (Pt–Co)	*COD (mg dm^{-3})*	*Cr (mg dm^{-3})*	*NH_4^+ (mg dm^{-3})*	*S^{2-} (mg dm^{-3})*	*ICR^c*
0	1124	506	4.3	106	1.9	—
1	448	413	3.1	43	1.5	0.183
2	365	354	2.2	28	0.9	0.3
3	250	288	1.1	18	0.5	0.43
4	160	224	0.98	7.9	0.38	0.557
%R	86	56	77	92.5	86	
Electrolysis Time (h)	*Ti/Pt*					
	I = 0.5A V = 27.6V		*5.56A dm^{-2a}*		*13.8 Wh*	
	Colour (Pt–Co)	*COD (mg dm^{-3})*	*Cr (mg dm^{-3})*	*NH_4^+ (mg dm^{-3})*	*S^{2-} (mg dm^{-3})*	*ICR^c*
0	1107	515	4.3	107	2	—
1	437	363	3	44	1.4	0.295
2	168	325	2.2	26	0.8	0.369
3	155	247	1.1	16	0.4	0.52
4	136	189	1	4.2	0.24	0.633
%R	88	62	76	96	88	

I: Current in amperes; V: Cell voltage in volts; a: Current density, A dm^{-2}; b: Power consumption; c: Normalized instantaneous COD removal (ICR) = $[COD_{t=0} - COD_t]/COD_{t=0}$.

ranged between 75 and 95 per cent, COD reduction between 50 and 75 per cent, NH_4^+ reduction between 48 and 96 per cent, S^{2-} reduction was in the range 60-84 per cent and reduction in total chromium was between 65 and 82 per cent depending upon the type of anode and the magnitude of applied current over a period of 4h. Colour

Table 47.3: Electrochemical Treatment of Tannery Wastewater using Ti/PbO$_2$ Anode

Electrolysis Time (h)	Ti/PbO$_2$					
	I = 0.2A V = 4.05V		0.67A dm^{-2a}		0.81 Wh^{b}	
	Colour (Pt–Co)	COD (mg dm^{-3})	Cr (mg dm^{-3})	NH_4^+ (mg dm^{-3})	S^{2-} (mg dm^{-3})	ICR^{c}
0	1137	536	4.8	104	1.98	—
1	678	404	3	85	1.51	0.246
2	585	296	2.3	71	1.23	0.448
3	150	240	1.8	59	1.01	0.552
4	126	192	1.6	54	0.76	0.642
%R	89	64.2	66.6	48.1	61.6	

Electrolysis Time (h)	Ti/PbO$_2$					
	I = 0.35A V = 5.71V		1.17A dm^{-2a}		2.0 Wh	
	Colour (Pt–Co)	COD (mg dm^{-3})	Cr (mg dm^{-3})	NH_4^+ (mg dm^{-3})	S^{2-} (mg dm^{-3})	ICR^{c}
0	1113	520	4.8	105	1.98	—
1	483	304	3	75	1.5	0.415
2	295	272	2.2	54	1.1	0.477
3	183	232	1.4	42	0.95	0.554
4	103	180	1	35	0.7	0.654
%R	91	65.4	79.1	66.6	54.6	

Electrolysis Time (h)	Ti/PbO$_2$					
	I = 0.50A V = 10.12V		1.67A dm^{-2a}		5.06 Wh	
	Colour (Pt–Co)	COD (mg dm^{-3})	Cr (mg dm^{-3})	NH_4^+ (mg dm^{-3})	S^{2-} (mg dm^{-3})	ICR^{c}
0	1122	544	4.8	102	1.99	—
1	634	280	3	71	1.45	0.485
2	252	232	2.4	45	1.21	0.573
3	132	184	1.2	31	0.95	0.661
4	124	156	0.9	22	0.7	0.713
%R	89	71.3	81.2	78.4	64.8	

I: Current in amperes; V: Cell voltage in volts; a: Current density, A dm^{-2}; b: Power consumption; c: Normalized instantaneous COD removal (ICR) = [COD$_{t=0}$ – COD$_t$]/COD$_{t=0}$.

removal was highest (dikka91 per cent) using a Ti/MnO$_2$ anode at the lowest applied current density of 0.2A dm^{-2}, whereas comparable colour removal could be obtained at current density >0.67A dm^{-2} using Ti/PbO$_2$ and >5.56A dm^{-2} using Ti/Pt anode.

Table 47.4: Electrochemical Treatment of Tannery Wastewater using Ti/MnO$_2$ Anode

Electrolysis Time (h)	Ti/MnO$_2$					
	I = 0.2A V = 4.06V		0.204A dm^{-2a}		0.812 Wh^b	
	Colour (Pt–Co)	COD (mg dm^{-3})	Cr (mg dm^{-3})	NH_4^+ (mg dm^{-3})	S^{2-} (mg dm^{-3})	ICR^c
0	1114	480	4.81	104	1.9	—
1	319	372	3.02	81	1.61	0.225
2	284	306	2.01	63	1.42	0.362
3	99	256	1.3	47	1.18	0.466
4	99	220	0.89	35	0.9	0.542
%R	91	54.2	81.5	66.3	52.6	
Electrolysis Time (h)	Ti/MnO$_2$					
	I = 0.4A V = 12.9V		0.408A dm^{-2a}		5.16 Wh	
	Colour (Pt–Co)	COD (mg dm^{-3})	Cr (mg dm^{-3})	NH_4^+ (mg dm^{-3})	S^{2-} (mg dm^{-3})	ICR^c
0	1114	480	4.81	104	1.91	—
1	554	356	2.9	78	1.6	0.258
2	217	290	1.9	57	1.38	0.396
3	162	238	1.1	38	1.05	0.504
4	118	184	0.89	28	0.7	0.616
%R	89.4	61.6	81.5	73	63.3	
Electrolysis Time (h)	Ti/MnO$_2$					
	I = 0.60A V = 14.15V		0.612A dm^{-2a}		8.49 Wh	
	Colour (Pt–Co)	COD (mg dm^{-3})	Cr (mg dm^{-3})	NH_4^+ (mg dm^{-3})	S^{2-} (mg dm^{-3})	ICR^c
0	1134	480	4.81	102	1.93	—
1	558	280	2.7	65	1.55	0.416
2	349	208	1.61	51	1.21	0.566
3	235	164	1.05	28	0.91	0.658
4	69	116	0.87	10	0.58	0.758
%R	94	75.8	81.9	90	69.9	

I: Current in amperes; V: Cell voltage in volts; a: Current density, A dm^{-2}; b: Power consumption; c: Normalized instantaneous COD removal (ICR) = [$COD_{t=0}$ – COD_t]/$COD_{t=0}$.

The reduction in COD and its current dependence using the chosen anodes is depicted in Figure 47.2. In the same way, NH_4^+ reduction and its dependence on the current density are shown in Figure 47.3.

In general, higher applied current densities resulted in greater per cent removals of COD as well as NH_4^+. However, the removal of NH_4^+ at Ti/Pt showed very little dependence on the current density [see Figure 47.3 (a)]. The first-order rate constants derived for COD and NH_4^+ removals are presented in Table 47.5.

Table 47.5: Rate Constants for COD and NH_4^+ Removals using Various Anodes

Sample No.		Ti/Pt			
	I (A dm^{-2})	COD k x 10^{-3} min^{-1}	R^2	NH_4^+ k x 10^{-3} min^{-1}	R^2
1	1.00	1.17	0.97	3.94	0.94
2	2.22	1.41	0.98	4.64	0.97
3	5.56	1.80	0.98	5.40	0.96
Sample No.		Ti/PbO_2			
	I (A dm^{-2})	COD k x 10^{-3} min^{-1}	R^2	NH_4^+ k x 10^{-3} min^{-1}	R^2
1	0.67	1.93	0.99	1.27	0.98
2	1.17	2.06	0.87	2.12	0.98
3	1.67	2.56	0.84	2.82	0.99
Sample No.		Ti/MnO_2			
	I (A dm^{-2})	COD k x 10^{-3} min^{-1}	R^2	NH_4^+ k x 10^{-3} min^{-1}	R^2
1	0.204	1.49	0.98	1.93	0.99
2	0.408	1.75	0.99	2.35	0.99
3	0.612	2.68	0.96	3.62	0.92

I: Current density.

Although the current density at these anodes was raised 2.5-5.5 times, the change in the rate constants is observed to be small. Thus, for 1.00, 2.22 and 5.56A dm^{-2} using the Ti/Pt anode, the rate constants for COD removal are found to be 1.17×10^{-3}, 1.41×10^{-3} and 1.80×10^{-3} min^{-1}. An increase in *k* of 1.20 and 1.54 times, relative to that at 1.00A dm^{-2}, is found at 2.22 and 5.56A dm^{-2} respectively. The *k* values for COD removals using the Ti/PbO_2 anode are: 1.93×10^{-3}, 2.05×10^{-3} and 2.56×10^{-3} min^{-1}, respectively at 0.67, 1.17 and 1.67A dm^{-2}. The *k* values for COD removal using the Ti/MnO_2 anode are 1.49×10^{-3}, 1.75×10^{-3} and 2.68×10^{-3} min^{-1} respectively at 0.204, 0.408 and 0.612A dm^{-2}. Similarly, for 1.00, 2.22 and 5.56A dm^{-2} using Ti/Pi anode, the rate constants for NH_4^+ removal are: 3.94×10^{-3}, 4.64×10^{-3} and 5.40×10^{-3} min^{-1}; for 0.67, 1.17 and 1.67A dm^{-2} using the Ti/PbO_2 anode the *k* values are 1.27×10^{-3}, 2.12×10^{-3} and 2.82×10^{-3} min^{-1}; for 0.204, 0.408 and 0.612A dm^{-2} using the Ti/MnO_2 anode, the *k* values are 1.93×10^{-3}, 2.35×10^{-3} and 3.62×10^{-3} min^{-1}. It is clear that ammonia removal is more facile using the Pt/Ti electrode.

The efficiency of Ti/Pt, Ti/PbO_2 and Ti/MnO_2 anodes and energy consumption at various current densities are presented in Table 47.6. The anode efficiency has been calculated in kilograms of COD or NH_4^+ removed per hour per ampere and per square metre (kgCOD or NH_4^+ h^{-1} A^{-1} m^{-2}). The Ti/Pt anode showed the highest anode efficiency for the reduction of COD as well as NH_4^+. The efficiency of the Ti/Pt anode was 0.802 kgCOD h^{-1} A^{-1} m^{-2} and 0.270 kgNH_4^+ h^{-1} A^{-1} m^{-2}. At the lowest current applied, the Ti/PbO_2 anode showed a five times lower efficiency for COD removal and a 13 times lower efficiency for NH_4 removal. Similarly, the efficiency of Ti/MnO_2 was 24 times (for COD) and 30 times (for NH_4^+) lower when compared with that of Ti/Pt. Hence, the order of anode efficiency may be written as Ti/Pt » Ti/PbO_2 > Ti/MnO_2. This is in apparent contradiction to the somewhat higher percentage COD removal and higher rate constants found with Ti/PbO_2 and Ti/MnO_2 anodes (Tables 47.2–47.4). It may be noted that the calculation of anode efficiency, unlike that of percentage COD reduction and rate constants, takes into account the anode surface area and applied current. Therefore, anode efficiency values allow an appropriate comparison of performance of anodes. The energy consumption was 5.77 kWh kg^{-1} COD and 16.63 kWhkg^{-1} of NH_4^+ using Ti/Pt. The energy consumption per kgCOD removal at the lowest current density is about 1.6 and 2.2 times higher for Ti/PbO_2 and Ti/MnO_2 anodes, respectively, when compared with that for a Ti/Pt anode. The energy consumption per kgNH_4^+ removal using Ti/PbO_2 and Ti/MnO_2 is about three to four times more when compared with that for Ti/Pt. Application of higher current densities actually decreased the anode efficiency and increased the energy consumed per kg of COD of NH_4^+ removed. This may be due to the competing water electrolysis reaction that becomes dominant at higher current densities. In a similar context, Szpyrkowicz *et al*[12] reported that 1.32 kWh is required to remove 1 kg of NH_4^+ from biologically treated tannery wastewater using a graphite anode at a current efficiency of 19.1 per cent. This is eight, two and three times higher than the energy requirement

Table 47.6: Comparison of Energy Consumption Data of different Anode Materials at different Current Densities

Sample No.	*Anode Type*	*Current Density*	*Rate[a] (mg dm^{-3} h^{-1})*		*Anote Efficiency[b]*		*Actual Power Consumed (Wh)*	*Energy Consumption*	
			COD	*NH_4^+*	*KgCOD h^{-1} A^{-1} m^{-2}*	*KgNH_4^+ h^{-1} A^{-1} m^{-2}*		*kWh kg^{-1} COD*	*kWh kg^{-1} NH_4^+*
1	Ti/Pt	1.00	65.0	22.25	0.802	0.270	0.37	5.77	16.63
		2.22	69.0	24.52	0.383	0.136	2.00	29.00	81.56
		5.56	76.8	25.70	0.171	0.057	13.80	179.80	537.00
2	Ti/PbO_2	0.67	86.0	12.50	0.142	0.021	0.81	9.42	64.80
		1.17	85.0	17.50	0.081	0.017	2.00	23.53	114.28
		1.67	97.0	20.0	0.065	0.013	5.06	52.16	253.00
3	Ti/MnO_2	0.204	65.0	17.25	0.033	0.009	0.81	12.46	46.95
		0.408	74.0	19.00	0.019	0.005	5.16	69.73	271.60
		0.612	91.0	23.00	0.015	0.004	8.49	93.30	369.00

a: 4-h average of net COD reduction observed in each case; b: Anode efficiency estimated as kgCOD or kgNH_4^+ removed per hour per ampere and per square metre.

for Ti/Pt, Ti/PbO_2 and Ti/MnO_2 anodes respectively in the present study. With regard to NH_4^+ removal from aerobically treated tannery wastewater, Szpyrkowicz *et al* [13] reported a first-order rate constant of 0.045 min^{-1}, 25.8 per cent current efficiency and 87.2 kWh kg^{-1} NH_4^+ removed using Ti/Pt anode at 2.0 Adm^{-2}. Similar energy consumption for ammonia removal at a comparable current density using Ti/Pt was obtained in the present study also (81.56kWh kg^{-1} NH_4^+, Table 47.6). Vlyssides *et al* [9] reported the use of a Ti/Pt anode, although referring to electrochemical oxidation of a textile dye wastewater. They report the anode efficiency was about 0.170 kgCOD h^{-1} A^{-1} m^{-2} and the mean energy consumption was 21 kWh kg^{-1} COD. The presence of reactive azo dyes in this dye wastewater may be responsible for lower anode efficiency and higher energy consumption. A titanium alloy covered with 0.25 mm thick Pt alloy foil was used as anode in these studies.

The details of compounds identified through GC-MS analysis are presented in Table 47.7. The raw wastewater showed the presence of various organic compounds of which the prominent ones were: phenol, 1,6-anhydro-3-4-dideoxy-ß-d-glucohexapyranose, naphthalene, pelletierine, 5-benzoyloxy benzoic acid, 2,4-dioctylphenol, 1-methoxy-4- (2-phenyl-ethenyl)-benzene, 1,2-benzene dicarboxylic acid di-isooctylester, 17α-ethnyl-17ß- hydroxy-5ß-estran-3-one and dibutyl phthalate. Fewer compounds were found in the liquid phase after the electrochemical treatment. The treatment resulted in disappearance of peaks due to phenol, naphthalene and steroid from the liquid phase, while the peaks corresponding to some other compounds showed a reduction in their concentrations. Interestingly, the liquid phase after electro-oxidation contained some substituted phenolic compounds, *viz.* 2,5-bis (1,1-dimethylethyl) phenol and 2,4,6-tris (1,1-dimethylethyl) phenol, indicating that some chemical transformations occur. The foam sample showed peaks attributable to 1,1,2,2-tetrachloroethane, phenol, naphthalene, 2-hydroxybenzaldehyde, butylated hydroxytoluene, diphenylamine and sulfur. The reaction of anodically-generated free chlorine with organic pollutants can lead to the formation of chloro-organic compounds [12]. In the present study, the only chloro-organic compound detected is tetrachloroethane. Due to its higher volatility, it appears to have transferred into the vapour phase and become trapped in the foam. Many other lower molecular weight compounds such as phenol, naphthalene, hydroxy benzaldehyde, etc. are also carried into the foam. Molecular sulfur, probably a product of electrochemical oxidation of sulfide, was also identified in the foam. Rajalo and Petrovskaya [19] have earlier demonstrated electro-chemical oxidation of sulfides to sulfur in tannery wastewater.

The electrochemical reactions that take place during electrolysis of wastewater are complex. They are frequently interpreted as being based on the formation of products such as •O, •OH, O_2, •O_2H, H_2O_2, •Cl, Cl_2, ClOH•, ClO_2 and H_2 at electrodes. The radicals among them, owing to their higher oxidation potential, oxidize organic compounds directly. Alternatively, they may disintegrate into more stable oxidants. The more stable oxidants diffuse into the solution phase and react with organic compounds. This is referred to as indirect oxidation. The direct oxidation of organic compounds is mainly influenced by the catalytic activity of the anode, the diffusion rates of the organic compounds towards the anode surface and the applied current density. On the other hand, the rate of indirect electro-oxidation is dependent on the

Table 47.7: Chemical Compounds Found in Tannery Wastewater before and after Electrochemical Oxidation Treatment (GC-MS data)

Before Electrochemical Oxidation		
RT (min)	*Chemical Name*	*Formula*
4.71m	1,6-Anhydro-3-4-dideoxy-β-o-glucohexapyranose	$C_5H_{10}O_2$
4.91s	Pelletierine	—
5.56m	Phenol	C_6H_6O
8.41s	Naphthalene	$C_{10}H_8$
15.33s	5-Benzoyloxy benzoic acid	$C_{12}H_{14}O_3$
20.99s	1-Methoxy-4- (2-phenyl-ethenyl)-benzene	$C_{15}H_{14}O$
17.04s20.81s	Dibutyl phthalate	$C_{16}H_{22}O_4$
16.61s18.91s	2,4-Dioctylphenol	$C_{22}H_{38}O$
22.91s	1m2-benzene dicarboxylic acid di-iso-octylester	$C_{24}H_{38}O_4$
23.56vs	17α-ethynyl-17-β-hydroxy-5β-estran-3-one	$C_{24}h_{39}NO_2S$

After Electrochemical oxidation					
Liquid Phase			*Foam*		
RT (min)	*Chemical Name*	*Formula*	*RT (min)*	*Chemical Name*	*Formula*
4.71m	1,6-Anhydro-3-4-dideoxy-β-o-	$C_5H_{10}O_2$	4.10s	**1,1,2,2-Tetrachloro-ethane**	$C_2H_2Cl_4$
4.88m	glucohexapyranose		5.56s	Phenol	C_6H_6O
15.35s	**5- (Benzyloxy)-pentanal**	$C_{12}H_{14}O_{3}$-	6.29w	2-Hydroxybenzaldehyde	$C_7H_6O_2$
16.17m	**2,5-Bis (1,1-dimethyl-ethyl)phenol**	$C_{14}H_{22}O$	8.33s	Naphthalene	$C_{10}H_8$
			12.71m	*Sulfur*	S_8
17.05m	Dibutyl phthalate	$C_{16}H_{22}O_4$	18.69m		
18.00m	Bis (Butylglycolate)phthalate	$C_{18}H_{24}O_6$	14.03m	Diethyl phthalate	$C_{12}H_{14}O_4$
			15.33m	**5- (Benzyloxy)-pentanal**	$C_{12}H_{14}O_3$
18.17m	Butyl-2-ethylhexylester,Phthalic acid	$C_{20}H_{30}O_4$	14.34m	*Diphenylamine*	$C_{12}H_{11}N$
			12.90	**Butylated hydroxy-tolune**	$C_{15}H_{24}O$
16.62m	**2,4,6-Tris (1,1-dimethyl-ethyl) phenol**	$C_{18}H_{30}O$	20.99m	1-Methoxy-4- (2-phenyl-ethenyl) benzene	$C_{15}H_{14}O$

s: Strong; m: Medium; w: WeakNew compounds are shown in **bold**.

diffusion of stable oxidants in solution, the temperature and the pH. In the electrochemical oxidation of tannery wastewater, typically containing chloride in the range of 4500-4600 mg dm^{-3}, one would predict both direct and indirect oxidation routes to play a role. In the present study, the removal of colour, COD and sulfide may have occurred through the direct oxidation route. This is also supported by the GC-MS observation that very few chloro-organic products are formed and that the COD removal increased with increasing current density. The removal of ammonia may have occurred through the indirect oxidation route[13]–, whereby the ammonium ions are oxidized using *in-situ* formed Cl_2 or OCl^- into N_2 according to the overall reaction given below:

$$2NH_4^+ + Cl_2 \rightarrow N_2 + 2HCl$$

Conclusions

The results suggest that tannery wastewater can be effectively treated by an electro-oxidation method and a final effluent with substantially reduced pollution load can be obtained. The efficiency of Ti/Pt was 0.802 kgCOD h^{-1} A^{-1} m^{-2} and 0.270 kgNH_4 h^{-1} A^{-1} m^{-2}, while the energy consumption was 5.77 kWh kg^{-1} COD and 16.63 kWh kg^{-1} of NH_4^+. The efficiency of Ti/PbO_2 and Ti/MnO_2 was found to be low and energy consumption was high. The order of efficiency of anodes was found to be Ti/Pt » Ti/PbO_2 > Ti/MnO_2 [20].

References

1. Smith de Sucre V., and Watkinson J.P.; Anodic oxidation of phenol for wastewater treatment, *Canadian J. Chem. Engg.*, 59, 52-59 (1981).
2. Comninellis Ch., and Pulgarin C.; Anodic oxidation of phenol for wastewater treatment, *J. Appl. Electrochem.*, 21, 703-708 (1991).
3. Comninellis Ch., and Pulgarin C.; Electrochemical oxidation of phenol for wastewater treatment using SnO_2 anodes, *J. Appl. Electrochem*, 23, 108-112 (1993).
4. Rodgers J.D., Jedral W., and Bunce N.J.; Electrochemical oxidation of chlorinated phenols, *Environ Sci. Technol.*, 33, 1453-1457 (1999).
5. Pulgarin C., Adler N., Peringer P., and Comninellis Ch.; Electrochemical detoxification of 1,4-benzoquinone solution in wastewater treatment, *Wat. Res.*, 28 (4), 887-893 (1994).
6. Burke L.D., and Murphy O.J.; The electro-oxidation of methanol and related compounds at ruthenium dioxide coated electrodes, *J. Electroanal. Chem.*, 101, 351-361 (1979).
7. Murphy O.J., Hitchens G.D., Kaba I., and Verostko C.E.; Direct electrochemical oxidation of organics for wastewater treatment, *Wat. Res.*, 26, 443-451 (1992).
8. Gedye R.N., Sadana Y.N., Enmonds A.C.E., and Langlois M.L.; Anodic oxidation of organic waste from the manufacture of trinitrotoluene, *J. Appl. Electrochem.*, 17, 731-736 (1987).

9. Vlyssides A.G., Loizidou M., Karlis P.K., Zorpas A.A., and Papaioannou D.; Electrochemical oxidation of textile dye wastewater using Pt/Ti electrode, *J. Hazard Mater*, B70, 41-52 (1999).

10. Vlyssides A.G., and Israilides C.J.; Electrochemical Oxidation of a textile dye and finishing wastewater using a Pt/Ti electrode, *J. Environ. Scui. Health*, A33 (5), 847-862 (1998).

11. Allen S.J., Khader K.Y.H., and Bino M.; Electro-oxidation of dyestuffs in wastewaters, *J.Chem. Technol. Biotechnol.*, 62, 111-117 (1995).

12. Szpyrkowicz L., Naumczyk J., and Zilio-Grandi F.; Application of electrochemical processes for tannery wastewater treatment, *Toxic and Environ. Chem.*, 44, 189-202 (1994).

13. Szpyrkowicz L., Naumczyk J., and Zilio-Grandi F.; Electrochemical treatment of tannery wastewater using Ti/Pt and Ti/Pt/Ir electrodes, *Wat. Res.*, 29 (2), 517-524 (1995).

14. Berchmans L.J., and Vijayavalli R.; Electrochemical oxidation as a tool for pollution control: effect of anodic oxidation on the treatment of distillery effluent, *Ind. J. Environ. Health*, 31 (4), 309-313 (1989).

15. Cossu R., Polcaro A.M., Lavagnali M.C., and Palmas S.; Treatment of MSW landfill leachate by electrochemical oxidation, in *Proc. Sardinia 97 Sixth Int Landfill Symp, S Margherita di Pula*, Cagliari, Italy, 13-17 October (1997).

16. Scott K.; Electrochemical aspects of chemical waste minimization, in *Chemistry of Waste Minimization*, Ed. by Clarke J.H., Blackie Academic and Professional, London, pp.360-416 (1995).

17. Feng J., and Jonson D.; Electrocatalysis of anodic-oxygen transfer reaction. Titanium substrates for pure and doped lead oxide film, *J. Electrochem Soc*, 138 (11), 3328-3337 (1991).

18. APHA, AWWA, WPCF; Standard *Methods for the Examination of Water and Wastewater*, 17th Edn., American Public Health Association, Washington, DC (1989).

19. Rajalo G., and Petrovskaya T.; Selective electrochemical oxidation of sulphides in tannery wastewater, *Environ Technol*, 17, 605-612 (1996).

20. Rao N.N., Somasekhar K.M., Kaul S.N., and Szpyrkowicz L.; Electrochemical Oxidation of Tannery Wastewater, *J. Chem Technol and Biotechnol*, 76, 1124-1131 (2001).

Chapter 48

Recovery of Quality Water from Tannery Wastewater Adopting Photochemical Oxidation and Membrane System

S. Rajamani, E. Ravindranath, R. Suthanthararajan, B. Umamaheswari and K. Chitra

CSIR-CLRI, Chennai

Introduction

Conventional effluent treatment system with physio-chemical and biological components could remove the organic and inorganic pollutants such as Suspended Solids (SS), Biological Oxygen Demand (BOD), Chemical Oxygen Demand (COD), heavy metals such as chromium etc., but not the Total Dissolved Solids (TDS) and chloride. Due to the inherent nature, the effluent from tanneries contains high TDS and other non-degradable matter. The ions that contribute towards TDS are sodium, chloride, and sulphate etc. Due to inadequate water source and poor quality of the available groundwater, many tanneries in South India transport water through tankers at a high cost. The treated effluent from Effluent Treatment Plant (ETP) is not suitable for reuse in the process due to the presence of high TDS. Therefore to recover water from treated effluent from conventional treatment plants and reuse it in tanning process a pilot integrated tertiary treatment system including photochemical oxidation, two stage Reverse Osmosis (RO) system etc. has been designed and installed for a capacity of 20,000 litres/day, in one of the tanneries in Melvisharam, Ranipet, India. The quality of recovered water meets the drinking water standard and is used in wet finishing operations in the tannery. This integrated plant, which is first of its kind in the world, is implemented with technical and financial support of United Nations Industrial Development Organisation (UNIDO), Chennai.

Wastewater from the tanneries is treated by physio-chemical and biological treatment methods for removal of organic and inorganic pollutants. The treated wastewater from effluent treatment plant contain high amount of TDS in the range of 5000 – 12000 mg/L. The ions that contribute towards TDS are Sodium, calcium, chloride and sulfate. The treated effluent from conventional physio-chemical and biological treatment water is not suitable for reuse in the leather finishing process due to the presence of high TDS with residual COD and other non-degradable constituents. Therefore to recover water from treated tannery wastewater and reuse it in tanning process a pilot scale RO system integrated with tertiary treatment units such as photo chemical oxidation, micro filter etc. was installed, operated and studied for a period of nearly two years in one of the commercial tanneries in Melvisharam, Ranipet, India.

Objective of the Study

The objective of the study is to investigate the suitability of applying Reverse Osmosis (RO) system with other tertiary treatment units such as photochemical oxidation, micro filter etc. required prior to RO unit for the removal of residual suspended organic and inorganic matter, for recovery and reuse of quality water in tanning process from biologically treated tannery effluent.

Unit Operations

The important units of the selected system are collection tank (CT), pressure sand filter (PSF), photo-chemical oxidizer (PCO), activated carbon filter (ACF), softener, microfilter, dosing units, Intermediate storage tanks and two stage RO units. The schematic process flow diagram is shown in Figure 48.1. The integrated system with treatment capacity of 20,000 litres per day was designed to operate with the influent characteristics is as given in Table 48.1.

Table 48.1: Basic Design Parameters of Integrated Membrane System

Sl.No.	*Parameter*	*Value*
1.	pH	7.0 – 9.0
2.	COD	< 400 ppm
3.	SS	< 150 ppm
4.	TDS	5600 – 8000 ppm
5.	Silica	< 20 ppm
6.	Permeate Recovery	80 – 85 %
7.	TDS in Permeate	< 700 ppm

Operation of the System

Treated effluent from the conventional effluent treatment plant was taken up for the study. Wastewater from the secondary settling tank is pumped to the collection

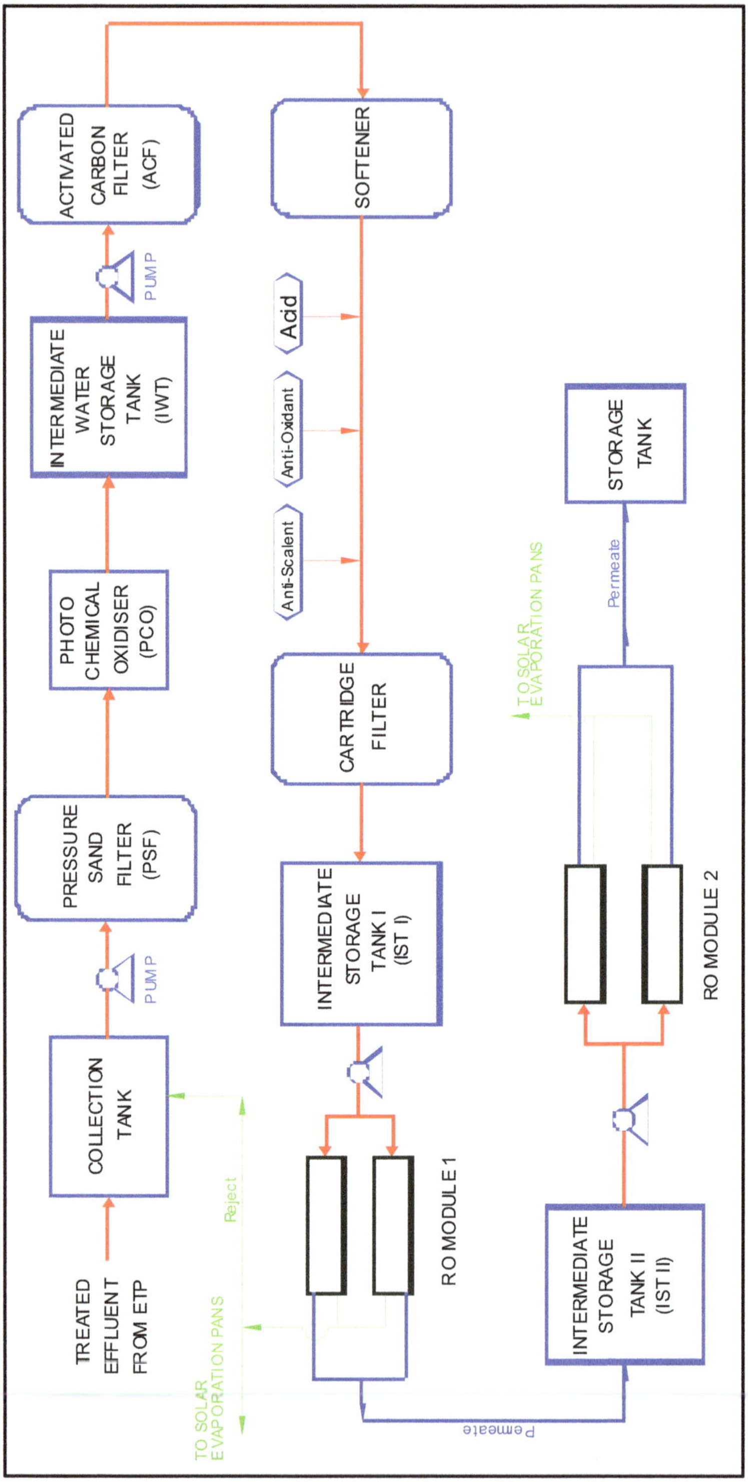

Figure 48.1: Flow Diagram of Photochemical Oxidiser System and Reverse Osmosis Plant.

tank of the RO system. Wastewater from the collection tank is pumped to Intermediate Water Tank (IWT) through Pressure Sand Filtration and Photo Chemical Oxidation. From IWT wastewater is pumped into Intermediate Storage Tank (IST-I) through Activated Carbon Filtration, softener and cartridge filter. Anti-scalent, anti-oxidant and acid were added to softener using online dosing pumps before cartridge filter.

All the units were operated at ambient temperature of 28 – 35°C on batch wise upto two shifts in a day. Wastewater from IST-I is pumped using high pressure pump with a hydraulic pressure of 10-14 psi to RO module I. Effluent from IST-II is pumped using high pressure pump with a hydraulic pressure of 12-16 psi to RO module II. Part of the reject-I from RO module I continuously flows to collection tank. 100 per cent Reject-II from RO module II is mixed with wastewater from PCO in IWT.

Performance of Photochemical Oxidation System

Prior to Photo Chemical oxidation system, wastewater was passed through pressure sand filter for removal of suspended solids. Photo Chemical Oxidizer system with UV lamp and hydrogen peroxide dosing was installed for destruction of organics prior to first stage membrane system. The dosing of hydrogen peroxide was monitored closely to avoid excess free oxygen radical getting into the subsequent treatment units, especially the first stage membrane system. From the analytical results, it was observed that COD concentration reduced from 200 mg/L to 100 mg/L after Photo Chemical Oxidation unit. Though higher reduction of organics could have been achieved by increasing the dosage this was not done to protect the membrane system.

Performance of Reverse Osmosis System

Samples from various tertiary treatment units integrated with RO system were analysed for the parameters including pH, suspended solids, total dissolved solids, chloride, sulfate, sodium, silica, calcium, magnesium and COD by procedures recommended in the Standard Methods for the Examination of Water and Wastewater (APHA-AWWA-WPCF, 19th Edition, 1995).

Table 48.2: Overall Performance of Reverse Osmosis (RO) System

Component	*Feed Concentration (mg/l)*			*Removal Efficiency(%)*		
	Min.	*Max.*	*Ave.*	*Min.*	*Max.*	*Ave*
Total Diss. Solids	3370	7190	4965	89.0	98.3	95.1
Chloride as Cl^-	760	2560	1390	84.0	99.1	93.0
Sulfate as SO_4^{2-}	810	2615	1645	99.7	100	99.9
Sodium as Na^+	535	2180	1185	85.6	99.2	93.8
Calcium as Ca^{++}	52	440	240	96.5	100	98.9
Magnesium as Mg^{++}	7	38	23	90.9	100	97.0
Silica as SiO_2	5	25	12	81.2	100	93.4

Portable WTW pH meter was used for pH measurement. COD was estimated using Thermoreactor 300. Sodium, calcium, magnesium and silica were determined using Atomic Absorption Spectrophotometer. The overall removal efficiency observed during the study period is given in Table 48.2.

Part of the reject from RO Module I was discharged into solar evaporation pan. The flow was controlled manually. The characteristics of concentrate discharged into solar pan is given in Table 48.3.

Increase in concentrate volume was observed due to fouling in the RO module and this resulted in increasing the hydraulic pressure of RO module I. This resulted in slow filling up of permeate in IST II.

Table 48.3: Characteristics of Concentrate

Parameters	*Reject*		
	Min.	*Max.*	*Ave.*
Total Dissolved Solids (TDS)	7960	27830	12830
Chloride as Cl^-	1540	4285	2670
Sulfate as SO_4^{2-}	1855	13640	6060
Sodium as Na^+	1100	5920	2750
Calcium as Ca^{++}	8	950	360
Magnesium as Mg^{++}	6	285	86
Silica as SiO_2		6	100 33

Note: All values are expressed in mg/L.

Water Recovery

The Total Dissolved Solids concentrations in the treated tannery effluent during the period was in the range of 3300-7200 mg/L. TDS concentration in the permeate was about 88-440 mg/L. The average recovery of permeate was 60 per cent. Characteristics of permeate collected during the study period is given in Table 48.4.

Recovered water from RO module II is collected in syntex tank and reused in wet finishing operations in tannery.

Table 48.4: Characteristics of Permeate

Parameters	*Permeate Characteristics*		
	Min.	*Max.*	*Ave.*
Total Dissolved Solids (TDS)	88	440	225
Chloride as Cl^-	17	190	82
Sulfate as SO_4^{2-}	BDL	8.2	2
Sodium as Na^+	15	122	60
Calcium as Ca^{++}	BDL	11	3
Magnesium as Mg^{++}	BDL	1.1	0.5
Silica as SiO_2	BDL	3.0	1.0

Note: BDL: Below Detectable Limit; All values are expressed in mg/L.

Clearing of Tertiary Treatment Units

At the end of the day's operation, almost all the units in the RO system were washed. PSF and ACF are back washed with the wastewater in IWT tank to remove the suspended solids. Both the RO modules are flushed with permeate water. Softener was regenerated with about 15-20 kg of commercial sodium chloride everyday. Brine liquor from softener after regeneration is discharged into solar evaporation pan.

Cost Analysis

- ☆ Total capital cost of the system: Rs. 1.2 millions (or say 30,000 USD)
- ☆ Fixed operating cost per 1000m^3 of water: Rs. 20,000 (or say 450 USD)
- ☆ Variable operating cost per 1000m^3 of water: Rs. 20,000 (or say 450 USD)
- ☆ Total recovery cost per 1000m^3 of water: Rs. 40,000 (or say 900 USD)

Conclusion

Ionic removal efficiency of reverse osmosis membranes from treated tannery wastewater is more than 93 per cent and the recovered water met the drinking water standards. Average water recovery limited to 60 per cent because of the dye present in the treated tannery wastewater and membrane fouling. Recovered water was used for wet finishing operations in tannery regularly.

References

1. Ravindranath E., Suthanthararajan R., Umamaheswari B., Chitra, K., and Rajamani, S.; *Recovery and reuse of tannery wastewater adopting commercial scale RO system*, in Third Annual Conference of the Indian Environmental Association, Enviro Vision 2000, on Industrial Wastewater Recycle and Reuse at Mumbai, May 5-6, 2000.
2. Rajamani, S., Jakov Buljan, UNIDO, *Technology Package–System for recovery and reuse of chromium from spent tanning liquor using magnesium oxide and sulphuric acid*, TECHPACK/UNIDO/RePO/1, 1996.
3. Shweta Singh, Rajamani S., and Ramasami T.; *Sustainability of Leather Tanning Small and Medium Enterprises (SMEs) through Environmental Compliance in Developing Countries*–XXVI IULTCS Congress, Cape Town, 7-10 March 2001.
4. *Sampling, Testing and Evaluation of the pilot Membrane System (RO) unit for processing treated tannery effluent* at A.T.H Leder Fabrik, Melvisharam, March 2000.
5. *Indian Journal of Environmental Health*, Vol. 41, October 1999.
6. *Membrane Filtration*, R G Gutman, 1987.
7. Rajamani S., Ravindranath E., Suthanthararajan R., Umamaheswari B., and Chitra K.; *Recovery of Quality Water from Tannery Wastewater adopting Photochemical Oxidation and Membrane System*. Proc. of the Indo-Italian Workshop on Emerging Technologies for Industrial Wastewater and Environment, NEERI, Nagpur, September 2-4, 2002, pp. 339-342.

Chapter 49

Glucose-6 Phosphatase an Indicator for Pollution Stress for Sulphur Dioxide

V.I. Pandit, S. Rawalgaonkar and S. Hasan

CSIR-NEERI, Nagpur

ABSTRACT

The physiological and biochemical changes are known to occur in plants as a response to air pollution though the injury is not externally apparent. Photosynthesis, respiration as well as other metabolic processes are affected adversely due to stress conditions caused by air pollutants. Sulphur dioxide is the most important gaseous pollutant present in the ambient atmosphere which damages vegetation. The plant species of beans (Vigna catjang and Dolichos lablab), cotton (Gossypium hirsutum), tomato (Lycopersion esculentum) and wheat (Triticum aestivum) were exposed to different levels of sulphur dioxide. Wheat and cotton plants are sensitive to SO_2 while tomato and bean plants are resistant to SO_2. The plant leaves were analysed for sugar content and the enzyme activity glucose-6-phosphatase while photosynthetic rate and transpiration rates were measured in situ by portable photosynthesis system (LiCOR-600 model). The results showed significant reduction in photosynthesis as well as glucose-6-phosphatase enzyme activity. Transpiration, on the other hand, in all the plants increased while sugar content indicated mixed response. The decline in enzyme activity glucose-6-phosphatase was observed with different concentration of sulphur dioxide treatments in all the plants makes it a suitable indicator for sulphur dioxide pollution.

Keywords: *Pollution stress, Static exposure chamber, Bioindicator, Glucose-6 phosphatase, Biochemical alterations, Physiological responses.*

Introduction

Vegetation is the most important natural resource. Sulphur dioxide is one of the major air pollutants in many parts of the world [1]. Its effect on vegetation have been well reviewed in terms of foliar injury, physiological and biochemical alternations [2,3]. The lower pollution concentration cause injury to plants which are less severe and no loner externally apparent [4]. In the absence of visible injury, the changes in the activity of various nzymes due to influence of pollution has been used as an indicator of pollution stress.Experiments were carried out to study the effect of SO_2 on activity of enzyme glucese-6-phosphatase, transpiration rate, sugar content and photosynthesis in general on different plant species. An attempt was made to find out dose-response relationship.

Methods

To evaluate the responses of plants to a single pollutant demands exposure chamber studies. The studies were carried out using a simple static exposure chamber (Figure 49.1). The plants selected include wheat (*Triticum aestivum*) and cotton (*Gossypium hirsutum*) sensitive to SO_2 while tomato (*Lyopersicon esculentum*) and bean (*Vigna catjang* and *Dolichos lablalb*) resistant to SO_2 have been selected. The SO_2 was generated by bubbling air in an aqueous solution of metabisulphite [5]. The experiments were carried out with four weeks old plants potted in earthen pots having soil mixture of manure: sand: soil (1: 1: 2). The plants were exposed to 0.25, 0.5 and 1.0 ppm SO_2 for 8 hrs daily while the control pants were not exposed to SO_2.

The position of the potted plants were changed daily so as to avoid positional effect. The leaves of the control as well as exposed plant were harvested after 10 days

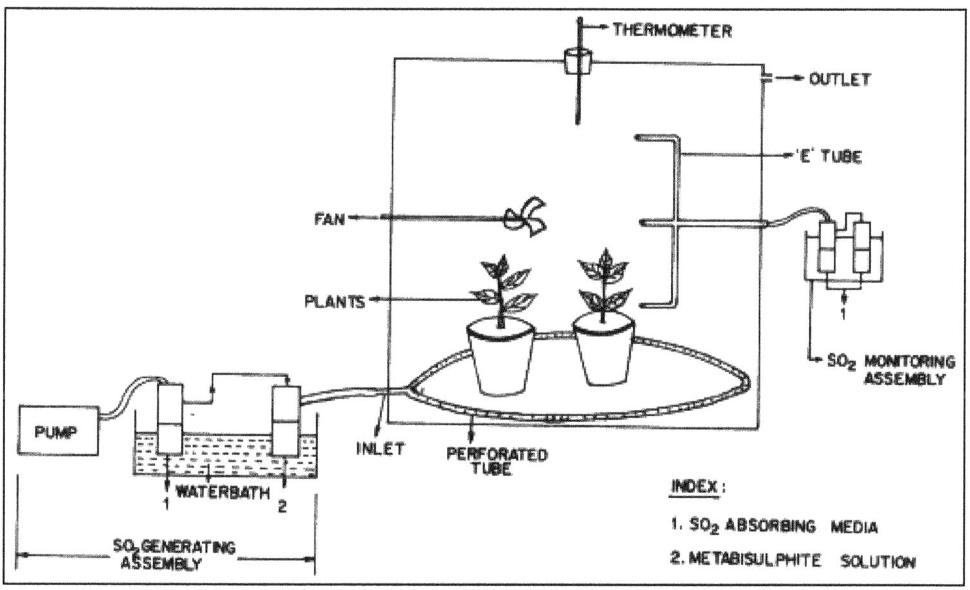

Figure 49.1: Schematic Experimental Setup.

and analysed for biochemical assay. The activity of enzyme blucose-6 phosphatase was estimated by method of Harper [6] and the sugar content by phenol-sulphuric acid method [7]. The photosynthesis and transpiration rates were measured by use of portable photogymetesis system (Licor-6000).

Results

Bean, tomato and cotton plants exposed to 0.5 and 1.0 pm SO_2 exhibited foliar injury. The effect of different doses of SO_2 on photosynthesis and transpiration rate of control and treated plants are presented (Tables 49.1 and 49.2). The reduction in photosynthesis rate (8-45 per cent) and increase in transpiration rate (5-50 per cent) was the common response observed in all the treatments for different plants studied (Figures 49.2 and 49.3).

Table 49.1: Effects of different Levels of Sulphur Dioxide on Photosynthetic Rate of Plants

Sl.No.	Plant Species	Photosynthesis Rate* (mg $CO_2/m^2/s$)					
		Control	Exposed to 0.25 ppm	Control	Exposed to 0.5 ppm	Control	Exposed to 1.0 ppm
1.	*Vigna catjang*	1.2315 ±0.102	0.9565 ±0.44	0.3513 ±0.06	0.2981 ±0.14	0.3598 ±0.06	0.2749 ±0.04
2.	*Gossypium hirsutum*	1.746 ±0.39	0.9604 ±0.66	0.2809 ±0.04	0.2094 ±0.08	0.3268 ±0.14	0.2655 ±0.054
3.	*Dolichos lablab*	0.5697 ±0.47	0.3424 ±0.21	0.2933 ±0.03	0.1766 ±0.07	0.1834 ±0.032	0.1395 ±0.042
4.	*Lycopersicon esculentumum*	3.28 ±0.388	3.011 ±0.28	0.8124 ±0.13	0.6843 ±0.096	0.3150 ±0.047	0.2241 ±0.12
5.	*Triticum aestivum*	2.305 ±0.09	2.009 ±0.24	1.2692 ±0.27	0.937 ±0.32	1.483 ±0.32	0.9629 ±0.11

* Each value is mean of five observations.

Decrease of glucose-6-phosphatase activity was observed in all the treated plants when compared to control (Tables 49.3 and 49.4). Over 60 per cent reduction in enzyme activity was observed in Dolichos lablab and cotton, while in other plants reduction ranged 10-40 per cent (Figure 49.4). The sugar content increased with 0.25 ppm exposure while no definate pattern was observed for treatment with 0.5 and 1.0 ppm SO_2 (Figure 49.5).

Discussion

Reduction in photosynthesis rate has been observed as a general response to SO_2 pollution stress [8-11]. Enhancement of transpiration rate was also observed and is of not much concern as it is less responsive. Thus reduce photosynthesis indicate reduction in the productivity. Glucose-6-phosphatase is an enzyme of reversal of glycolysis which carries out the hydrolysis of glucose-6-phosphate to glucose which is then made available in cellular environment. Evans and Miller [12] and Materna [13] pointed out that before appearance of any visible injuries physiological injuries

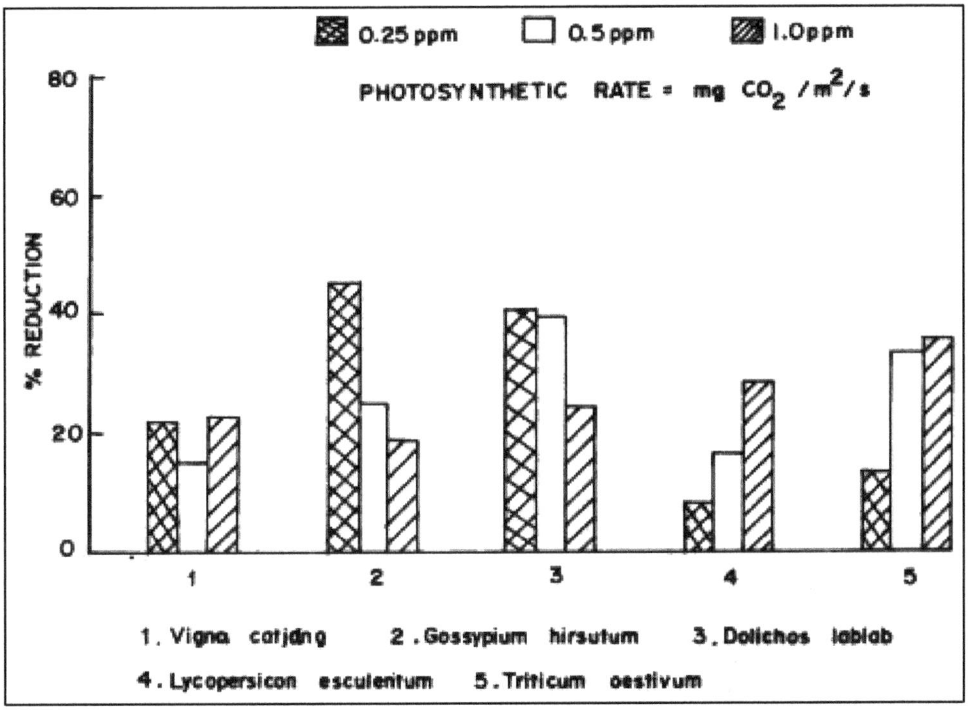

Figure 49.2: Per cent Reduction in Photosynthetic Rate of Plants after Exposure to different Levels of SO_2.

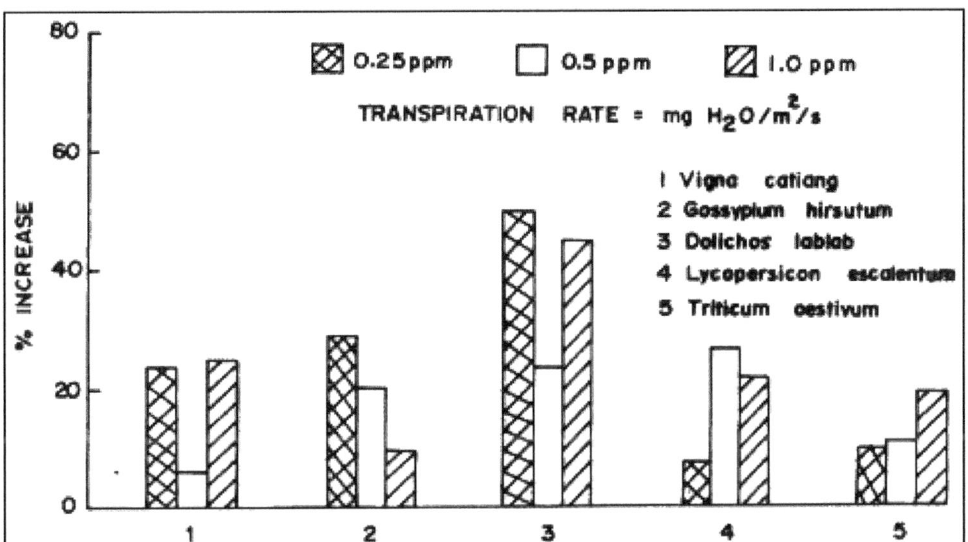

Figure 49.3: Per cent Increase in Transpiration Rate of Plants after Exposure to different Levels of SO_2.

Table 49.3: Effects of different Levels of Sulphur Oioxide on Enzyme Activity Glucose-6-Phosphatase of Plants

Sl.No.	Plant Species	Glucose-6-Phosphatase* (μg phosphate liberated/min/g)					
		Control	Exposed to 0.25 ppm	Control	Exposed to 0.5 ppm	Control	Exposed to 1.0 ppm
1.	*Vigna catjang*	2.3942± 0.33	1.3627± 0.28	3.7595± 1.6512	2.3826± 1.42	0.92± 0.34	0.626± 0.04
2.	*Gossypium hirsutum*	1.2473± 0.53	0.8624± 0.15	1.964± 0.77	1.3846± 1.02	0.892± 0.6	0.8± 0.4
3.	*Dolichos lablab*	3.724± 0.69	2.632± 0.43	4.7895± 3.17	1.7253± 0.81	1.544± 0.07	1.13± 0.23
4.	*Lycopersicon esculentum*	2.6536± 0.37	1.926± 0.21	6.342± 3.91	2.376± 0.68	1.372± 0.07	0.927± 0.27
5.	*Triticum aestivum*	3.2791± 0.75	2.379± 0.72	3.174± 0.35	2.1± 0.82	1.696± 0.11	1.085± 0.04

* Each value is mean of five observations.

Table 49.4: Effects of different Levels of Sulphur Dioxide on Sugar Content of Plants

Sl.No.	Plant Species	Sugar Content * (mg/g dry wt)					
		Control	Exposed to 0.25 ppm	Control	Exposed to 0.5 ppm	Control	Exposed to 1.0 ppm
1.	*Vigna catjang*	93.94 ±32.4	99.105 ±13.77	36.94 ±6.62	33.86 ±12.08	70.78 ±7.38	73.10 ±29.2
2.	*Gossypium hirsutum*	36.24 ±15.71	38.53 ±10.67	23.64 ±5.84	13.14 ±5.12	33.15 ±3.89	27.7 ±6.08
3.	*Dolichos lablab*	77.29 ±9.83	85.67 ±7.58	38.35 ±2.65	36.34 ±8.80	20.96 ±3.43	28.78 ±8.24
4.	*Lycopersicon esculentum*	38.29 ±10.38	46.32 ±6.53	25.23 ±3.37	35.62 ±6.35	40.37 ±6.97	35.69 ±5.62
5.	*Triticum aestivum*	62.60 ±26.46	71.75 ±12.57	23.08 ±6.30	17.73 ±3.15	30.56 ±9.03	35.82 ±11.86

* Each value is mean of five observations.

occur to plants which can be used as an indicator of pollution levels. Rabe and Kreeb [14] expresed their views regarding the formulation of physiological or biochemical indicator. The best physiological indicator, according to them, is the one with strongest reaction. They further suggested that for the experiments exposing one species of plant is not enough but comparison of two or more plants exposed simultaneously gives a better picture and this is better if these plants differ in their resistance as studied in present experiments belongs to sensitive class *viz.* wheat and beans as well as relatively resistant species *viz.* cotton and tomato. In all the analysis mentioned a comparison with standard is indispensable. This standard is provided by control

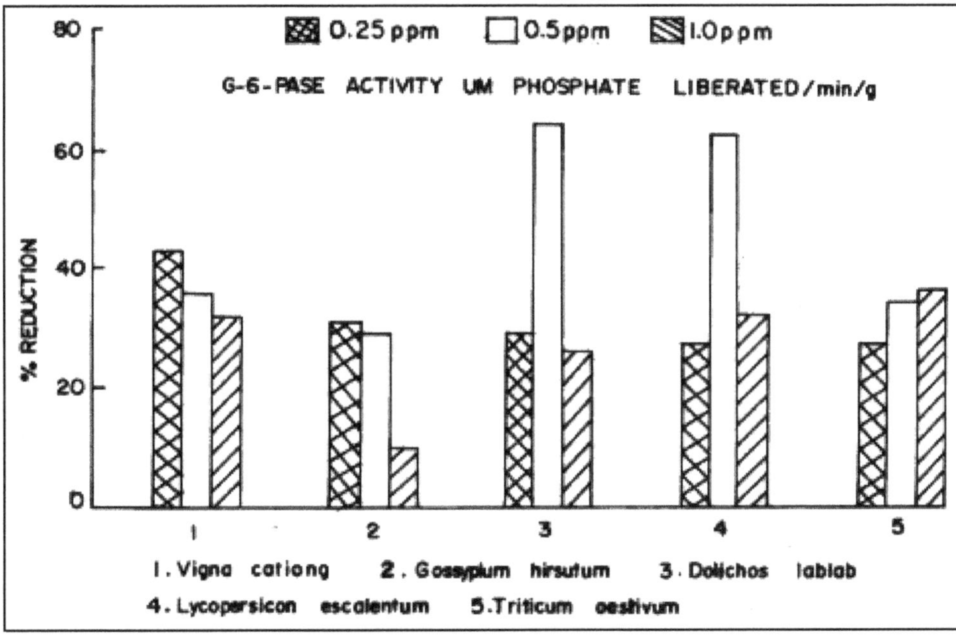

Figure 49.4: Per cent Reduction in Glucose-6-Phosphatase Activity after Exposure to different Levels of SO_2.

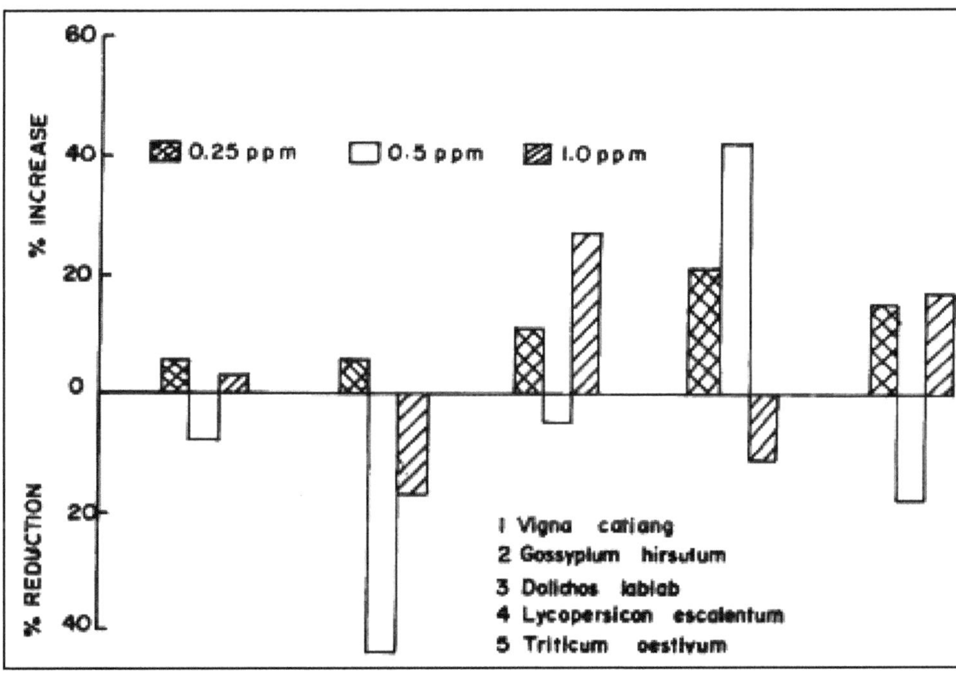

Figure 49.5: Per cent Change in Sugar Content of Plants after Exposure to different Levels of SO_2.

plants which are exposed under identical conditions except pollutant supply. It has been reported that in stress condition sugars accumulate as a response to increasing pollution load [15]. In the present studies the same was confirmed for 0.25 ppm SO_2 exposure while the depletion in sugars for some plants may be taken as an adjustment towards the tolerance mechanism shown by the plant to combat the stress situation.

Conclusion

In all exposure studies the reduction in photosynthesis rate and increase in the transpiration rate was observed as the common response.

The activity of glucose-6-phosphatase significantly reduced in comparison with the respective control plants. Hence, there is a clear connection between level of pollution the lowering of this enzyme activity may be considered as an indicator of SO_2 pollution. However, further research is desirable for lower exposure concentration normally recorded in polluted urban atmosphere. The lowering of glucose-6-phosphatase activity in all the treatments to different plant species both sensitive and resistant studied make it a suitable indicator of pollution stress.

In all treatments the sugar content increased with 0.25 ppm SO_2 exposure but no defiant pattern was observed for 0.5 and 1.00 ppm SO_2 treatment. These results confirm that at lower exposure the sugars accumulate as a response of pollution load.

References

1. Katainen HS, Makinen E Jokinen J, Karjalainen R, Kellomaki S. Effects of SO_2 on the photosynthetic and respiration rates in Scots pine seelings. *Environ Pollut.* 1987; **46;** 241-51.
2. Singh SN, Yanus M, Srivastava K, Kulshreshtha K, KJ Ahmed. Response of *Calendula officinalis* L. to long term fumigation with SO_2. *Environm. Pollut.* (Ser A) 1985; **39;** 17-25.
3. Ewald D, Schlee D. Biochemical effects of sulphur dioxide on proline metabolism in alga *Trebouxiz* sp. *New Phytol.* 1983; **94;** 235-40.
4. Farooq Mohammed, Akbar Masood, MU Beg. Effect of acute exposure of SO_2 on the metabolism of *Holoptelea integrifolia*. *Environ. Pollt.* (Ser A) 1985; **39;** 197-205.
5. Sharma HC, Thakre RA Generation and monitoring of gaseous pollutants. Ind. *J.Air Pollut. Contr.* 1984; **5;** 24-27.
6. Harper AE, *Glucose-6-phosphatase in Methods of Enzymatic Analysis* Ed by H.U. Bergmeyer. Verlag Chemie, GMBH Weinheim/Bergstr. Academic Press. New York and London. 1963; pp. 788-792.
7. Plummer DT, *An Introduction to Biochemistry* McGraw Hill Book Co., England 1971; p. 223.
8. Olszyk DM, Bytenerowicz A, Fox CA, Kats G, Dawson PJ, Wolf J. Injury and physiological responses *Larrea tridentala* (DC) (Coville) exposed *in situ* to SO_2 *Environ. Pollut.* 1987; **48,** 197-211.

9. Lamoraux RJ, Chang WR. Phosynthesis and transpiration of excised silver maple leaves exposed to cadmium and SO_2. *Environ. Pollut*. 1978; **17,** 257-68.
10. Bennet JH, AC Hill. acute inhibition apparent photosynthesis by phytotoxic air pollutants. *Am. Chem. Soc. Symp*. 1974; Series No. **3;** 115-27.
11. Tanaka *et al.*, Experimental studies on SO_2 injuries in higher plants. III. Inhibitory effect of sulphite ion on $14CO_2$ fixation. *Water, Air, Soil Pollut*. 1974; **3 (1);** 11-16.
12. Evans LS, Miller PR. Ozone damage to pondersoa pine: A histochemical appraisal. *Ann.J. Bot*. 1972; **59;** 297-304.
13. Materna J. EinfluB niedriger Schwefeldioxidk onzentrazionen auf die Fichte. In Wirkungen von Luftveruntereinigungen auf Waldbaume, VII. Int. Tagung Forstl. Rauchschadensachverstandiger Essen. 1970; 7-11.
14. Rabe R, Kreeb KH. Enzyme activities and chlorophyll and protein content in plants as indicators of air pollution. *Environ. Pollut*. 1979; **19,** 119-37.
15. Jager HJ, Bender J, Grunhage L. Metabolic responses of plants differing in SO_2 sensitivity towards SO_2 fumigation. *Environ. Pollut*. (Ser A), 1985; **39,** 317-35.

Chapter 50

Effluent Recycling through Application of Membrane Separation Processes for Zero Effluent Discharge in Textile Industry: A Case Study

T. Nandy[1], R.S. Dhodapkar[1], C.V. Deshpande[1], S.N. Kaul[1] and Lidia Szpyrkowicz[2]

[1]*CSIR-NEERI, Nagpur*

[2]*University of Venice, Venice, Italy*

ABSTRACT

The paper highlights the characteristics of wastewater generated from a small scale dyeing and printing synthetic textile industry. The adequacy assessment of full scale effluent treatment plant comprising chemical, biological and filtration processes are also presented. In addition, implementation of measures for discernable improvement in the performance of the existing units through effective operation and maintenance, and application of membrane separation processes leading to zero effluent discharge is also highlighted.

Keywords: *Textile industry, Dyeing, Printing, Treatment, Membrane process, Recycle/reuse.*

Introduction

Small scale textile processing industries have been set up in an unplanned manner in many cities in the country resulting in violation of discharge norms. These units even if they strongly desire to install an effective effluent treatment plant to protect the water environment and meet social obligations, are unable to do so because of various constraints (Nandy *et al.*, 1998).

The textile wastewaters are characterized with high suspended solids (SS), chemical oxygen demand (COD), biological oxygen demand (BOD) and total dissolved solids (TDS) is intensely of coloured which makes it very difficult to treat by any single conventional method. The common treatment methods implemented for textile wastewaters are the physico-chemical and biological aerobic routes (Hamza and Homodo, 1980; Brower and Reed, 1987; Paprowicz and Slodczyk, 1988; Kabdasali, *et al.*, 2000; Krull and Dopkens, 2004). However, these systems are inadequate for achieving discharge effluent quality (Sheng and Chi, 1966). Therefore, a combination of treatment processes that are reliable and effective in removing a wide range of pollutants are required to be implemented.

Today the water scarcity is posing severe problem for sustainable development, since the water demand is more, whether its human consumption, agricultural application or industrial use (Cicek, 2003). The textile industries accentuate this problem by their high water consumption and wide variety of pollutants discharge (Brik and Chaman, 2004). Therefore, in water intensive industries water management must be directed towards economisation and rationalization of the use of water resources. Waste management with recourse to effluent recycle/reuse is necessary to meet the ever increasing water demand leading to water conservation. The application of membrane separation processes do provide the industries with a technology to achieve water quality limits, and produce reusable water (Celenza, 2000; Frank, 2000), and has proved to be an effective process in concentrating the bulk of pollutant into small liquid volume for further disposal (Bilstad and Espedal, 1994).

The paper presents the detailed evaluation of an existing treatment facility in a small scale textile processing industry, and implementation of measures based on treatability studies for its effective performance thereby the improving the efficiency of the existing treatment plant. The application of membrane separation technology intended to produce treated effluent quality meeting the requirements for recycle as boiler feed water make-up is also highlighted. Implementation of the treatment scheme comprising chemical–biological–adsorption–membrane separation processes results in conservation of freshwater demand.

Industry Details

The industry studied is a small scale textile unit located in Southern India mainly engaged in dyeing and printing of manmade polyester fabric. The industry has a dyeing and printing processing capacity of 30000 meters per day. The polyester fabric processing involves scouring, bleaching, dyeing, printing, washing and finishing. The present total freshwater consumption of the industry is 135 m^3 per day including boiler feed water, the source of water being groundwater. The combined wastewater generation from various processing in the industry is 84 m^3 per day including boiler blowdown and sanitary wastewater.

Materials and Methods

Existing Effluent Treatment Plant

The industry has a full scale effluent treatment plant (ETP) treating the combined process wastewater generated. The combined wastewater emanating from the various process units is transported to the ETP through a closed conduit. The ETP comprising primary, secondary and tertiary treatment options has the following unit processes and operations:

- ☆ Preliminary Treatment
  - Screens
  - Oil and grease trap
  - Equalization basin
- ☆ Primary Treatment
  - Reaction tank
  - Flocculation channel
  - Primary settling tank
- ☆ Secondary Treatment
  - Aeration tank
  - Secondary settling tank
- ☆ Tertiary Treatment
  - Activated carbon column
- ☆ Sludge Dewatering
  - Sludge drying beds.

Assessment of Effluent Treatment Facilities

The performance evaluation of the effluent treatment plant was carried out through flow monitoring, and collection of wastewater samples at various stages of treatment, and its characterization. Hourly wastewater flow was monitored at the inlet to reaction tank, and the hourly flow range between 70.1 and 78.6 m^3/d.

Based on hourly samples collected and composited for twenty-four hours, the detailed analyses for major parameters at various stages of treatment are presented in Table 50.1. All analysis have been conducted as per Standard Methods (1998). The performance of ETP alongwith the stipulated Standards of the Statutory Board for discharge into Inland Surface Waters are presented in Table 50.2.

The chemical and biological sludges generated from the ETP fall under Hazardous Waste Category (MEF, 2000). The sludges after dewatering through sludge drying beds are stored in containers within the industry premises, and finally transported to a common facility for disposal through secured landfill.

Table 50.1: Assessment of Effluent Treatment Plant (ETP)

Sl.No.	*Parameters*	*Effluent from*			
		Equalization Basin	*Primary Settling Tank*	*Secondary Settling Tank*	*Carbon Column*
1.	Appearance	Coloured and turbid	Turbid white	Turbid	Clear
2.	pH	7.3-7.5	7.0-7.6	7.1-7.5	7.1-7.6
3.	Alkalinity	495	322	332	332
4.	Suspended Solids	600	200	100	50
5.	Total Dissolved Solids	2594	2410	2337	2250
6.	COD	2486	1418	621	310
7.	BOD	1044	588	235	90
8.	Total Kjeldahl Nitrogen	20	11	2	1.7
9.	Phosphates	5.2	3.0	2.0	1.7
10.	Sulphides	18	13	10	7.8
11.	Oil and Grease	11.2	5.6	3.6	2.6
12.	Chlorides	595	580	575	560
13.	Calcium	498	460	446	392
14.	Magnesium	205	192	184	172
15.	Sulphates	464	420	405	390
16.	Sodium	364	342	329	322
17.	Potassium	18	19	19	20
18.	Heavy Metals				
	Zinc	0.054	0.014	0.008	0.004
	Lead	0.098	0.048	0.042	0.015
	Nickel	0.856	0.826	0.754	0.714
	Cadmium	0.012	0.014	0.012	0.012
	Cobalt	0.025	0.017	0.011	0.011
	Manganese	0.110	0.055	0.055	0.057
	Iron	0.802	0.539	0.424	0.214
	Chromium	0.480	0.139	0.094	0.058
	Copper	ND	ND	ND	ND

Hourly samples collected and composited for 24 hours for 5 days.

Concentration values are average of five sets of data.

All values are expressed in mg/L except pH.

ND – Not Detectable.

Table 50.2: Performance of Effluent Treatment Plant (ETP)

Sl.No.	*Parameters*	*Effluent*		*Standards for Effluent Discharge into Inland Surface Waters*
		Raw	*Tertiary Treated*	
1.	Appearance	Coloured and turbid	Clear	*
2.	pH	7.1-7.5	7.1-7.6	5.5-9.0
3.	Alkalinity	540	332	–
4.	Suspended Solids	770	50	100
5.	Total Dissolved Solids	2600	2250	2100
6.	COD	2878	310	250
7.	BOD	1295	90	30
8.	Total Kjeldahl Nitrogen	28	1.7	100
9.	Phosphates	5.5	1.7	5.0
10.	Sulphides	21	7.8	2.0
11.	Oil and Grease	14.4	2.6	10
12.	Chlorides	570	560	1000
13.	Calcium	486	392	–
14.	Magnesium	220	172	–
15.	Sulphates	405	390	1000
16.	Sodium	400	322	–
17.	Potassium	26.5	20	–
18.	Heavy Metals			
	Zinc	0.082	0.004	5.0
	Lead	0.064	0.015	0.1
	Nickel	0.742	0.714	3.0
	Cadmium	0.024	0.012	2.0
	Cobalt	0.020	0.011	–
	Manganese	0.114	0.057	–
	Iron	0.900	0.214	–
	Chromium	0.340	0.058	2.0
	Copper	ND	ND	3.0

Hourly samples collected and composited for 24 hours for 5 days.

Concentration values are average of five sets of data.

All values are expressed in mg/L except pH.

ND – Not Detectable.

*All efforts should be made to remove colour and unpleasant odour as far as practicable.

Table 50.3: Assessment of ETP after Implementation of Improvement Measures

Sl.No.	*Parameter*	*Effluent from*				*Standards for Effluent Discharge into Inland Surface Waters*
		Equalization Basin	*Primary Settling Tank*	*Secondary Settling Tank*	*Carbon Column*	
1.	Appearance	Coloured and turbid	Turbid white	Turbid	Clear	*
2.	pH	7.4-7.6	7.5-7.9	7.2-7.4	7.0-7.6	5.5-9.0
3.	Alkalinity	510	332	335	335	–
4.	Suspended Solids	620	143	44	18	100
5.	Total Dissolved Solids	2540	2393	2310	2220	2100
6.	COD	2432	1026	269	118	250
7.	BOD	1023	392	45	16	30
8.	Total Kjeldahl Nitrogen	24	12	2	1.6	100
9.	Phosphates	5	2.5	1.4	1.1	5.0
10.	Sulphides	20	12	5	1.8	2.0
11.	Oil and Grease	13	5	2.4	1.4	10
12.	Chlorides	580	551	540	506	1000
13.	Calcium	478	430	404	334	–
14.	Magnesium	225	198	179	160	–
15.	Sulphates	475	402	368	334	1000
16.	Sodium	358	319	304	291	–
17.	Potassium	22	21	18	14	–
18.	Heavy Metals					
	Zinc	0.064	0.011	0.0056	0.0023	5.0
	Lead	0.084	0.034	0.027	0.0081	0.1
	Nickel	0.856	0.770	0.654	0.602	3.0
	Cadmium	0.016	0.015	0.012	0.011	2.0
	Cobalt	0.019	0.012	0.0072	0.0069	–
	Manganese	0.110	0.044	0.044	0.044	–
	Iron	0.710	0.426	0.298	0.134	–
	Chromium	0.440	0.088	0.053	0.029	2.0
	Copper	ND	ND	ND	ND	3.0

Hourly samples collected and composited for 24 hours for 5 days.

Concentration values are average of five sets of data.

All values are expressed in mg/L except pH.

ND – Not Detectable.

*All efforts should be made to remove colour and unpleasant odour as far as practicable.

Table 50.4: Assessment of Membrane Separation Processes (ATP)

Sl.No.	*Parameters*	*Feed**	*Permeate from*		*Concentrate from Reverse Osmosis*
			Ultrafiltration	*Reverse Osmosis*	
1.	Appearance	Clear	Clear	Clear	Muddy
2.	pH	7.7	7.7	6.0	7.8
3.	Alkalinity	345	325	12	1100
4.	Suspended Solids	22	ND	ND	ND
5.	Total Dissolved Solids	2234	2196	40	7584
6.	COD	120	20	ND	68
7.	BOD	20	3	ND	11
8.	Total Kjeldahl Nitrogen	ND	ND	ND	ND
9.	Phosphates	1.1	0.066	ND	0.22
10.	Sulphides	1.6	ND	ND	ND
11.	Oil and Grease	1.4	<1	ND	3.3
12.	Chlorides	494	483	12	1653
13	Calcium	330	325	3	1125
14.	Magnesium	164	143	2	493
15.	Sulphates	350	307	ND	1070
16.	Sodium	289	264	3	917
17.	Potassium	15	<1	<1	1
18.	Heavy Metals				
	Zinc	0.0024	ND	ND	ND
	Lead	0.0079	0.0079	ND	0.028
	Nickel	0.600	0.206	ND	0.721
	Cadmium	0.012	0.0049	ND	0.017
	Cobalt	0.0072	0.0059	0.002	0.019
	Manganese	0.050	0.010	0.003	0.028
	Iron	0.132	0.117	0.018	0.114
	Chromium	0.025	0.018	ND	0.063
	Copper	ND	ND	ND	ND

*Tertiary treated effluent from feed tank of ATP

Concentration values are average of five sets of data

All values are expressed in mg/L except pH

ND – Not Detectable.

Results and Discussion

The performance of various unit processes and operations of the ETP were not at their optimal level of performances mainly attributed to the following reasons:

- ☆ Improper equalization of wastewater flow and characteristics
- ☆ Non-optimal coagulation dosage and improper dosing
- ☆ Irregular desludging from primary clarifier
- ☆ Low MLSS concentration in the aeration tank of the activated sludge process, and
- ☆ No regular replacement of spent carbon in the activated carbon column.

The secondary and tertiary treated effluent does not conform to the stipulated Standard of the State Board for discharge into Inland Surface Waters with respect to BOD, COD, TDS and sulphide parameters.

Proper Operation and Maintenance

Based on the adequacy assessment of the ETP, measures were implemented to improve the plant performance which included:

- ☆ Proper wastewater equalization through uniform mixing provided in the equalization basin
- ☆ Optimization of coagulants dosage based on treatability studies
- ☆ Proper dosing by providing dosing tanks instead of manual operation
- ☆ Regular desludging from primary clarifier
- ☆ Build-up of design MLSS concentration in the aeration tank of biological treatment process, and
- ☆ Replacement of spent activated carbon in the tertiary treatment column.

Optimization of Coagulants Dosage

Treatability studies for treatment of combined wastewater were conducted to determine the optimum coagulant dosage for lime and alum using commercial grade chemicals already in use at the plant site. Initially, lime was used in various doses ranging from 400 to 1000 mg/L. After optimizing the lime dose, alum was used in concentration ranging from 400 to 800 mg/L in combination with optimum lime dose. The results indicated an optimum dosage of lime: alum as 800: 600 mg/L as against 950: 700 mg/L, achieving SS, COD and BOD removals of 80.0, 60.6 and 63.5 per cent, respectively with treated effluent pH of 7.7. The sludge generated was 150 mg/L.

Based on scientific treatability studies, optimum dosage of chemical coagulants was practiced for improvement in the performance of chemical treatment. In addition, optimum dosing of coagulants facilitated reduction in chemical consumption, and subsequent chemical sludge generation as compared to random addition of coagulants.

Optimization of Activated Sludge System

Performance of the activated sludge system was poor as compared to design criteria as the solid retention time (SRT) in the aeration tank was estimated to be in the range of 65-70 days, which is high as compared to the design value of 10 days and the oxygen uptake rate (OUR) of biomass (MLVSS) was 26.7 mg/L.h indicating low OUR of biomass. Sludge volume index (SVI) was 163 ml/g indicating poor sludge settleability.

Measures implemented to improve the performance of activated sludge system were inoculation of mixed liquor for volatile biomass (MLVSS) build-up, and regular sludge wastage from sludge recycle line to maintain SRT of around 10 days. After biomass build-up in the aeration tank, with MLVSS concentration in the range of 2245 to 2680 mg/L, the SVI and OUR values observed were 105 ml/g and 41.2 mg/L.h, respectively. OUR of 41.2 mg/L.h as compared to 26.7 mg/L.h as observed during earlier monitoring indicate build-up of active volatile biomass (MLVSS) in the aeration system. The average concentration of COD in the secondary treated effluent was 269 mg/L as compared to 621 mg/l as observed during the earlier monitoring. The SS, COD and BOD removals of 69.2, 73.8 and 88.5 per cent, respectively were achieved as against 50.0, 56.2 and 60.1 per cent, respectively under the earlier operating conditions.

Spent Activated Carbon Replacement

Efforts were made to reduce the concentration of major polluting parameters well below the stipulated Standards. The spent carbon in the tertiary treatment carbon column was replaced with fresh carbon. The characteristics of treated effluent at various stages of treatment as monitored after implementation of the measures for improvement in the performance of the ETP alongwith the discharge Standard are presented in Table 50.3. The combined wastewater flow during evaluation varied in the range of 70.5-96.3 m^3/d. The results indicate improvement achieved in the effluent characteristics through proper operation and maintenance of the various unit operations and processes of the plant. The tertiary treated effluent from the ETP conformed to the prescribed limits for Inland Surface Water Discharge except for the TDS parameter.

Implementation of Membrane Processes

In compliance with the directives of the Board for zero effluent discharge, application of membrane separation processes comprising ultrafiltration (UF) and reverse osmosis (RO) were initiated and commissioned. A multigrade filter has been provided as a pretreatment facility. Figure 50.1 present the schematics of the effluent treatment facilities including membrane processes.

The advanced treatment facilities (ATP) were evaluated after implementation of the improvement measures in the chemical, biological, and tertiary unit processes. The performance evaluation of the UF and RO systems were carried out by monitoring flow, and collection of samples of permeate from UF and RO, and rejects (concentrate) from the latter. Based on the samples collected from the aforereferred sources, the detailed analyses for major parameters are presented in Table 50.4. The UF was operated at feed flow of 8.0 m^3/hr and the RO system at a feed flow of 7.0 m^3/hr and pressure of 13 kg/cm^2, achieved permeate recovery of 5.0 m^3/hr (71.4 per cent) and rejects of 2.0 m^3/hr (28.6 per cent). ATP (UF and RO) implemented achieved 100 per cent removal of SS, BOD and COD, and 98.2 per cent TDS removal.

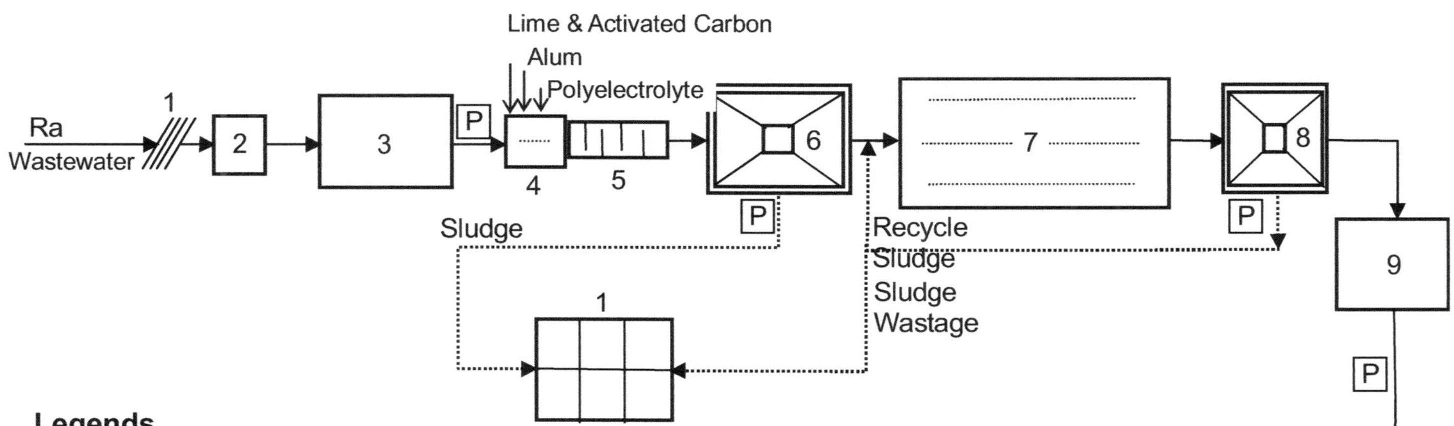

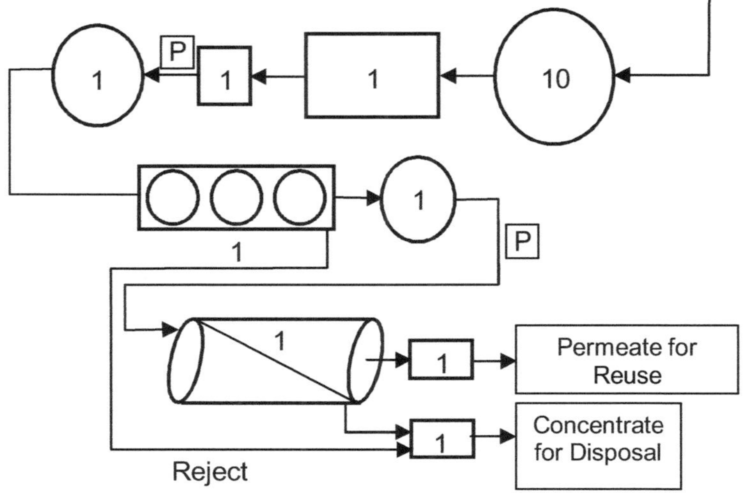

Legends

1 Screens
2 Oil and Grease Trap: 5x2.3x2.8 m
3 Equalization Basin: 6.0x6.3x3.2 m
4 Reaction Tank: 3.0x1.6x1.3 m
5 Flocculation Channel: 1.3x1.6x1.3 m
6 Primary Settling Tank: 6.7x5.8x3.5 m
7 Aeration Tank: 11.2x7.2x4.6 m
8 Secondary Settling Tank: 3.9x3.9x4.9 m
9 Collection Sump: 2.4x2.3x3.4 m
10 Activated Carbon Column: Dia. 2.0 m; Height: 2.5 m
11 Collection Tank: 6.0x4.0x4.0 m
12 MGF Feed Tank: 1.25x1.25x2.8 m
13 Multigrade Filter: Dia. 1.1 m; Height: 2.2 m
14 Ultrafiltration Unit
15 Collection Tank: Vol. 5 m^3

Figure 50.1: Schematics of the Effluent Treatment Facilities including Membrane Processes.

The rejects (concentrate) generated from the reverse osmosis unit is collected in a collection sump, and finally transported through tankers to a Common Effluent Treatment Plant (CETP) for treatment through an evaporation system.

Improvement in the performance of the existing facilities, and implementation of advance treatment processes facilitated reuse/recycle of 64 m^3/d of tertiary treated effluent for process use, washing of screens, blankets, and winch machines, and boiler feed water makeup including ash quenching. The permeate (boiler feed water) recovered from the reverse osmosis system (10 m^3/d) is reused as boiler feed water make-up. The entire treated effluent after tertiary and advanced treatment are recycled/reused within the industry for various purposes, and no effluent is discharged from the industry, thereby attaining zero liquid effluent discharge.

Conclusion

The various unit processes and operations of the ETP designed to treat combined wastewater from a small scale textile industry under existing operating conditions were not functioning at optimal levels of performance as per design. This was mainly attributed to poor operation and maintenance of the plant. The performance of the ETP was improved by optimizing the operational parameters in the individual units. Based on the adequacy assessment, efforts were made to improve the performance of the plant through optimization of commercial grade chemical coagulants dosages and build-up of active biomass in the activated sludge system. In addition, proper maintenance activity comprising regular pumping schedule of various pumps, regular coagulant dosing and desludging, continuous operation of mixers and surface aerators were also initiated.

The existing treatment facility was further upgraded through implementation of membrane separation processes comprising ultrafiltration and reverse osmosis units to produced recyclable effluent quality. The advanced treatment facility achieved reusable effluent characteristics with concentrations of SS, BOD and COD below detectable limits and TDS concentration of 40 mg/L (average). The characteristics of treated effluent after proper operation and maintenance, optimization of chemical and biological treatment system and upgradation of the ETP produced boiler feed quality water. The treated effluent recycle within the plant not only achieved zero effluent discharge, but also results in conservation of 74 m^3 per day (around 35 per cent) of freshwater requirement.

The chemical and biological sludges generated from the ETP are disposed through a common secure landfill facility. The concentrate generated from the RO system is treated through an evaporation system operated under the management of a common effluent treatment facility.

References

1. Bilstad T. and Espedal E. (1994). Membrane Separation of Wool Scour Effluent. *Water Science and Technology*, 29, 251-256.
2. Brik M. and Chaman B. (2004). Effect of Ozone Chlorine and Hydrogen Peroxide on the Elimination of Colour in Treated Textile wastewater by MBR, *Water Science and Technology*, 49, 299-303.

3. Brower G.R. and Reed G.D. (1987). Economic Pre-treatment for Colour Removal from Textile Dye Waste, *Proc. 41st Purdue Ind. Waste Conf.*, Lafayette, Indiana, 612-620.
4. Celenza, G. J. (2000). *Industrial Waste Treatment Process Engineering*, Vol III, Technomic Publication, Pennsy Lvania, USA.
5. Cicek N. (2003). A Review of Membrane Bioreactor and their Potential Application in the Treatment of Agriculture Wastewater. *Biosystem Engineering*, 45, 6.37-6.44.
6. Frank J. J. (2000). Introduction to Membrane Technologies for Industrial and Municipal Wastewater Treatment and Reuse. *Proc. Membrane Technologies for Industrial and Municipal Wastewater Treatment and Reuse*, Water Environment Federation, 2-15.
7. Hamza A. and Hamodo M.F. (1980). Multi-process Treatment of Textile Wastewater, *Proc. 35th Purdue Ind. Waste Conf. Lafayette*, Indiana, 151-159.
8. Hazardous Waste Category, (2000). MEF Notification No. 465 Gaz, India Part II Section 3, Subsection II.
9. Kabdasali J., Gorel H. and Tunay O. (2000). Characterization and Treatment of Textile Printing Wastewaters, *Env. Technol.*, 21, 1147-1155.
10. Krull R. and Dopkens (2004). Recycling of Dyehouse Effluents by Biological and Chemical Treatment. *Water Science and Technology*, 49, 311-317.
11. Nandy Tapas, Kaul S. N. and Shastry S. (1998). Wastewater Management for Cluster of Small Scale Dyeing and Printing Synthetic Textile Industry, *Envi. Pollution Control Journal*, 3, 5-15.
12. Nicolaou M. and Hadjivassilis I. (1992). Treatment of Wastewater from the Textile Industry. *Water Science and Technology*, 25, 31-35.
13. Paprowicz J. and Slodczyk S. (1988). Application of Biologically Activated Sorptive Column for Textile Wastewater Treatment, *Environmental Technology*, 9, 271-280.
14. Sheng and Chi (1996). Continuous Treatment of Textile Wastewater by Combined Coagulation, Electro-chemical Oxidation and Activated Sludge, *Water Resources*, 30, 587-592.
15. *Standard Methods for the Examination of Water and Wastewater* (1998). APHA, AWWA and WPCF, 20th Edition.

Chapter 51

Georaphical Information System and its Applications in Wastewater Management

V.K. Kondawar and S.M. Kashyap

CSIR-NEERI, Nagapur

Introduction

Geographical Information System (GIS) can be defined as organized collection of computer hardware, software, geographic data, and personnel designed to efficiently capture, store, update, manipulate, analyze and display/present all forms of geographically referenced information. In the most general sense, a GIS need not be computerized, but should be an integrated set of procedures for the input, manipulation, database management and output of geographic information. The system is concerned with spatial objects, their locations, and collateral data on specified scales of measurements referenced to some coordinate system of earth surface. The geographic data can be about natural resources, environmental status, infrastructure, demography, socio-economic parameters, etc. The geographical database can be queried for extraction of specific and meaningful information from a diverse collection of data, which is possible because the data are organized into a 'model' of the real world. The GIS can be employed for predictive modeling, planning/ decision making, implementation and finally for long term monitoring of impacts of plan/decisions on the environment.

Generally information systems can be non-spatial (as Management Information System) or spatial. Spatial systems can be Non-geographic (CAD/CAM) where the data is not 'geocoded' or Geographic, where each point, line or polygon is assigned geographic co-ordinates. Geographic Information Systems are often referred based on 'what it does?' such as

- ☆ Land Information System
- ☆ Land Resources Information System
- ☆ Resources Information System
- ☆ Natural Resource Management Information System
- ☆ Geographically Referenced Information System
- ☆ Spatial Information System
- ☆ Spatial Data Management and Analysis System
- ☆ Planning Information System
- ☆ AM/FM (Automated Mapping and Facilities Management)

Why GIS

General objectives for the application of GIS are

- ☆ Spatial predictive modeling
- ☆ Optimized resource management
- ☆ Cost minimization
- ☆ Ease and transparency in planning and decision making efficient data processing, communication and presentation
- ☆ Easy availability of good quality (accuracy-position and classification) data Precision, resolution, currency and completeness

Spatial Objects: Maps

In order to represent a portion of earth's surface as map, the area must be reduced. The extent of this reduction is expressed as a ratio called Map Scale- the ratio of map distance to ground distance. Generally small scale (1: 250,000) maps depict large ground area but they have low spatial resolution whereas large-scale maps (1: 1,200) depict small ground area with higher spatial resolution.

Maps are flat, but the earth's surface is curved (ellipsoid). A geographic coordinate system is a reference system used to locate and measure geographic features. The spherical geographic coordinates system is based on a three-dimensional sphere. Real-world locations in it are measured in degrees of latitude and longitude. The Cartesian coordinate system is based on a flat surface wherein real-world locations are measured using x- and y- coordinate values from a specified point of origin. In order lo represent three-dimensional earth's surface in two-dimensional map, a mathematical transformation called a projection is performed. Projection formulas convert data from a geographic location (latitude and longitude) on the earth to a representative location on a flat surface. The projection process will distort one or more of the following four spatial properties *viz.* Shape, Area Distance, Direction. Because these properties are often used to make decision, anyone who uses map should know which projection distorts which properties and to what extent'.

Components

Geographical Information systems have three important components–computer hardware, sets of system and application software and a proper organizational arrangements These three components need to be in balance if the system is to function satisfactorily. The main hardware components of a GIS include central processing unit, mass storage units such as hard disk/tape drives, input and output devices/ peripherals such as digitizer, scanner, printer or plotter and Visual Display Unit (VDU). The software component of GIS is related to various function and operation of GIS.

Data input, database management

Data entry, editing, verification, storage, retrieval

Data transformation, manipulation, analysis, modeling

Raster: Local, Zonal, Focal, and Global

Vector: Spatial overlay. Buffering, Map generalization. Feature extraction, Feature reduction, and theme co-ordinate transformation.

Data output, presentation and Reporting

Zooming, colour assignment and change operation, view. Text writing, Annotation, Graphic User Interface (GUI).

Till recently most GIS applications were based on "map geometry" (only x, y co-ordinate) in 2-dimensional topology. Users' perception is three-dimensional. Therefore currently, more efficient GIS products (2.5 D or 3 D) are offered by vendors which integrate the third-dimension using the Digital Terrain Model (DTM). The fourth dimension is introduced through the time for dynamic systems leading to temporal GIS systems.

Data Types and Structure

GIS database includes land related data showing spatial relationship (graphical data) and attribute of the land parcels itself.

Graphical data are digital representative of spatial entities. The graphical component of land related data can be readily captured and stored in graphical data files in the form of various spatial elements. Spatial objects in the real world can be thought of as four easily identifiable objects.

Map scale dictates the type of map feature used to represent real world features.

- ✰ Point: Represent location feature
- ✰ Lines: Represent linear feature
- ✰ Polygons: Represents area features
- ✰ Surface: Adding the dimension of height (elevation) to features and records surfaces as "Three dimensional object"

Points define discrete locations of geographic features, which are too small to be depicted as polygons. Lines represent the shape of geographic objects too narrow to depict as polygons such as streets and streams. Polygons are closed figures that represent the shape and location of homogeneous, real-world features such as district, parcels, soil types or land use zones. Symbols and labels area used to describe features.

Non-spatial attribute data consists of textural description on properties which may be associated with graphical entities, for example, name of location, width of road, diameter of pipeline, owner of land parcel, land use category, vegetation characteristic, etc. Many of the attributes data can be linked together through the use of a common identifiers or keys.

There are two main ways to organize the spatial data: Raster Data Structure and Vector Data Structure. The choice of a particular spatial data structure is one of the important early decisions in designing a geographic information system. Comparison of raster and vector models is given below:

Raster Model	*Vector Model*
Simple data structure Efficient overlay operation Handles high spatial variability Easy handling of remotely sensed images	Compact data structure Efficient encoding of topology Easy network and geometrical analysis Easy handling of line maps

Data Processing and Analysis

Main data processing operations used frequently to generate a database are

- ☆ Digitization or scanning of data
- ☆ Topology construction, data reduction and generalization
- ☆ Errors detection and editing
- ☆ Merging of points and edge matching
- ☆ Rectification/registration
- ☆ Format conversion

Spatial relationships between geographic features are based on the location of features. Some of the relationships can be enumerated as:

- ☆ Connectivity: Describes the link (s) between different map features
- ☆ Adjacency: Describes what features are adjacent to each other
- ☆ Proximity: Describes what features are near each other
- ☆ Intersection: Describes which features intersect (cross) each other
- ☆ Containment: Describes what features are located within an area
- ☆ Relative Position: Describes the comparative position between features
- ☆ Elevation differences: Describe the height difference between features.

GIS tools have to efficiently integrate tools of relational attribute database with spatial database characterized by above relationships. They range from simple database query system to complex analysis and decision support system. The simple analysis tools provided by GIS include:

- ☆ User defined views involving reclassification, aggregation, overlay, logical and arithmetic operations, etc
- ☆ Difference and connecting measurements, network analysis
- ☆ Neighborhood characterization including buffering
- ☆ 3 D models including estimation of point to point visibility, shadow region, slope and aspect and flow models.

GIS analysis can help to develop relations between geographic features and their associated attributes drawn from different sources. Integration analysis is employed to identify and extract the useful hidden relationship in multi-thematic datasets. Integration can be for passive enquiry, probing possibilities and active planning. It involves conjunctive analysis of appropriate multi-thematic data and weighted integration to generate composite information sets. The relationships can be expressed as statement or tables. Determining weightages requires human judgment supplemented by mathematical/statistical tools.

Digital models can be linked to some layers of information with conventional GIS functionality. However full integration of physical models with GIS is presently not possible.

Some applications of GIS that can interest engineers for wastewater collection and treatment can be:

- ☆ Number of road cuttings for particular segment of pipeline.
- ☆ Find elevation difference between source of effluent and treatment plant site
- ☆ Discover the best route between the source of effluent and treatment plant.
- ☆ Determine a suitable site for a waste-water treatment plant, which cannot be located near a stream or residential area.
- ☆ Ascertain which land use is contained within a given region
- ☆ Overlays of landuse and groundwater table

Some common GIS packages used frequently in India are ARC/INFO (ESRI, USA), PAMAP (PAMP graphics Limited, Canada), SPANS (TYDAC Technologies, USA), Geomedia, MGE (Intergraph System, The Netherlands), TNTlite (MicroImages Inc, USA), Geomatica (PCI Geomatics, USA), IDRISI (Clark University, USA), GRASS, MAPINFO, Auto-CAD, etc. Various departments of Government of India have developed GIS packages suitable for specific applications in mind such as ISROGIS, GEOSPACE, INGIS, GRAM, etc. However their utilization remained restricted.

Applications

The power of a GIS is to organize and display large land-use areas on a computer screen with zoom-in and zoom-out capabilities. Maps of selected areas can be printed on almost any scale and with variable amounts of detail. Output maps can be color-coded for viewing on-screen or printing. This capability is ideal for environmental planning when land uses or resources change with time. Another advantage of a GIS over conventional analyses is an increase in efficiency. A GIS is not just a relational database management system since in addition to the storage capabilities it includes spatial operators, which permit the manipulation of spatial data.

GIS in the past have been used primarily for the production of digital maps for resources management and management of land information systems. However, in the last 10 years with the wide availability of digital map data and the emergence of desktop mapping systems, GIS are applied in many different fields ranging from marketing to environmental planning and from city planning to water resources management.

In the initial years of GIS, the system was visualized as an extension of computer aided graphic/design tools with very limited spatial analysis capabilities. It was essentially addressing the needs of the geographical community. Therefore, the traditional geological community was skeptical about its usage in solving serious geological problems and it still preferred specialized mining software or geostatistical packages for their applications. However, the GIS developers soon realized the need for incorporating multivariate geostatistical modules and powerful 3-D analysis and these components work often considered as advanced components and become the selling point of GIS packages.

GIS application in water resources is itself very broad. However, much of this work in water resources has been in the area of large-scale, river basin hydrology. GIS has a long history of use in the water resources field, this is due in large part to the early availability of remotely sensed data suited for this purpose. Work in natural areas tends to focus on grid, or raster-based, hydrology, whereas in urban areas– because of their complex hydraulic and hydrologic regimes- tends to use more complex models that are vector based. Raster based approaches use rectangles as their fundamental unit within which hydrologic characteristics are uniform. Vector-based models use coordinate geometry to define unique boundaries of hydrologic characteristics.

Recently, a new trend is emerging where GIS systems are combined with analytical, mathematical models to form spatial decision support toolkits. In these integrated systems the GIS is used to handle the spatial data and provide a user friendly interface to the complex analytical models used to simulate our knowledge about events in the real world. Such toolkits are transferable, permit users to operate the models without having a complete knowledge of the intricacies of the model or the data structures and could be used for guiding policy makers in evaluating the impact of alternative decisions.

GISs have been used in groundwater contamination studies to organize site-specific data, plot concentration contours, handle complex geological data, and run transport models linked to mapped output for hazardous waste sites. By using GIS applications in remote-sensing technology, the pipeline industry can benefit from computer imaging for planning, maintaining, and controlling pipeline systems. A prototype least-cost analysis was considered using remote sensing data and GIS to determine a least-cost routing for a small section of a proposed pipeline for the transport of crude oil. The proposed model was designed to include pipeline length, topography, geology, land use, and stream, wetland road, and railroad crossings. Custer County in South Dakota used GPS and GIS technology to develop a master street address guide to every structure and facility. Dobson[1] has discussed effectiveness of GIS as a tool for decision-makers. It supports decisions ranging from matters of state and environment to routine matters, such as siting facilities, expanding residential subdivisions, and routing transportation networks. The GIS developed for Brookline, Massachusetts was seen as a way of improving the efficiency of the maintenance of the city's water and wastewater infrastructure. The use of GIS for analysis and display in the temporal domain will help industry because a long-range view encourages investment in physical infrastructure and environmental protection.

A load allocation module allows multiyear distribution system studies in a GIS context to best assist utility planners in deciding when and where customers will grow and how to expand the system facilities to meet the demand growth. Officials in Clark County, Nevada, planned to include the GIS technology in its latest master plan designed to improve the design process on more than $1 billion in flood control projects. It is claimed that GIS will give the Regional Flood Control District's engineers and developers valuable new planning tools for the rapidly growing greater Las Vegas, Nevada area.

GIS has increasingly provided a foundation for integration and analyses of water environment data in context with other geographically related data. The degree of data diversity within a geographic framework employed in many water environment applications today extend to include economic, ecological, political and hydrological, to name a few examples. Technical tools, such as global positioning system (GPS) technology, satellite remote sensing and modeling integration, has allowed GIS to continue to expand beyond digital mapping to embrace a growing number of water resource protection and management domains.

Wastewater Management

Holliman[2] *et al.*, described the development of an GIS to support the management, maintenance and improvement of the Long Beach reclaimed water distribution systems and wastewater collection systems. By generating databases that combine requirements from a range of organization activities, wastewater utility managers can more easily automate their operation and maintenance activities. An integrated data environment with GIS as a central component allows utilities to respond to changing user requirements cost effectively. The affordability of a GIS by municipalities was discussed by Mitschele R[3.] in terms of sharing the data, costs, and benefits.

Fienberg and Uhrick[4] discuss integrating an infrastructure database in Broward County, Florida, with a GIS and water distribution and wastewater models. The HydroWorks model is used to simulate wastewater collection system, with close integration with the database of infrastructure characteristics and the GIS. Refsgaard[5] *et al.*, describes the evaluation the Danish Hydraulic Institute's (DHI) land-process hydrologic model SHE (Systeme Hydrologique European) and its extensive use of

GIS. Ribeiro[6] describes the use of raster-based GIS to interface the HSPF (Hydrologic Simulation Program-Fortran) to analyze the effects of basin urbanization. Hellweger[7] developed an ArcView application (ArcView popular GIS software) using the Avenue scripting language to perform calculations of an SCS (Soil Conservation Service) based model TR-55. Makropoulas[8] *et al.*, demonstrate how a raster-based GIS (IDRISI) can be used to integrate theoretical concepts and site-specific spatial characteristics. The strength of GIS can be used as a common ground between specialists and non-specialists to help them communicate effectively.

GIS have been used extensively for visualizing the projected developments in metropolitan areas, in terms of changes in population, industrial patterns, infrastructural utilities, etc. One very interesting aspect in this context is the expected future load and behavior of sewer systems, which convey the municipal and industrial wastewater flow to sewage treatment plants. Large sewer systems are structurally very complex networks with many nodes, pumps and weirs, often, with overflow chambers and loops. The hydraulic performance and capacity can only be evaluated by use of comprehensive mathematical modeling tools. MOUSE[9] (DHI, Denmark) is a professional engineering software package for the simulation of surface runoff, flows, water quality and sediment transport in urban catchments and sewer systems. The program can be equally successfully applied in any other type of pipe networks with alternating free surface and low pressurized flows. It is a dynamic, user-friendly modeling tool for the analysis, design, management and operation of both simple and large pipe networks. It provides a complete and effective working environment for sophisticated urban drainage and sewer engineering. As one step towards an efficient integration between the MOUSE modeling system and GIS technologies, DHI has developed the MOUSE Model Editor application using the ArcView and Avenue from Environmental System Research Institute (ESRI). This application offers the user an interactive tool for the definition of sewer network systems to be used for simulation in the MOUSE package. The MOUSE simulation results (overflow volumes, average and extreme filling of pipes etc.) are available to the MOUSE Model Editor application for graphical display and further analyses.

By applying MOUSE it is possible to obtain answers to questions such as:

1. Which parts of existing sewer systems will be overloaded under different conditions?
2. What are the main causes of overloading–backwater or insufficient pipe capacity?
3. What are the implications of replacing critical sewers, installing new basins, weirs, etc.?

Sewer asset and management database systems stores high level of details, the level of detail required for hydraulic modeling is much simpler, including every single minor component will not improve the accuracy of the model due primarily to the uncertainty in the hydraulic input data (rainfall, foul flow, etc.). Hence there is a need to simplify the network from the database level to the model network. MOUSE, an ArcView application allows the user the possibility to extract data from the asset management system, condense it semi-automatically–while monitoring consistency in the description of the physical system–and then store the data as a model description for use.

The MOUSE Model Editor includes a wide range of facilities:

- ☆ Display the entire sewer system network based on the external database.
- ☆ Interactive selecting of relevant nodes, pumps, weirs, etc. to be included in the final MOUSE sewer network.
- ☆ Editing facilities for adding or removing nodes, pumps and weirs to the sewer network.
- ☆ Editing attributes related to the nodes, pumps, weirs or pipes through a dialogue box.
- ☆ Facilities for viewing and analyzing the MOUSE simulation results, such as
  - Inflow to the system
  - Overflow volumes
  - Average filling of pipes
  - Extreme filling of pipes
  - External information- other hydraulic information

The package consist of various modules such as an advanced hydrological model for continuous time series, analyses of the potential of Real-Time Control (RTC) applications, computational models for the simulation of sediment transport and water quality in urban catchments and sewer systems, computation and statistical analysis of combined sewer overflows.

The modular structure offers great flexibility:

1. Each module can be handled separately
2. Data transfer between modules is automatic
3. Coupling of physical processes (*e.g.* sediment deposition–hydraulic conveyance) is facilitated
4. Updating or expansion of existing installations with renewed or additional modules is simple

The Pipe Flow Model performs the computations of unsteady flow in pipe networks. The computations are founded on an implicit, finite difference numerical solution of the 1D, free surface flow equations (Saint Venant). The implemented algorithm provides efficient and accurate solutions in multiply connected, branched and looped pipe networks.

The computational scheme is applicable to vertically homogeneous flow conditions, which occur in pipes ranging from small- profile collectors for detailed urban drainage, to low-lying, often-pressurized sewer mains, affected by a varying outlet water level. Both sub critical and supercritical flows are treated by means of the same numerical scheme, which is adaptive according to the local flow conditions.

Naturally, flow features like backwater effects and surcharges are simulated. Pressurized flow computations are facilitated through the implementation of a narrow 'slot', as a vertical extension of a closed pipe cross section. The complete non-linear flow equations can be solved for user-specified or automatically supplied boundary conditions. In addition to the fully dynamic description, simplified flow descriptions are also available. Within the Pipe Flow Model, advanced computational formulations enable the description of a variety of pipe network elements and flow phenomena including:

- ☆ Flexible cross-section database, including standard shapes
- ☆ Circular manholes
- ☆ Detention basins
- ☆ Overflow weirs
- ☆ Pump operations
- ☆ Flow regulation
- ☆ Constant or time variable outlet water level/inflows into the network
- ☆ Non-standard headlosses at manholes and basins
- ☆ Depth-variable friction coefficients

With such a description, the Pipe Flow Model can predict the detailed hydraulic conditions throughout the pipe network. Depending on the purpose of the study, the user may specially concentrate on the following aspects:

- ☆ The treatment plant loads (outflow hydrographs)
- ☆ Combined sewer overflows
- ☆ Pump activity
- ☆ Filling of detention basins
- ☆ Local surface flooding, etc.

There is no question that GIS approaches will revolutionize the ability to manage and visualize large environmental and infrastructure datasets, but more importantly, they will help us understand the linkages between monitoring data and the processes being analyzed.

Future

Developments in hardware, software, manpower and networking, as also their social acceptability is increasing due to various advantages of GIS listed earlier. The increased penetration of GIS into nodal agencies, which act as infrastructure planners and information providers is likely to increase availability of digital data. If the data exchange standards are universal, data gathering would be cheaper and easier, further accelerating growth of GIS.

The diversity of analysis and simulation tools that would conceivably be needed is enormous. A more open GIS architecture or 'toolbox' approach wherein additional functions would be grafted on easily or existing functions modified, might allow unusual analysis requirements to be carried out in GIS environment easily and efficiently. Future likely developments in GIS, useful to public health engineers, can be in following areas:

- Integration of Modeling in a GIS
- Simulation model, Stochastic modeling, Fuzzy and probabilistic models, Optimization, Genetic Algorithm
- Expert System, Neural Nets and Intelligent GIS
- Dynamic Modeling and animated Visualization
- Object Oriented GIS
- Advanced user interface in GIS
- Integrated use of the technologies to solve real life problems
- Multi-dimensional GIS
- Information packet for decision makers, mangers and citizens.

References

1. Dobson, J.E. (1997). GIS supports Decision Making Time after Time, *GIS World*, 10, 9, 34
2. Holliman, T.R. *et al.* (1995). AM/FM/GIS Implementation for Water and Wastewater Systems, Proc. of the 1995 Annual Urban and Regional Info. Syst. Assoc. (1), 373.
3. Mitschele, R. (1996). Share and Share Alike: Creating A Cost-Effective GIS, *Am. City County*, III, 3, 24
4. Fienberg,D., and Uhrick, S.W. (1997). "Integrating GIS with Water Wastewater Hydraulic Models" Proc., 1997 ESRI User Conf. (http: //www.esri.com/library/ userconf/proc97/PROC97/TO200/PAP1999/P199.htm)
5. Refsgaard, J.C., Storm, B., and Refsgaard, A. (1995). "Recent developments of the Systeme Hydrologique European (SHE) towards the MIKE SHE". *Modeling and management of sustainable basin-scale water resource systems*, S.Simonovic, Z. Kundzewicz, D. Rosbjerg, and K.Takeuchi, eds., International Association of Hydrologic Sciences Pubs. No. 231, IAHS Press, Wallingford, U.K. 427-434.
6. Rebeiro, C.T. (1996). "Impact of land use on water resources: Integrating HSPF and a raster-vector GIS", Proc. HydroGIS '96: Applic. of Geographic Info. Sys. In Hydro and Water Resour. Mngmt., K. Kovar and H.P.Nachmebel, eds., International Association of Hydrologic Sciences Publ. No 235, IAHS Press, Wallingford, U.K., 349-356.
7. Hellweger, R (1996). "TABHYD-TR55 Tabular hydrograph method in Arc View " Ctr. For Research in Water Resources, University of Texas at Austin, Texas, USA (http: //www.crwr.utexas.edu/gis/gishyd98/runoff/webfiles/ TABHYD/TABHYD.htm)

8. Markropoulous, C., Butler, D., and Maksimovic, C. (1998). "A GIS-based methodology for the evaluation of suitability of urban areas for source control application", UDM Proc., ' 98 4th *Int. Conf. On Devel. In Urban Drain Modeling*, D.Butler and C. Maksimovic, eds., Imperial college of Science,Technology and Medicine, London, 59-66.

9. Henrik G. Muller, Sten Lindberg, Danish Hydraulic Institute Agern Alle 5, DK-2970 Horsholm, Denmark.

Chapter 52

Environmental Impacts of Persistent Organic Pollutants and its Remediation

A. Juwlakar
CSIR-NEERI, Nagpur

The world's environment has enjoyed a huge prominence in political and social way of thinking in recent years. This has allowed the concept of sustainable development to flourish and many countries are embracing this goal with great zeal. Rapid industrialization all over the world has led to the introduction of numerous xenobiotic chemicals into the global ecosystem which in turn leads to the environmental contamination. Over 1,00,000 chemicals are currently in use world-wide and the number and amount of chemicals used is still increasing. Many of these are hazardous chemicals. Of particular concern are the persistent organic pollutants (POPs) since last three decades. Although many different chemicals, both natural and anthropogenic *(i.e.* produced by man), may be defined as POPs, 12 of them, all chlorine containing organic compounds, have been chosen initially as priority pollutants by the United Nations Environment Programme (UNEP) for their impact on human health and environment. The initial 12 POPs are also termed as the "***dirty dozen***" include many of the first generation organochlorine insecticides, *e.g.* DDT, aldrin, industrial chemical products such as polychlorinated biphenyls (PCBs) and unwanted by-products such as dioxins and furans (Stockholm Convention on POPs, 2001). Later on, 10 more chemicals (*e.g.* Chlordecone, Lindane, Polybrominated diphenyl ethers etc) were added as POPs which are commonly known as new POPs.

POPs are organic compounds that, to a varying degree, resist photolytic, biological and chemical degradation. POPs are often halogenated and characterized by low water solubility and high lipid solubility leading to their bioaccumulation in fatty tissues (Breivik *et al.*, 2004). POPs are also noted for their semi-volatility; that property of their physico-chemical characteristics that permit these compounds to occur either in the vapour phase or adsorbed on atmospheric particles, thereby facilitating their long range transport through the atmosphere. They are detected in regions where they have never been used or produced (AMAP, 1998). Greatest attention to date has focused, understandably, on POPs; the general problem of the widespread contamination in the environment with persistent chemicals extends across other chemical groups. During the last two decades much attention has been given to this group of substances at the international level after it became apparent that they are transported through the environment across borders. Individual countries alone are unable to control the environmental pollution from such border-crossing substances and critical concentrations have been reached in some regions, even in places where they have never been produced or used. Negotiations on a global, legally binding instrument to reduce and/or eliminate releases of POPs started in Montreal, Canada in 1998 under the auspices of UNEP. In May 2001, 126 countries and the EU agreed and adopted the text of this global treaty, referred to as the Stockholm Convention on POPs. To date, 178 countries have signed the Convention.

Brief Introduction POPs

As mentioned above, POPs chemicals includes pesticides, industrial chemicals and unintentionally produced by-products. The dirty dozen and new POPs are described below;

The Dirty Dozen

1. Aldrin

Aldrin is a pesticide used to control soil insects such as termites, corn rootworm, wireworms, rice water weevil, and grasshoppers. It is white, odourless crystals when pure; technical grades are tan to dark brown with a mild chemical odour. It has been widely used to protect crops such as corn and potatoes, and has been effective to protect wooden structures from termites. Aldrin is readily metabolized to dieldrin by both plants and animals. As a result, aldrin residues are rarely found in foods and animals, and then only in small amounts.

2. Chlordane

Chlordane is a broad spectrum contact insecticide that has been used on agricultural crops including vegetables, small grains, maize, other oilseeds, potatoes, sugarcane, sugar beets, fruits, nuts, cotton and jute. It has also been used extensively in the control of termites. Chlordane is highly insoluble in water, and is soluble in organic solvents. It is colourless to yellowish-brown viscous liquid with an aromatic, pungent odour similar to chlorine. It is semi-volatile and can be expected to partition into the atmosphere as a result.

3. DDT

Dichlorodiphenyl trichloroethane (DDT) is a pesticide that was very widely used in the past, because of its very low acute toxicity. It is odourless to slightly fragrant colourless crystals or white powder. DDT was widely used during the Second World War to protect the troops and civilians from the spread of malaria, typhus and other vector borne diseases. After the war, DDT was widely used on a variety of agricultural crops and for the control of disease vectors as well. It is still being produced and used for vector control. DDT is highly insoluble in water and is soluble in most organic solvents. It is semi-volatile and can be expected to partition into the atmosphere as a result. It is lipophilic and partitions readily into the fat of all living organisms and has been demonstrated to bioconcentrate and biomagnify. The breakdown products of DDT are DDD and DDE which are also present virtually everywhere in the environment and are more persistent than the parent compound.

4. Dieldrin

Dieldrin has been used in agriculture for the control of soil insects, termites and several insect vectors of disease but this latter use has been banned in a number of countries due to environmental and human health concerns. Dieldrin binds strongly to soil particles and hence is very resistant to leaching into groundwater. Volatilization is an important mechanism of loss from the soil and, because of its persistent nature and hydrophobicity, dieldrin is known to bioconcentrate. The half life of dieldrin in temperate soils is approximately 5 years. This persistence, combined with high lipid solubility, provides the necessary conditions for dieldrin to bioconcentrate and biomagnify in organisms.

5. Polychlorinated Dibenzo-p-Dioxins and Furans

Polychlorinated dibenzo-*para*-dioxins (dioxins) and polychlorinated dibenzofurans (furans) are two groups of planar tricyclic compounds that have very similar chemical structures and properties. Their properties vary with the number of chlorine atoms present. They are generally very insoluble in water, are lipophilic and are very persistent. Neither dioxins nor furans are produced commercially, and they have no known use. They are by-products resulting from the production of other chemicals. Dioxins may be released into the environment through the production of pesticides and other chlorinated substances. Furans are a major contaminant of PCBs. Both dioxins and furans are related to a variety of incineration reactions, and the synthesis and use of a variety of chemical products. Dioxins and furans have been detected in emissions from the incineration of hospital waste, municipal waste, hazardous waste, cars, and the incineration of coal, peat and wood. Of the 210 dioxins and furans, 17 contribute most significantly to the toxicity of mixtures. Dioxins and furans are considered to be very stable and persistent. This persistence, combined with high partition coefficients provides the necessary conditions for these compounds to bioconcentrate in organisms. The chemical properties of dioxins and furans (low water solubility, high stability and semi-volatility) favour their long range transport and these compounds have been detected in arctic organisms.

6. Endrin

Endrin is a foliar insecticide used mainly on field crops such as cotton and grains. It has also been used as a rodenticide to control mice and voles. It is white, odourless, crystalline solid when pure; light tan colour with faint chemical odour for technical grade. It is rapidly metabolized by animals and does not accumulate in fat to the same extent as other compounds with similar structures. It can enter the atmosphere by volatilization, and can contaminate surface water from soil run-off. The half life of endrin in soil may be up to 12 years, depending on local conditions. The chemical properties of endrin (low water solubility, high stability in the environment, and semi-volatility) favour its long range transport, and it has been detected in arctic freshwater.

7. Hexachlorobenzene

Hexachlorobenzene (HCB) is a fungicide that was first introduced in 1945 for seed treatment, especially for control of bunt of wheat. HCB is also a byproduct of the manufacture of industrial chemicals including carbon tetrachloride, perchlorethylene, trichloroethylene and pentachlorbenzene. It is a known impurity in several pesticide formulations, including pentachlorophenol and dicloram and may be present as an impurity in others. HCB is highly insoluble in water, and is soluble in organic solvents. It is quite volatile and can be expected to partition into the atmosphere as a result. HCB is very persistent and its estimated half lives in soil from aerobic and anaerobic degradation range from 2.7 to 22.9 years.

8. Heptachlor

Heptachlor is a non-systemic stomach and contact insecticide, used primarily against soil insects and termites. It has also been used against cotton insects, grasshoppers, some crop pests and to combat malaria. It is white to light tan, waxy solid or crystals with a camphor-like odour. Heptachlor is highly insoluble in water, and is soluble in organic solvents. It is quite volatile and can be expected to partition into the atmosphere as a result. It binds readily to aquatic sediments and bioconcentrates in the fat of living organisms. The half life of heptachlor in temperate soil is up to 2 years. Heptachlor is highly insoluble in water, and is soluble in organic solvents. It binds readily to aquatic sediments and bioconcentrates in the fat of living organisms.

9. Mirex

Mirex is a stomach insecticide with little contact activity. Its main use was against fire ants in the southeastern United States, but it has also been used to combat leaf cutters in South America, harvester termites in South Africa, Western harvester ants in the USA, mealybug of pineapple in Hawaii and has been investigated for possible use against yellow jacket wasps in the USA. It has also been used as a fire retardant in plastics, rubber, paint paper and electrical goods. Mirex is very resistant to breakdown, is very insoluble in water and has been shown to bioaccumulate and biomagnify. Due to its insolubility, mirex binds strongly to aquatic sediments. Mirex is considered to be one of the most stable and persistent pesticides, with a half-life of up to 10 years. This persistence, combined with lipophilicity, provides the conditions necessary for mirex to bioconcentrate in organisms.

10. Polychlorinated Biphenyls

Polychlorinated biphenyls (PCBs) are mixtures of chlorinated hydrocarbons that have been used extensively since 1930 in a variety of industrial uses, including as dielectrics in transformers and large capacitors, as heat exchange fluids, as paint additives, in carbonless copy paper and in plastics. The value of PCBs for industrial applications is related to their chemical inertness, resistance to heat, non-flammability, low vapour pressure and high dielectric constant. There are 209 possible PCBs, from three monochlorinated isomers to the fully chlorinated decachlorobiphenyl isomer. Generally, the water solubility and vapour pressure decrease as the degree of substitution increases, and the lipid solubility increases with increasing chlorine substitution. PCBs in the environment may be expected to associate with the organic components of soils, sediments, and biological tissues, or with dissolved organic carbon in aquatic systems, rather than being in solution in water. PCBs volatilize from water surfaces in spite of their low vapour pressure, and partly as a result of their hydrophobicity; atmospheric transport may therefore be a significant pathway for the distribution of PCBs in the environment. Half-lives for PCBs undergoing photodegradation range from approximately 10 days for a monochlorobiphenyl to 1.5 years for a heptachlorobiphenyl.

11. Toxaphene

Toxaphene is a nonsystemic and contact insecticide that was used primarily on cotton; cereal grains fruits, nuts and vegetables. It has also been used to control ticks and mites in livestock. It is yellow, waxy solid with a chlorine/terpene-like odour.Toxaphene is highly insoluble in water, and has a half life in soil of up to 12 years. It has been shown to bioconcentrate in aquatic organisms and is known to undergo atmospheric transport.

The New POPs

1. Alpha hexachlorocyclohexane

Alpha-HCH by itself is neither intentionally produced nor placed on the market but produced as the main constituent of technical HCH which is used as organochlorine insecticide or chemical intermediate to manufacture enriched gamma-HCH (Lindane). Alpha-HCH is highly persistent in water in colder regions and may bioaccumulate and biomagnify in biota and arctic food webs. This chemical is subject to long-range transport, is classified as potentially carcinogenic to humans and adversely affects wildlife and human health in contaminated regions.

2. Beta hexachlorocyclohexane

Very similar to alpha-HCH, beta-HCH by itself is neither intentionally produced nor placed on the market. It is produced as constituent of technical HCH used as organochlorine insecticide or chemical intermediate to manufacture enriched HCH (lindane). Beta-HCH is highly persistent in water in colder regions and may bioaccumulate and biomagnify in biota and arctic food webs. This chemical is subject to long-range transport, is classified as potentially carcinogenic to humans and adversely affects wildlife and human health in contaminated regions.

3. Chlordecone

Chlordecone is closely related chemically to mirex, a pesticide which is already listed under the Stockholm Convention. The chemical structure of chlordecone differs from mirex in that the oxygen of the keto group in chlordecone is replaced by two chlorine atoms in mirex. Chlordecone is not anymore produced or used. Chlordecone has been used in various parts of the world for the control of a wide range of pests. Chlordecone is highly persistent in the environment, has a high potential for bioaccumulation and biomagnification and based on physico-chemical properties and modelling data, chlordecone can be transported for long distances.

4. Hexabromobiphenyl

Hexabromobiphenyl belongs to a wider group of polybrominated biphenyls (PBBs). The term "polybrominated biphenyls" or "polybromobiphenyls" refers to a group of brominated hydrocarbons formed by substituting hydrogen with bromine in biphenyl. The hexabromo congeners exist as 42 possible isomeric forms. Hexabromobiphenyl has been used as a fire retardant in acrylonitrilebutadiene-styrene (ABS) thermoplastics for constructing business, machine housings and in industrial and electrical products and in polyurethane foam for auto upholstery. The chemical is highly persistent in the environment, highly bioaccumulative and has a strong possibility for long-range environmental transport.

5. Hexabromodiphenyl Ether and Heptabromodiphenyl Ether (commercial octabromodiphenyl ether)

Commercial octabromodiphenyl ether (c-octaBDE) is a mixture of several polybrominated diphenyl ethers and congeners and commonly called as polybrominated diphenyl ethers (PBDE). These synthetic brominated compounds have mainly been used as flame retardants. In addition to octaBDE isomers, c-octaBDE contains significant amounts of other component groups [such as pentabromodiphenyl (pentaBDE) and hexabromodiphenyl ethers (hexaBDE)]. Commercial mixture of octaBDE is highly persistent, has a high potential for bioaccumulation and food-web biomagnification, as well as for long-range transport.

6. Lindane

Lindane is the common name for the gamma hexachlorocyclohexane (gamma HCH). Lindane has been used as a broad-spectrum insecticide for seed and soil treatment, foliar applications, tree and wood treatment and against ectoparasites in both veterinary and human applications. The production of lindane has decreased rapidly in the last few years and only few countries are still known to produce lindane. Lindane is persistent, bioaccumulates easily in the food chain and bioconcentrates rapidly.

7. Pentachlorobenzene

Pentachlorobenzene (PeCB) was used in PCB products, in dyestuff carriers, as a fungicide, a flame retardant and as a chemical intermediate *e.g.* previously for the production of quintozene. PeCB might still be used as an intermediate. PeCB is also produced unintentionally during combustion, thermal and industrial processes. It also present as impurities in products such as solvents or pesticides. PeCB is persistent in the environment, highly bioaccumulative and has a potential for long-range environmental transport. It is moderately toxic to humans and very toxic to aquatic organisms.

8. Perfluorooctane Sulfonic Acid (PFOS), its Salts and Perfluorooctane Sulfonyl Fluoride (PFOS-F)

PFOS is a fully fluorinated anion, which is commonly used as a salt or incorporated into larger polymers. PFOS and its closely related compounds, which may contain PFOS impurities or substances that can result in PFOS, are members of the large family of perfluoroalkyl sulfonate substances. PFOS is both intentionally produced and an unintended degradation product of related anthropogenic chemicals. The current intentional use of PFOS is widespread and includes: electric and electronic parts, fire fighting foam, photo imaging, hydraulic fluids and textiles. PFOS is still produced in several countries. PFOS is extremely persistent and has substantial bioaccumulating and biomagnifying properties, although it does not follow the classic pattern of other POPs by partitioning into fatty tissues but instead binds to proteins in the blood and the liver.

9. Endosulfan

Endosulfan is the youngest member of POPs group. It occurs as two isomers; alpha and beta endosulfan and used as broad spectrum insecticide for control of aphids, thrips, beetles, foliar feeding larvae, mites, borers, cutworms, bollworms, whiteflies, and leafhoppers. Historically it was used to control termites and tsetse fly. Endosulfan is persistent in atmosphere, sediments and water, it bioaccumulates and has potential for long-range transport. It is toxic to humans and has been shown to have adverse effects on a wide range of aquatic and terrestrial organisms.

The chemical structures of dirty dozen and new POPs are presented in Tables 52.1 and 52.2 respectively.

Sources and Uses

Most POPs originate almost entirely from anthropogenic sources associated largely with the manufacture, use and disposition of certain organic chemicals. In contrast, HCB, dioxins and furans are formed unintentionally in a wide range of manufacturing and combustion processes. Environmental contamination of POPs can be attributed to several sources and activities. The identified sources of POPs include production units, illegal imports as well as stockpiles of obsolete pesticide stocks (IPEN report 2006). Except for DDT, which continues to be used in vector control, seven other pesticide POPs listed initially in the Stockholm Convention are banned for manufacture and use. However, stockpiles of unused and date expired POPs are a cause of concern mainly because in many places they remain unidentified. POPs are often found in dangerous storage conditions, leaking from rusted containers, contaminating groundwater and soils and poisoning the health and environment of people everywhere. It is very difficult to estimate of PCBs in capacitors and

Table 52.1: Chemical Structures of Dirty Dozen

Name of Chemical	*Chemical Structure*
Aldrin	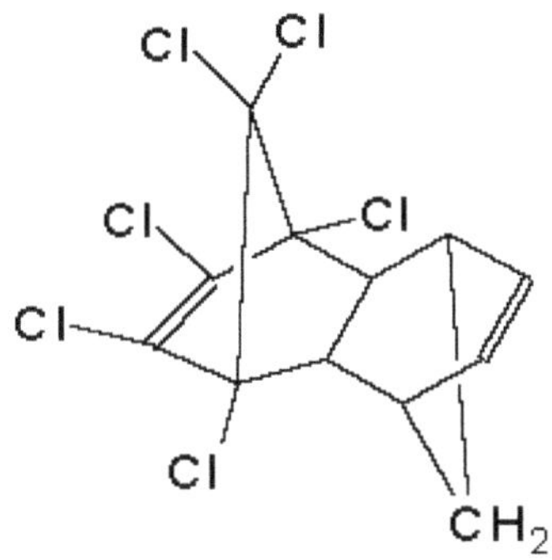
Chlordane	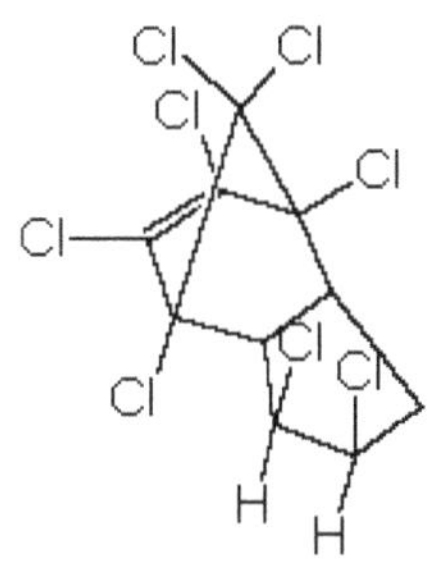 Cis 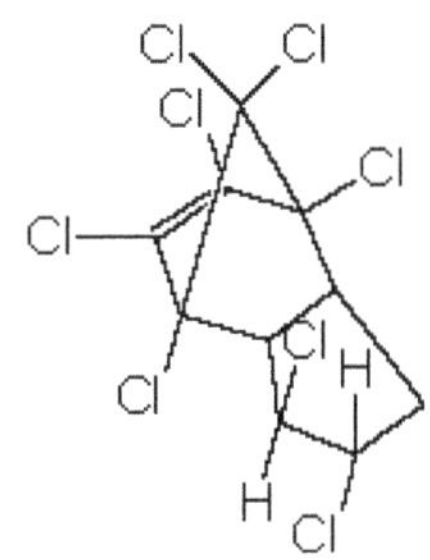 Trans
DDT	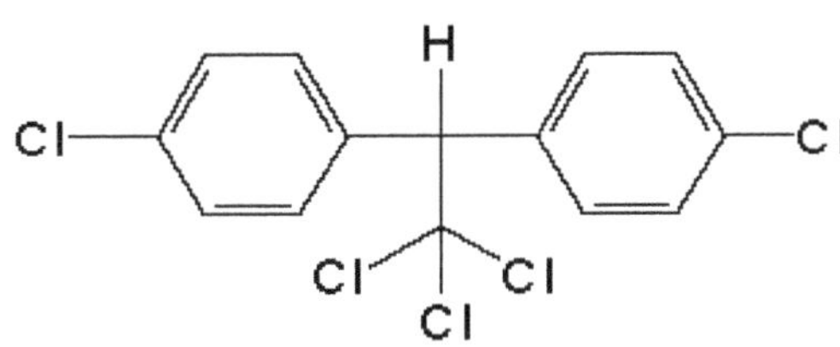
Dieldrin	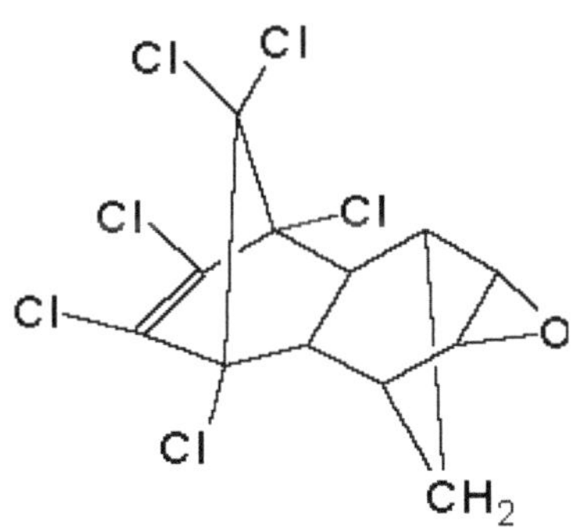

Contd...

Table 52.1–*Contd...*

Name of Chemical	*Chemical Structure*	
Dioxins and Furans	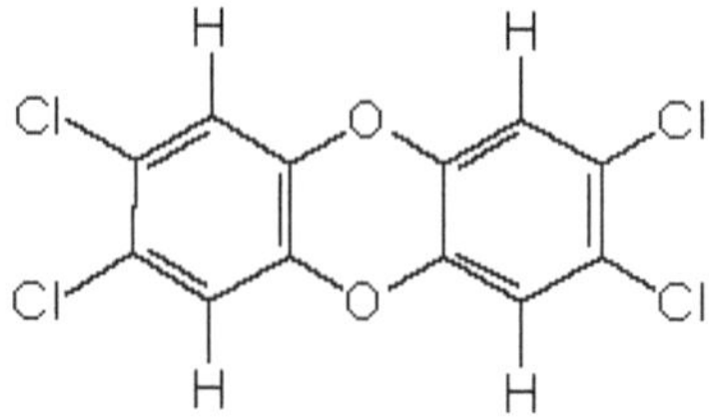 2,3,7,8-TCDD	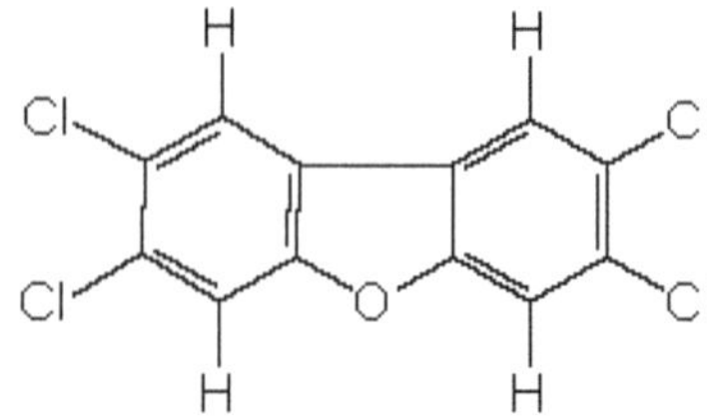 2,3,7,8-TCDF
Endrin	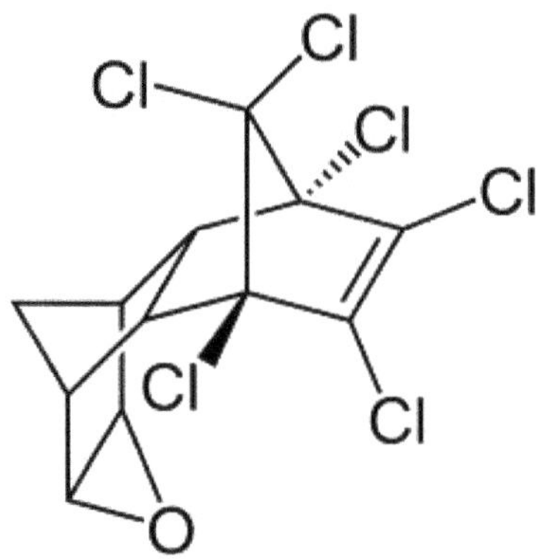	
Hexachlorobenzene	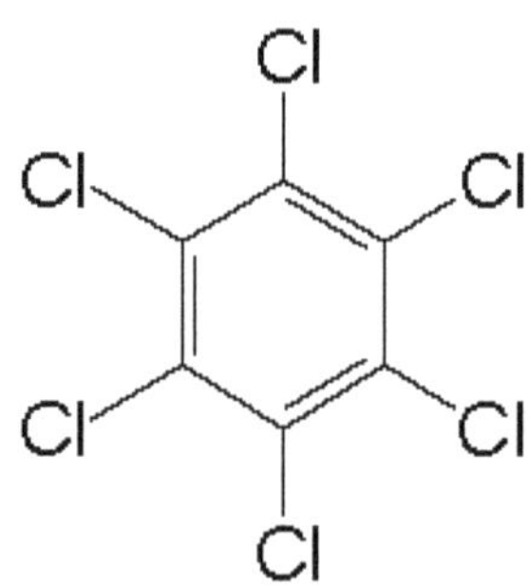	
Heptachlor	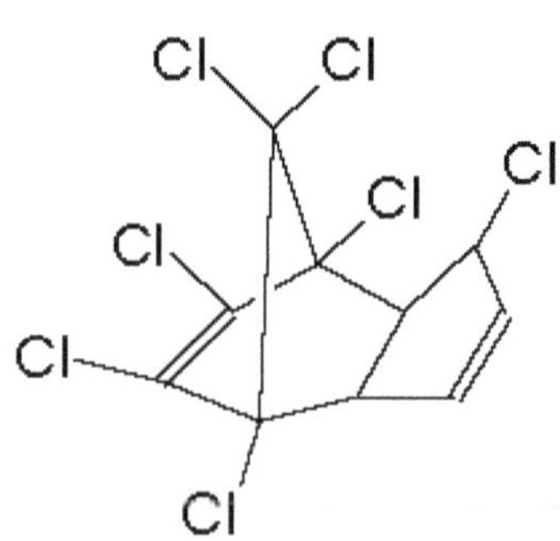	

Contd...

Table 52.1–*Contd...*

Name of Chemical	*Chemical Structure*
Mirex	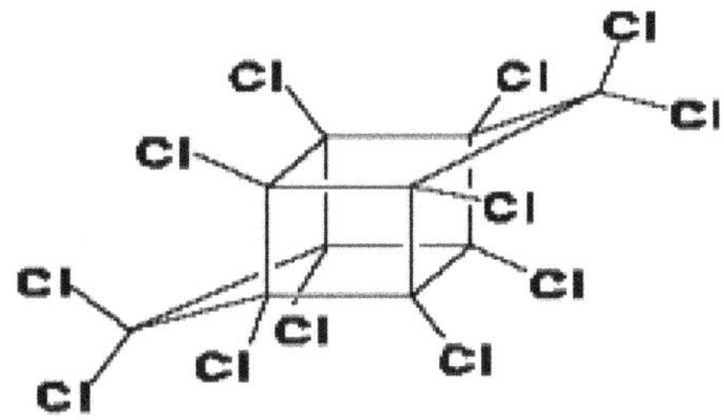
Polychlorinated Biphenyls	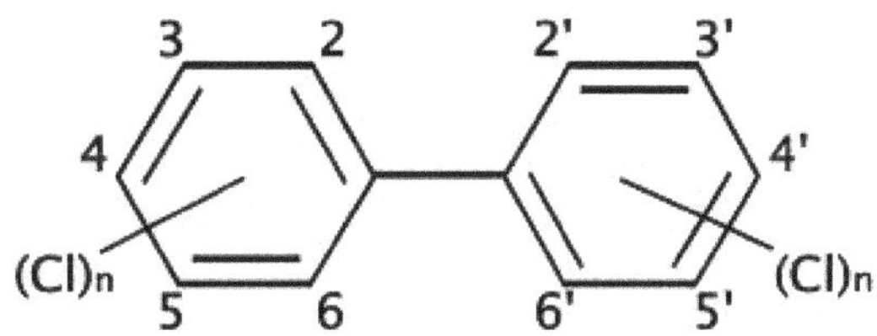
Toxaphene	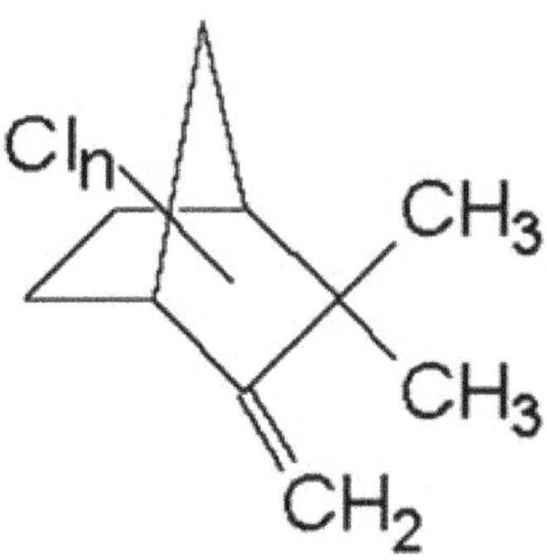

Source: http: //www.chem.unep.ch/pops/alts02.html.

transformers. Oil in old transformers is recycled into new transformers and may contain PCB's. Another critical issue is that of dioxins and furans, the sources of which are numerous like incinerations, cement factories, PVC units, biomass burning, open burning etc. Table 52.3 represents various sources of POPs contamination. Figure 52.1 shows the probable sources/pathways of POPs contamination and its transmission into the food chain.

Health Risks of POPs

The long biological half life of POPs facilitates the accumulation of seemingly small unit concentrations over extended periods of time. Serious health consequences are associated with POPs impacting humans, wildlife and the environment. International studies have revealed that wild life exposed to POPs have showed high rates of malformed genitalia, aberrant mating behaviour, sterility, cancer and immune

Table 52.2: Chemical Structures of New POPs

Name of Chemical	Chemical Structure
Alpha hexachlorocyclohexane	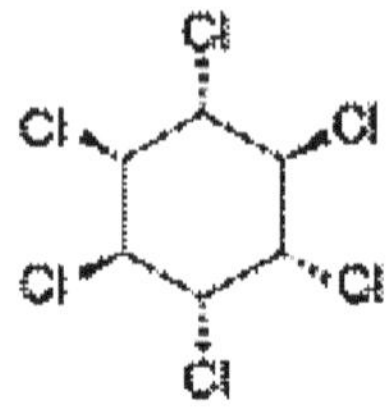
Beta hexachlorocyclohexane	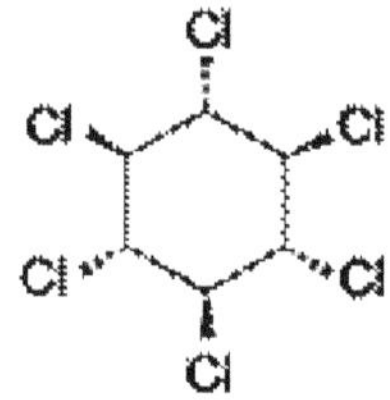
Chlordecone	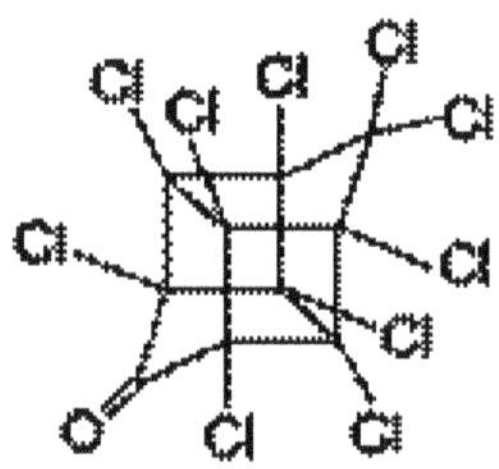
Hexabromobiphenyl	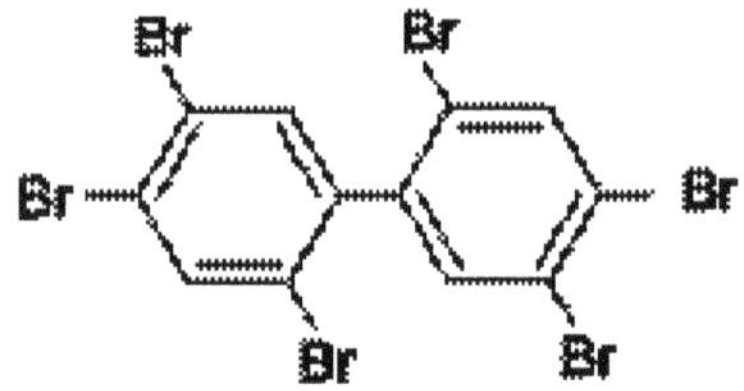
Hexabromodiphenyl ether and heptabromodiphenyl ether (commercial octabromodiphenyl ether)	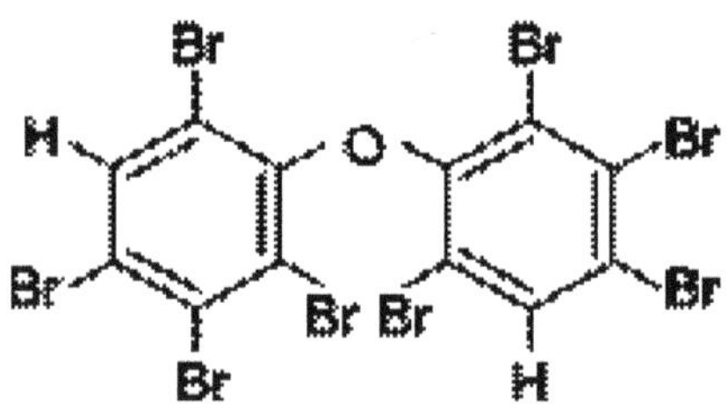
Lindane	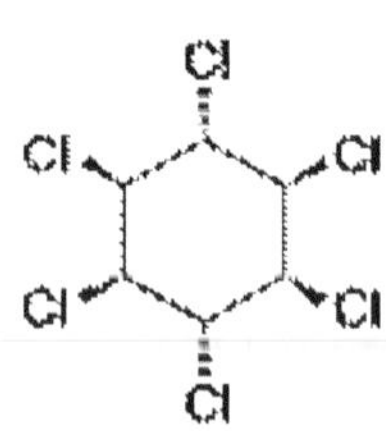

Contd...

Table 52.1–*Contd...*

Name of Chemical	*Chemical Structure*
Pentachlorobenzene	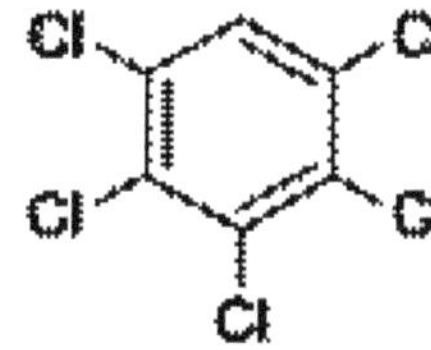
Perfluorooctane sulfonic acid, its salts andperfluorooctane sulfonyl fluoride	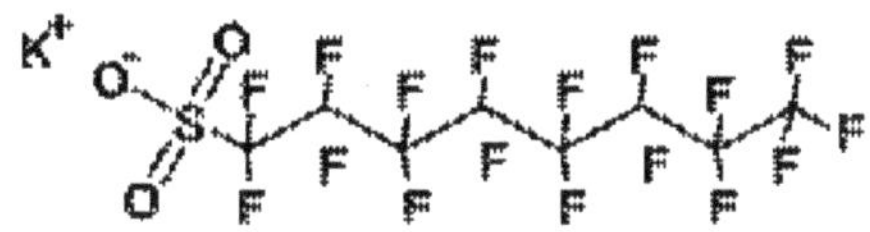
Endosulfan	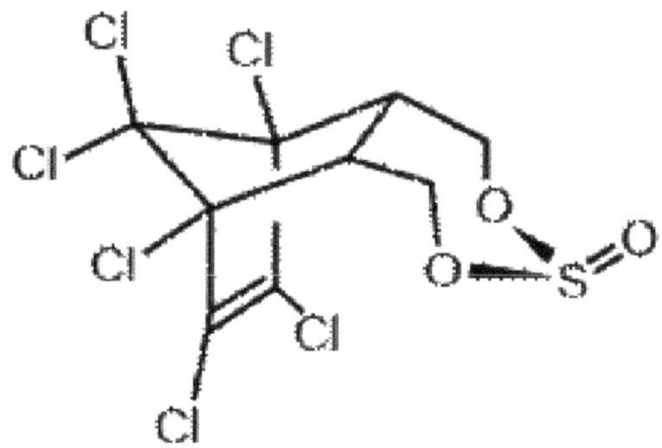

Source: http: //chm.pops.int/Convention/ThePOPs/TheNewPOPs/tabid/2511/Default.aspx.

system and thyroid dysfunction which might be the reason for decline in population of bottle-nosed dolphin, and beluga, common seals etc. Other known effects are thinning of the eggshells of fish-eating birds due to DDT exposure (ATSDR, 2002). These studies are suggestive of the fact that humans would experience similar effects as well. Scientific evidences suggest that some POPs have the potential to cause significant adverse effects to human health, at local level, at regional and at global level through long-range transport. Studies in India have shown high blood pressure, hormonal dysfunction, intrauterine growth retardation etc. in people with high levels of DDT. Some researchers have suggested a possible relationship between exposure to some POPs and human disease and reproductive dysfunction. Also, associations have been made between certain organochlorines contaminants and cancer in human populations. Therefore on a broad sense, it can be inferred that POPs are associated with a range of health complications in humans as well in other organisms. More recently some POPs have also been implicated in reduced immunity in infants and children, and the concomitant increase in infection, also with developmental abnormalities, cancer and tumor induction or promotion, sterility (ATSDR, 2005). In addition, studies have linked POP exposure to diseases and abnormalities in a number of wildlife species, including numerous species of fish, birds, and mammals (EPA, 2002).

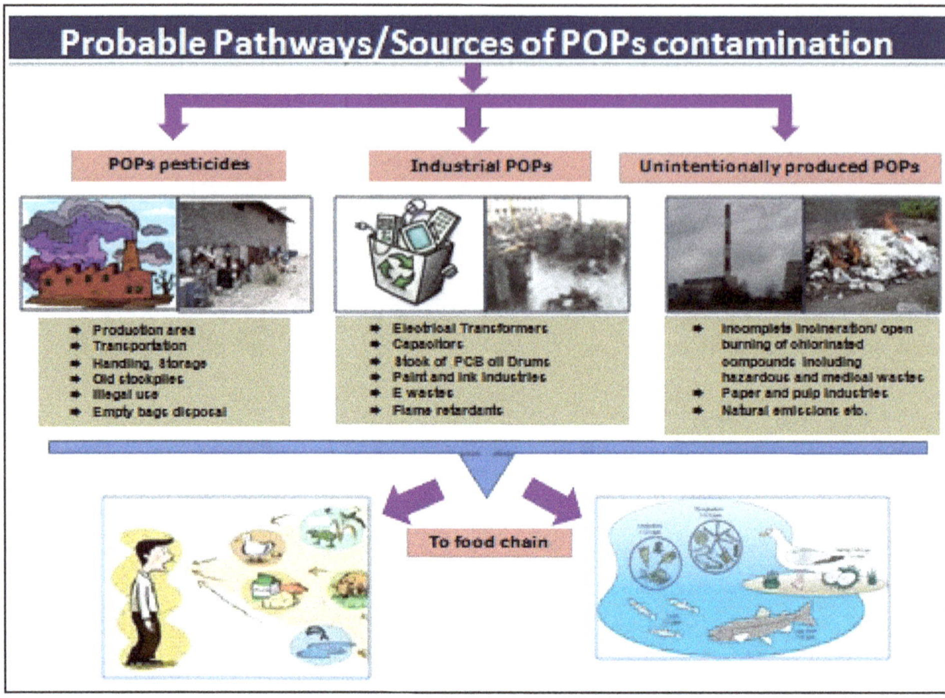

Figure 52.1: Probable Pathways/Sources of POPs Contamination and its Transmission in Food Chain.

Need for Remediation

As discussed above, POPs are toxic compounds that are chemically stable and do not easily degrade in the environment. Thus, they tend to accumulate and biomagnify as they move up through the food chain. Restrictions and bans on the use of POPs have resulted in a significant number of unusable stockpiles of POP-containing materials, internationally. In addition, deterioration of storage facilities used for the stockpiles, improper storage practices, and past production and use of POPs have resulted in contamination of soils around the world. Because of their chemical stability, tendency to bioaccumulate, adverse health effects associated with POPs, and widespread POP contamination, remediation technologies are needed to treat these pollutants.

Previously, POP-contaminated soil and stockpiles have been treated using technologies such as incineration that rely on combustion to destroy the contaminants. However, site owners and operators, remedial project managers, and other interested parties have expressed concern about the potential environmental and health effects associated with combustion of POPs. Combustion technologies can create dioxins and furans. Dioxins and furans have been characterized by EPA as human carcinogens and are associated with serious human health problems. Also, combustion technologies that have historically been used for the destruction of POPs may fail to

Table 52.3: Sources of POPs, Pathways of Contamination and Environmental Receptors

POPs Categories	*Sources of Release*	*Uses/Applications*	*Pathways of Contamination and Environmental Receptors*		
			Air	*Water*	*Soil*
Pesticides	✰ Stockpiles ✰ Manufacture of pesticides ✰ Imports	✰ Agricultural spraying for soiland crop pests ✰ Spraying/land application (*e.g.*, disease vector control, livestock)	✰ Stack emissions ✰ Industrial emissions ✰ Emissions during spraying	✰ Leakage from stockpiles ✰ Industrial effluents (*e.g.*, production waste) ✰ Agricultural and other soil run-off	✰ Leakages from stockpiles ✰ Industrial effluents (*e.g.*, production waste) ✰ Agricultural and other soil run-off
Industrial Chemicals (PCBs and HCB)	✰ Solid waste incineration ✰ Sewage sludge ✰ Ship breaking industry	✰ Raw material for industrial processes ✰ Electric equipment stocks	✰ Industrial emissions ✰ Emissions from solid waste and sewage sludge incineration ✰ Emissions from equipment and stocks ✰ Industrial emissions	✰ Industrial effluents ✰ Leakage from solid waste and sewage sludge dumps ✰ Leakage from equipment, stocks ✰ Industry (including ship-breaking) effluents	✰ Sewage sludge dumps ✰ Industrial effluents ✰ Leakage from equipment stocks ✰ Industry (including ship breaking) effluents
By-products (dioxins and furans, PCBs and HCB)	✰ By-products in manufacture of pesticides 5 and industrial chemicals ✰ Industrial, thermal and medical incineration processes 6 ✰ Transportation systems ✰ Biomass burning ✰ Forest fires/other wood combustion ✰ Sewage sludge		✰ Emissions from industrial, thermal and medical waste incineration processes ✰ Exhaust gas emissions from automobiles ✰ Releases in biomass/forest/woodburning ✰ Emissions from sewage sludge incineration/leakage from sewage sludge dumps	✰ Industrial effluents ✰ Leakage from sewage sludge dumps	✰ Industrial effluents ✰ Leakage from sewage sludge dumps ✰ Settling of air emissions from incineration proce-sses

Source: IPEN Report 2006.

meet the stringent environmental conditions or destruction and removal efficiency (DRE) requirements established for POPs. Because of these concerns and an ongoing desire to find more cost effective solutions, environmental professionals are examining the application of non- combustion technologies to remediate POPs in stockpiles and soil (UNEP 2004). Therefore, non combustion technologies are always recommended for the treatment and remediation of POPs and POPs contaminated sites. Certain non combustion technologies for POPs remediation are briefed below;

(1) Anaerobic Bioremediation Using Blood Meal

This technology was developed by EPA's Environmental Response Team (ERT). This technology is implemented at full scale to treat POPs mainly toxaphene and DDT contaminated sites. It uses biostimulation with amendments to promote degradation of toxaphene and DDT in soil and sediment by native anaerobic microorganisms (Harry., 2002).

(2) Bioremediation Using DARAMEND

DARAMEND is a proprietary technology provided by ART in Mississauga, Ontario, Canada. It is an amendment enhanced soil bioremediation technology for the treatment of POPs mainly DDT and toxaphene, using specialised and metallic amendments and water, followed by drying and tilling that involves the creation of sequential anoxic and oxic conditions (Philips *et al.,* 2001). This technology can be implemented in either *ex situ* or *in situ* where the treatment layer is 2 feet deep.

(3) Gas-Phase Chemical Reduction (GPCR)

GPCR technology was developed by Eco Logic International Inc. in Ontario, Canada. It is an *ex situ* technology, implemented at both full and pilot scales to treat solids and liquids contaminated with POPs such as HCB, DDT, PCBs, dioxins, and furans (Eco Logic, 2002).

(4) Geomelt Process

GeoMelt vitrification is a high-temperature technology that uses heat to destroy POPs and to permanently immobilize residual contaminants by incorporating them into the vitrified end product. It is a full-scale treatment technology used *in situ* and *ex situ* to treat high-strength wastes containing POPs such as dieldrin, chlordane, heptachlor, DDT, HCB, PCBs, dioxins, and furans (Geomelt., 2005).

(5) *In Situ* Thermal Desorption (ISTD)

ISTD technology is developed by Shell Oil Co. and Terra Therm This technology involves simultaneous application of heat and vacuum to subsurface soils to average temperatures of 1,000° F. It is used on full scale to treat PCBs, PAHs, dioxins, and chlorinated volatile organic compounds (CVOC) at volumes as low as a few hundred cubic yards to greater than 20,000 cubic yards in 6 to 9 months (EPA., 2004).

(6) Mechanochemical Dehalogenation (MCD)

MCD technology is developed by Environmental. Decontamination Ltd. (EDL). It occurs *ex situ* in an enclosed ball mill uses mechanical energy to promote reductive dehalogenation of contaminants. It is used to treat high-strength wastes containing POPs mainly DDT, aldrin and dieldrin, applicable to soil, sediments, and mixed solid-liquid phases. The MCD system exhibited a maximum treatment rate of 139 cubic meters per week (Volker., 2002).

(7) Xenorem

Xenorem technology is developed by Stauffer Management Company, a subsidiary of AstraZeneca Group PLC in Mississauga, Ontario, Canada. It is an *ex situ* bioremediation technology applicable to soil phase, used to treat low strength wastes containing chlordane, DDT, dieldrin, and toxaphene contamination. Organic amendments such as manure and wood chips are added to contaminated soil, which can increase the final amended soil volume by as much as 40 per cent (Gray *et al.*, 2002).

(8) Base-Catalyzed Decomposition (BCD)

BCD technology is developed by EPA's National Risk Management Research Laboratory in Cincinnati, Ohio. It is an *ex situ* technology used on pilot tests to treat high-strength waste containing POPs mainly PCBS, dioxins, and furans and is available in both transportable and fixed configurations (EPA., 2005).

(9) TDR-3R

TDR-3R is marketed by Terra Humana Clean Technology Engineering Ltd. in Hungary. It is a bench scale *ex situ* technology used to treat high and low-strength soils containing POPs contamination mainly HCB. The firm has developed pilot scale kilns that operate at a throughput of 0.1 tons per hour (EPA 2005).

(10) Sonic Technology

This technology is developed by Sonic Environmental Solutions Inc. in Vancouver, Canada. Sonic technology is a pilot scale *ex situ* technology that is used to treat low and high strength soils containing POPs. The sonic energy activates dechlorination of the PCBs in the solvent (EPA., 2005).

(11) Microbe Assisted Phytoremediation

Phytoremediation is an *ex situ* and *in situ* technology applicable in pilot scale to treat lowstrength soils, sludges, and sediments contaminated with POPs. It is a process that uses plants to remove, transfer, stabilize, or destroy contaminants in soil, sediment, and groundwater (Aken *et al.*, 2010). Microbe assisted phytoremediation (MAP) can be an eco-friendly and efficient technology for the remediation of contaminated sites (Juwarkar and Singh., 2010). Plants and associated bacteria constitute a strong symbiotic association for proficient and eco-friendly treatment of contaminated soil, sediments and water.

(12) Mediated Electrochemical Oxidation (CerOx)

This technology is developed by CerOx Corporation in Santa Maria, California. It is an *ex situ* electrochemical reaction technology used in pilot tests to treat low strength liquids containing POPs mainly chlordane, dioxins, and PCBs (UNEP). The systems range in size from modules with 25-gallon per day (gpd) treatment capacities to multimodular plants with 100,000-gpd treatment capacities. This technology was performed by UNR, CerOx Corporation, in May 2000. The medium treated was a pesticide-water emulsion in which 71 per cent by mass chlordane was mixed with water and fed to the gaseous-phase reactor. The system was reported to have achieved a chlordane destruction efficiency of 99.995 per cent (EPA., 2005).

(13) Plasma Arc (PLASCON)

PLASCON is a full scale *ex situ* technology potentially applicable to both low and high strength, solid and liquid wastes streams containing POPs contamination. This technology passes a direct current discharge through argon gas to create plasma with a temperature greater than 10,000°C. Organic compounds present in the waste get pyrolized. In this technology 1 to 3 tonnes/day of waste can be treated by a 150 kW unit. The residence time of the waste in the reaction chamber is very short (approximately 20- 50 milliseconds) with less than 0.5 g of waste being destroyed at any instant (EPA., 2005).

(14) Plasma Arc Centrifugal Treatment (PACT)

PACT is a full-scale *ex situ* technology that can be used on pilot scale to treat low and high strength wastes containing POPs contamination. This technology uses two treatments. Primary treatment occurs inside a centrifuge tank housing the plasma torch, which generates heat to melt and vitrify contaminated feed material to a temperature of about 1,650°C. Secondary treatment of gaseous contaminants occurs at a temperature of about 1,000°C, which ensures the destruction of products of incomplete combustion such as dioxins and furans (EPA., 2005).

(15) Supercritical Water Oxidation (SCWO)

SCWO is an *ex situ* technology applicable to both low and high strength solid and liquid wastes containing POPs contamination (Weber., 2004). This technology destroys toxic and hazardous organic wastes in a compact totally enclosed system, using an oxidant (*e.g.* oxygen or hydrogen peroxide) at temperatures and pressures above the critical point of water (374°C and 22.1 x 10^6 Pascal). During this process carbon dioxide, water and inorganic acids or salts are released as a by-products. SCWO systems operate continuously, use corrosion-resistant materials in their reactors and process only fluid influents. SCWO systems are limited to the treating of solids and liquids with a particle size of less than 200 microns suspended in a liquid. This technology is developed by General Atomics' Advanced Process Systems division was selected for use as an Assembled Chemical Weapons Assessment technology.

(16) Catalytic Hydrogenation

Catalytic Hydrogenation is a pilot plant technology applicable to low level wastes containing POPs contamination in liquid form. This process is based on the mild hydrogenation of spent oil in a packed bed catalytic reactor, operating at moderate temperatures and pressures. Under process conditions hydrogen reacts with hetero atoms in the oil itself, and also with any organochlorines present. During this process chlorinated species are converted in to hydrogen chloride and light hydrocarbons, released as byproduct. This technology is used for a commercial treatment of 3000 litres of PCB oil with concentration < 1000 ppm, also been tested on a wide range of POPs. This technology gives better than 99.999 per cent of destruction efficiency for PCB (UNEP).

(17) TiO_2-based V_2O_5/WO_3 catalysis

TiO2-based V_2O_5/WO_3 Catalysis technology is applicable to high strength wastes, containing POPs contamination such as PCBs. In this process, PCB were destroyed on a V_2O_5/WO_3 supported titanium catalyst at low temperature in the range of 150–300°C over 30 minutes. The efficiency of this process is >99.9 per cent.

Promising Eco-Friendly Technology for POPs Remediation

Chemo-Biological Approach

As POPs possess high chemical stability, super hydrophobicity and thermal resistivity, a cost effective, non combustion and eco-friendly technology is the demand of time. Therefore, bioremediation coupled with the use of chemical methodologies for the degradation of POPs can be proved as an effective tool for the degradation of POPs. The chemo-biological approach for the reductive dechlorination of PCBs has been studied by several researchers (Aronstein and Rice 1995). Chemical pretreatment can transform POPs to enhance their bioavailability for the microorganisms, which can result in their enhanced biodegradation (Aronstein *et al.*, 1995). Use of microorganisms coupled with chemical methods can be a novel approach to enhance the degradation of POPs and subsequent remediation of contaminated sites.

Conclusion

There is no doubt that certain excellent properties such as thermal resistivity, chemically stable nature, insecticidal properties etc., of few POPs chemicals make them beneficial in various uses. However, the persistence, bioaccumulation and toxic characteristic of POPs have put them among the most dangerous pollutants to the human being and the environment. However, there is a genuine lack of knowledge regarding their sources, impacts and remedial measures. Many advanced remediation techniques have been commercialized and adopted in industrialized and developed countries. This chapter covers an overview to the dirty dozen and new POPs, their source of contamination, health impacts and the best available technologies (BAT) for their remediation. However, due to the financial constraints, many advanced technologies are unlikely to be adopted by the developing countries. The success of finding suitable and affordable technologies is critical in solving problems in the developing countries.

References

1. Aken B.V., Correa P.A., Schnoor J.L. (2010). Phytoremediation of Polychloinated Biphenyls: New trends and promises. *Environmental Science and Technology* 44: 2767-2776.

2. AMAP, (1998). AMAP Assessment Report: Arctic Pollution Issues. *Arctic Monitoring and Assessment Programme (AMAP)*, Oslo, Norway. 859 ISBN 82-7655-061-4.
3. Aronstein B.A., Paterek J.R., Kelley R.L. and Rice L.E. (1995), The effect of chemical pretreatment on the aerobic microbial degradation of PCB congeners in aqueous systems. *Journal of Industrial Microbiology and Biotechnology*, 15: 55-59.
4. Aronstein B.N. and Rice L.E. (1995), Biological and integrated chemical–biological treatment of PCB congeners in soil/sediment-containing systems. *Journal of Chemical Technology and Biotechnology*, 63: 321–328.
5. Breivika K., Alcock R., Li Y.F., Bailey R.E., Fiedler H., Pacyna J.M. (2004). Primary sources of selected POPs: regional and global scale emission inventories. *Environmental Pollution* 128: 3–16.
6. Eco Logic. (2002). "Contaminated Soil and Sediment Treatment Using the GPCRTM/TORBED Combination". http: / /www.torftech.com / start.html.
7. EPA (2004). Field Evaluation of TerraTherm In Situ Thermal Destruction (ISTD) Treatment of Hexachlorocyclopentadiene. Office of Research and Development, Superfund Innovative Technology Evaluation Program. EPA/540/R-05/007. http: / /www.epa.gov / ORD / NRMRL / pubs / 540r05007 / 540R05007.pdf
8. EPA, 2002). Persistent Organic Pollutants, A Global Issue, A Global Response. EPA160-F-02-001.
9. EPA. (2005). Reference guide to non-combustiontechnologies for remediation of persistent organic pollutants in stockpiles and soil.
10. Geomelt technologies/AMEC earth and environmental, Inc., Richland, Washington. (2005). http: / /www.geomelt.com / technologies.
11. Gray N.C.C., Cline P.R., Gray A.L., Byod B., Moser G.P., Guiler H.A., and Gannon D.J. (2002). Bioremediation of a Pesticide Formulation Plant. *Proceedings of the Third International Conference on Remediation of Chlorinated and Recalcitrant Compounds. Monterey, California.*
12. Harry A. L. (2002). Anaerobic Bioremediation of Toxaphene Contaminated Soil – A Practical Solution. *17th World Congress of Soil Science*, Symposium No. 42, Paper No. 1509. Thailand.
13. IPEN (2006). Country Situation on Persistent Organic Pollutants (POPs) in India.
14. Juwarkar A.A. and Singh S.K. (2010). Microbe-assisted phytoremediation approach for ecological restoration of zinc mine spoil dump. *Int. J. Environment and Pollution*, 43: 1/2/3, 236–250.
15. Phillips T., Bell G., Raymond D., Shaw K., and Seech A. (2001). 6th International HCH and Pesticides Forum, Poznan, Poland. "DARAMEND Technology for In Situ Bioremediation of Soil Containing Organochlorine Pesticides."
16. Stockholm Convention on Persistent Organic Pollutants (POPs); Interim Secretariat for the Stockholm Convention, United Nations Environmental Programme (UNEP) Chemicals: Geneva, Switzerland, Oct 2001. http: / / www.pops.int.

17. Toxicological Profile: for DDT, DDE, and DDE. Agency for Toxic Substances and Disease Registry (2002).
18. U.S. Department of Health and Human Services, (2005). Toxicological Profile Information Sheet. Agency for Toxic Substances and Disease Registry (ATSDR). http: / /www.atsdr.cdc.gov / toxpro2.html.
19. UNEP Report. Review of emerging, innovative technologies for the destruction and decontamination of POPs and the identification of promising technologies for use in developing countries.
20. UNEP, (2004). STAP of the GEF. Review of emerging, innovative technologies for the destruction and decontamination of POPs and the identification of promising technologies for use in developing countries. http: / /www.unep.org/stapgef/ home/index.htm.
21. Volker B. (2002). Reductive Dehalogenation of Recalcitrant Polyhalogenated Pollutants Using Ball Milling. *Proceedings of the Third International Conference on Remediation of Chlorinated and Recalcitrant Compounds. Monterey, California.*
22. Weber R., 2004. Relevance of PCDD/PCDF formation for the evaluation of POPs destruction technologies. – PCB destruction by super critical water oxidation (SCWO). *Organohalogen Compd* 66: 1263–1269.

Chapter 53

Treatment of Textile Wastewater Using Aqueous Oxidation Agents and their Effect on Biodegradation

R.A. Pandey, K.V. Padoley, T.V. Subbarao, A.S. Rajvaidya, M.V. Lakshane, C.V. Deshpande, N.N. Rao, P.P. Pathe, T. Nandy, T. Chakarabarti and S.N. Kaul

CSIR-NEERI, Nagpur

ABSTRACT

The wastewater from a textile industry was characterized and results indicated that pH of the wastewater was alkaline (10.95) and contained average chemical oxygen demand (COD) of 1617mg/l, biochemical oxygen demand (BOD_5) of 688 mg/l, sulphide concentration of 21.6 mg/l and ammonia concentration 20.32 mg/l, alongwith intense colour. This investigation was designed to evaluate the extent of aqueous oxidative removal of organics and colour of the textile wastewater using Fenton's reagent and a combination of magnesium oxide and magnesium carbonate (magnesium peroxide) separately and also to assess the biodegradability of effluent generated after the abovesaid treatment options in comparison with the textile wastewater without any treatment, keeping same the COD concentration. The results indicate that under optimal conditions, pretreatment with Fenton's reagent could remove 69 per cent of COD, while pretreatment with magnesium peroxide could remove 62.23 per cent of COD. The wastewater was then subjected to biodegradation without any treatment using mixed biomass developed from bacterial cultures previously isolated on synthetic dyes and the results indicate a COD removal of 48 per cent within a period of 72 hours. Further, the biodegradation in case of Fenton treated wastewater resulted in 96 per cent removal of residual COD in 72 hours, while in case of magnesium peroxide treatment 93 per cent COD could be removed in the same time period compared to only 48 per cent COD removal for untreated wastewater within 72 hours. The chemical sludge produced during the fenton treatment was 0.6 g/l, while in case of magnesium peroxide treatment the sludge production was 3.3 g/l.

Keywords: *Textile wastewater, Pretreatment, Fenton oxidation, Magnesium peroxide, Biodegradation.*

Introduction

The textile wastewater poses a great challenge for treatment, since apart from being characterized with high pH, chemical oxygen demand (COD), biochemical oxygen demand (BOD), and total dissolved solids (TDS), it also contains high intensity of colour, which makes it very difficult to treat by conventional treatment methods. The untreated effluents if discharged into natural streams, can cause rapid depletion of dissolved oxygen of the receiving waterbodies, and the problem can further get aggravated due to the presence of suspended solids ultimately leading to subsequent decomposition of the sludge under anaerobic condition. The alkalinity of the wastewater and presence of chromium can severely affect the aquatic life and may also affect the biological treatment processes. Some of the dyes present in the wastewater are also reported to be toxic to aquatic life [1]. The colour often renders the water unfit for use for some industrial purposes in the downstream side. It has been reported that the presence of sodium hydroxide at a concentration of 25 ppm in the textile wastewater is lethal to fish life [1]. In view of the toxicity and adverse effects of the textile industry wastewater, it is essential to treat the wastewater so as to conform to the limits prescribed by the regulatory agencies.

Since the wastewaters generated from textile industry are toxic and resistant to biodegradation, the physico-chemical option in combination with biological methods has been reported in the literature by several workers. A combined approach for the treatment of textile wastewater using ozonation followed by biodegradation in activated sludge process has been reported [2].

The treatment of textile wastewater using ozone/ultraviolet light/hydrogen peroxide (O_3/UV/H_2O_2) treatment as a first step followed by biodegradation in activated sludge process and the results of their study indicate that ozonation was the most suitable pretreatment option for biodegradation, since the ozone pretreated wastewater did not show any inhibitory effect on subsequent treatment with microbial flora in the activated sludge system. [2]

The textile wastewater in the present investigation was subjected to treatment using Fenton's reaction and also magnesium peroxide. Further, the pretreated wastewater was subjected to biodegradation to assess the effect of these treatment options on amenability of the wastewater to biodegradation. The results of the study are presented and discussed in this paper.

Materials and Methods

The wastewater samples were collected in 50 liter capacity containers and were preserved at source using concentrated sulphuric acid and adjusting the pH to 4.0. Two aqueous oxidative agents *viz.* Fenton, and magnesium peroxide were employed for the pretreatment of textile wastewater. The pretreated wastewater was filtered through Whatman No.42 filter paper and was analyzed for COD, and also tested for further biodegradation. All the analyses were carried out on clarified samples as per Standard Methods [3].

Both pretreated (diluted/undiluted) wastewaters were subjected to static flask biodegradation experiments. Similarly, the wastewater without pretreatment was diluted so as to get the same COD magnitude as that of pretreated wastewater and was subjected to biodegradation in order to ascertain the effect of pretreatment on biodegradability of pretreated wastewater. For biodegradation, mixed biomass was developed using bacterial cultures previously isolated in the laboratory on synthetic dyes. Wastewater was supplemented with dipotassium hydrogen phosphate and potassium dihydrogen phosphate and ammonium sulphate, the medium pH was adjusted to 7.2 ± 0.02, and the flasks were incubated at 30±2°C.

Results

Evaluation of Fenton's Oxidation as a Pretreatment Option for Textile Wastewater

The textile wastewater was subjected to Fenton oxidation as one of the pretreatment options. The efficiency of the pretreatment was assessed on the basis of COD removal. Optimization of conditions for treatment of textile wastewater by Fenton's Reagent was carried out and details are given hereunder:

Optimization of pH of the Reaction

The optimization of pH was done using a constant dose of ferrous sulphate of 0.4 g/l and 2 ml/l of H_2O_2. The pH of the textile wastewater was varied in the range of 2.0, 2.5, 3.0, 3.5, and 4.0, using appropriate volume of sulphuric acid and a fixed contact time of four hours for the reactants and textile wastewater was maintained. The results on pH optimization are presented in Figure 53.1, which indicate that pH of 3.0 is optimal for the treatment of textile wastewater by Fenton's reagent, since maximum COD removal of 48.96 per cent is observed at this pH value.

Optimization of Contact Time

The optimization of contact time was carried out using a constant dose of ferrous sulphate 0.40 g/l and 2 ml/l of H_2O_2, reaction mixture at optimal of pH of 3.0. The contact time between Fenton's reagent and textile wastewater in the reaction mixture was varied from 60 minutes to 240 minutes (1 hr to 4 hrs) and the results presented in Figure 53.2, indicate that Fenton's reagent acts effectively at a contact time of 3 hours (180 minutes) with a maximum COD removal of 49 per cent.

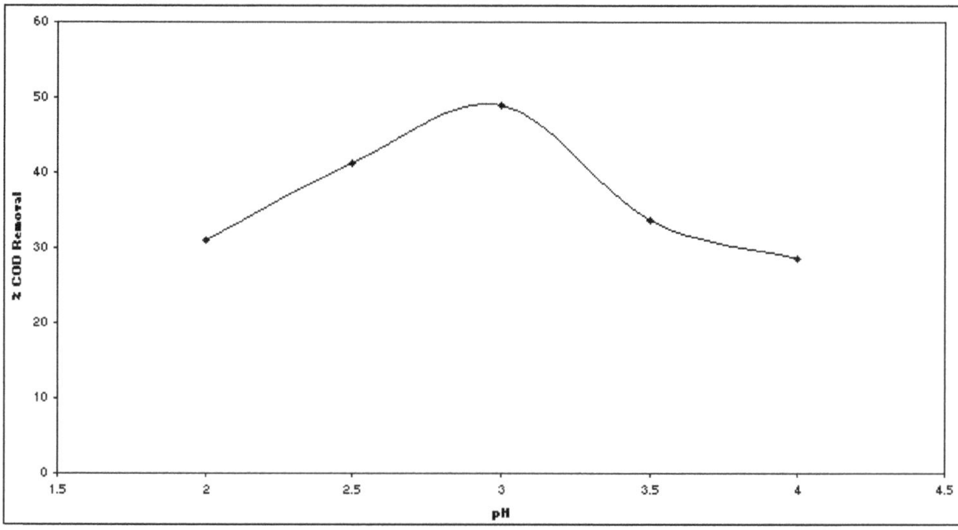

Figure 53.1: Optimization of pH of Textile Wastewater for Effective Treatment.

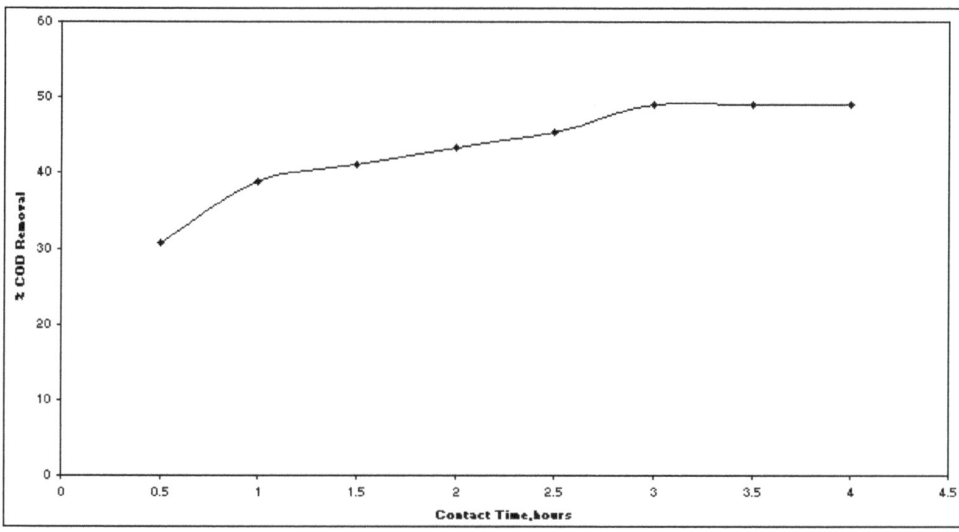

Figure 53.2: Optimization of Contact Time for Fenton Treatment of Textile Wastewater.

Optimization of Ferrous Sulphate Dose

Ferrous sulphate dose for the treatment of textile wastewater was varied from 0.05g/l to 0.90 g/l under optimal conditions of pH and contact time. The results of optimization of ferrous sulphate dose are presented in Figure 53.3, which indicate that a dose of 0.70 g/l ferrous sulphate is optimal for treatment of textile wastewater. This resulted in 54.64 per cent removal efficiency of COD. Further, increase in ferrous sulphate dose did not improve COD removal efficiency.

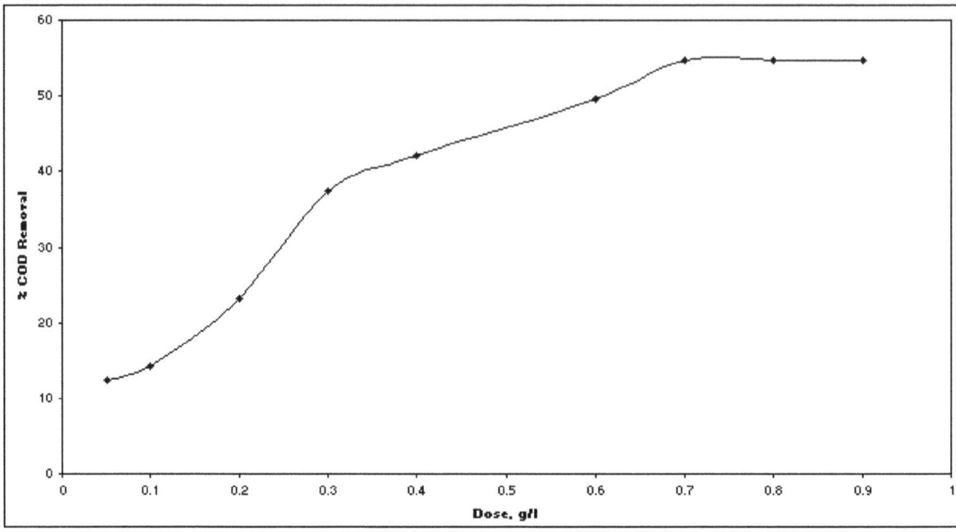

Figure 53.3: Optimization of Ferrous Sulphate Dose.

Optimization of Hydrogen Peroxide Dose

Hydrogen peroxide dose for the reaction was varied from 1 ml/l to 8 ml/l under optimal conditions of pH, contact time and ferrous sulphate dose and the results presented in Figure 53.4 indicate that a hydrogen peroxide dose of 4.0 ml/l is optimal for treatment of the textile wastewater. At optimal dose of H_2O_2, 68.69 per cent COD removal efficiency was achieved keeping other conditions constant. Further increase in the hydrogen peroxide dose did not lead to improvement in COD removal efficiency.

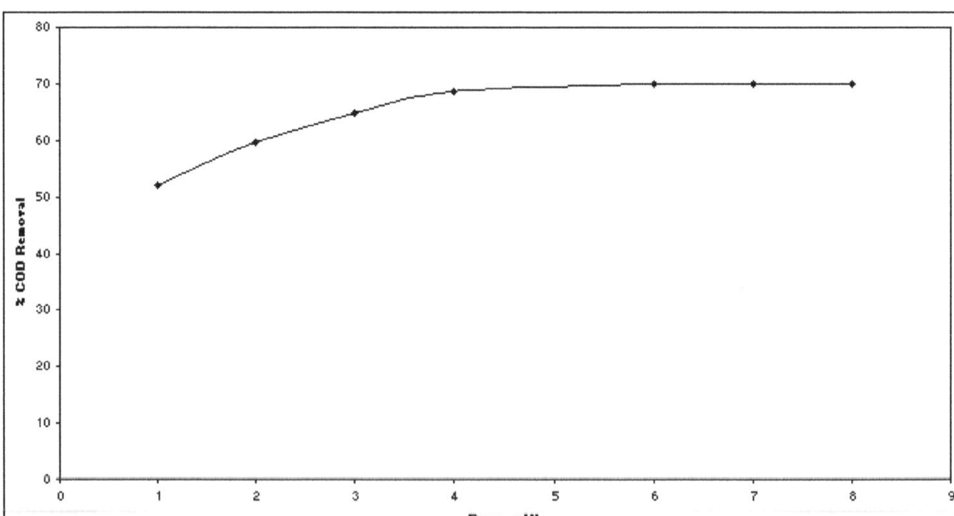

Figure 53.4: Optimization of Hydrogen Peroxide Dose.

Treatment of Textile Wastewater using Fenton's Reagent Under Optimal Conditions

After ascertaining all the optimal conditions viz. pH, contact time, and doses of ferrous sulphate and hydrogen peroxide, the textile wastewater was again subjected to treatment using Fenton's reagent. The pretreatment experiments were conducted five times under optimal conditions. The results indicate that under the optimal condition *i.e.* pH 3.0, ferrous sulphate dose of 0.70 g/l, contact time of 3 hours and hydrogen peroxide dose of 4 ml/l, the average COD removal for the textile wastewater using Fenton's reagent was 69 per cent. This pretreated textile wastewater was further subjected to biodegradation.

Evaluation of Magnesium Peroxide as a Pretreatment Option for the Textile Wastewater

The textile wastewater was subjected to pretreatment using another aqueous oxidant -a combination of magnesium oxide and magnesium carbonate alongwith hydrogen peroxide. The combination of these reagents yields magnesium peroxide slurry (as per the protocol given by Vol'nov and Latysheva, 1958). The efficiency of treatment was assessed on the basis of COD removal and other parameters *viz.* pH, contact time and slurry concentration were optimized subsequently for the treatment of textile wastewater.

Optimization of pH of the Reaction

The optimization of pH was carried out by varying the pH of the reaction mixture from 2.0 to 8.0. The contact time was maintained at one hour, the slurry concentration was maintained at 4 per cent, and the results are presented in Figure 53.5. The results on pH optimization indicate that a pH value of 3.0 is optimum for treatment of textile wastewater since maximum COD removal of 59.74 per cent was obtained at this pH value.

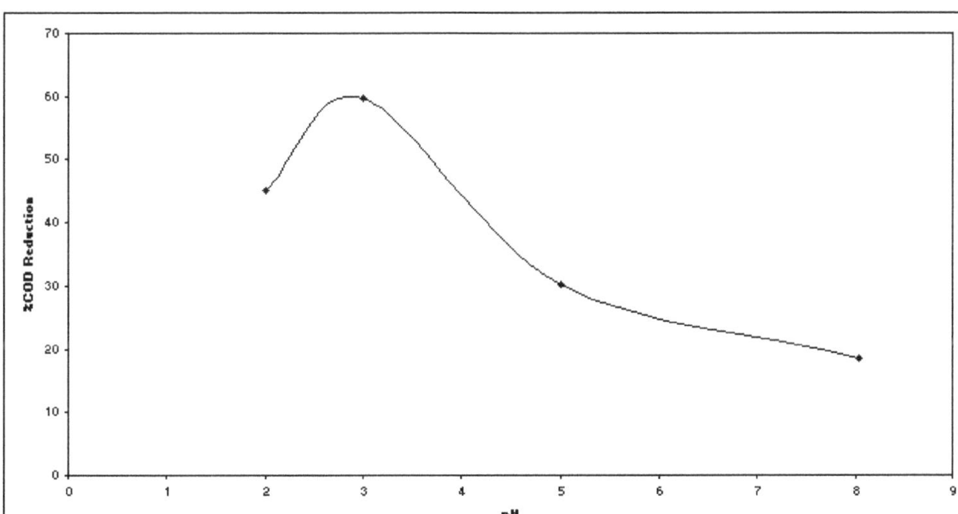

Figure 53.5: Optimization of pH for Magnesium Peroxide Treatment.

Optimization of Contact Time

The contact time for the treatment of textile wastewater was varied from 20 minutes to 100 minutes at pH 3.0 and with a slurry concentration of 4 per cent. The results presented in Figure 53.6 indicate that a contact time of 60 minutes (one hour) is optimum for the treatment of textile wastewater by magnesium peroxide, since maximum COD removal of 60.17 per cent was observed at this contact time.

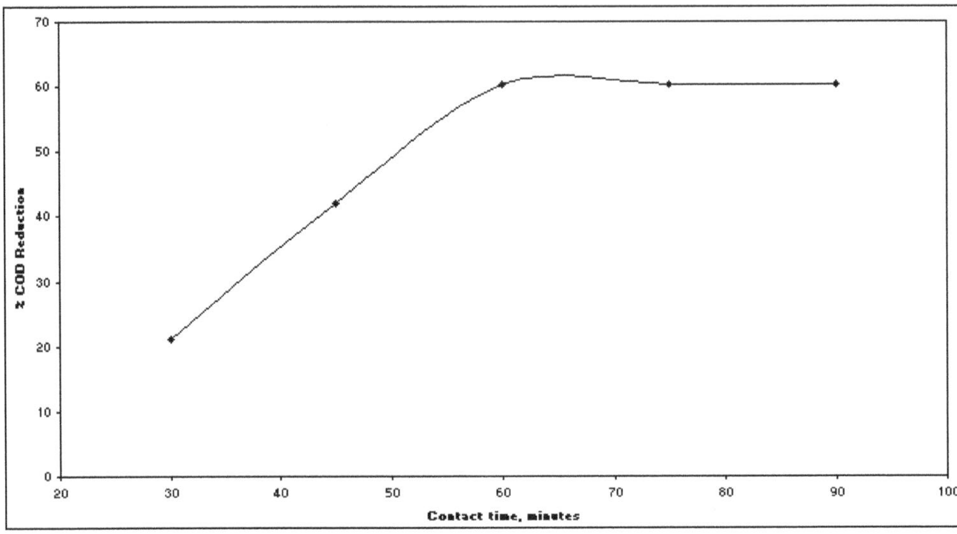

Figure 53.6: Optimization of Contact Time Magnesium Peroxide Treatment.

Optimization of Slurry Concentration

The magnesium oxide-magnesium carbonate mixture constituting the magnesium peroxide slurry was varied from 2 per cent to 14 per cent at a optimum pH of 3.0 and optimum contact time of 60 minutes. The results are presented in Figure 53.7 indicate that 4 per cent slurry concentration is optimal for the treatment of textile wastewater since it resulted in 60.27 per cent COD removal.

Treatment of Textile Wastewater Using Magnesium Peroxide under Optimal Conditions

After optimization of all the conditions *viz*. pH, contact time and per cent slurry concentration of magnesium peroxide, the jar test experiments were conducted with the textile wastewater and the results indicate that under optimal conditions 60.27 per cent of COD could be removed from the wastewater (average of five sets of experiments) using magnesium peroxide.

The sludge production in case of Fenton treatment system for textile wastewater was 0.6 g/l while in case of magnesium peroxide system was 3.3 g/l (Figure 53.8).

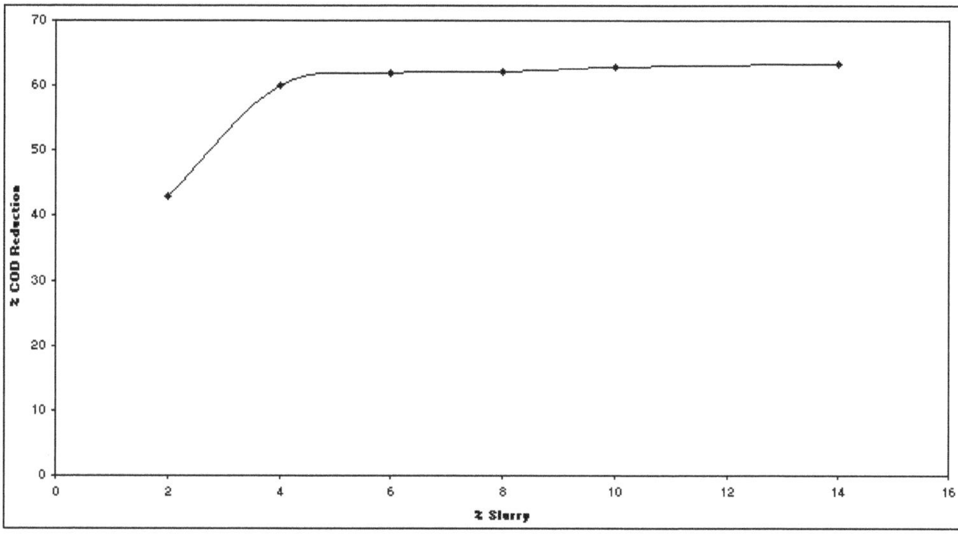

Figure 53.7: Optimization of Magnesium Peroxide Slurry Concentration.

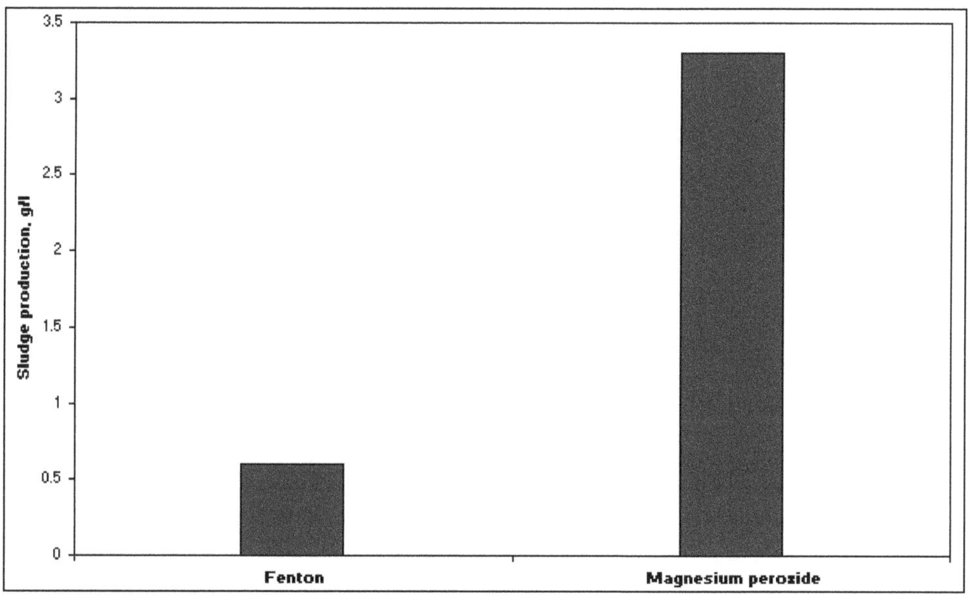

Figure 53.8: Sludge Production during Pretreatment of Textile Wastewater.

Comparative Evaluation of Biodegradative Characteristics of Textile Wastewater with and without Fenton's Treatment and Magnesium Peroxide Treatment

The treated textile wastewater generated after Fenton's treatment and magnesium peroxide treatment under optimal conditions was subjected to biodegradation under shake flask culture experiments using mixed biomass prepared by mixing bacterial cultures isolated on synthetic dyes, at a controlled temperature of incubation (30± 2°C) and shaking at 150 rpm. At the same time, the original textile wastewater without Fenton and magnesium peroxide treatment was diluted so as to bring down the COD value almost equal to that obtained by treatment with Fenton's reagent and magnesium peroxide. During biodegradation, the COD of the wastewater was assessed at fixed intervals of time and the results are presented in Figure 53.9. The results of these experiments indicate that Fenton treated textile wastewater offers better amenability to biodegradation, since within 72 hours 96 per cent of the residual COD gets removed in the textile wastewater as compared to only 48.0 per cent removal of COD in case of untreated wastewater in the same time period, while the magnesium peroxide treated wastewater could lead to a COD reduction of 93 per cent in the same time period. Thus, the results confirm that the biodegradability of Fenton's treated textile wastewater is more than magnesium peroxide treated and untreated, diluted textile wastewater.

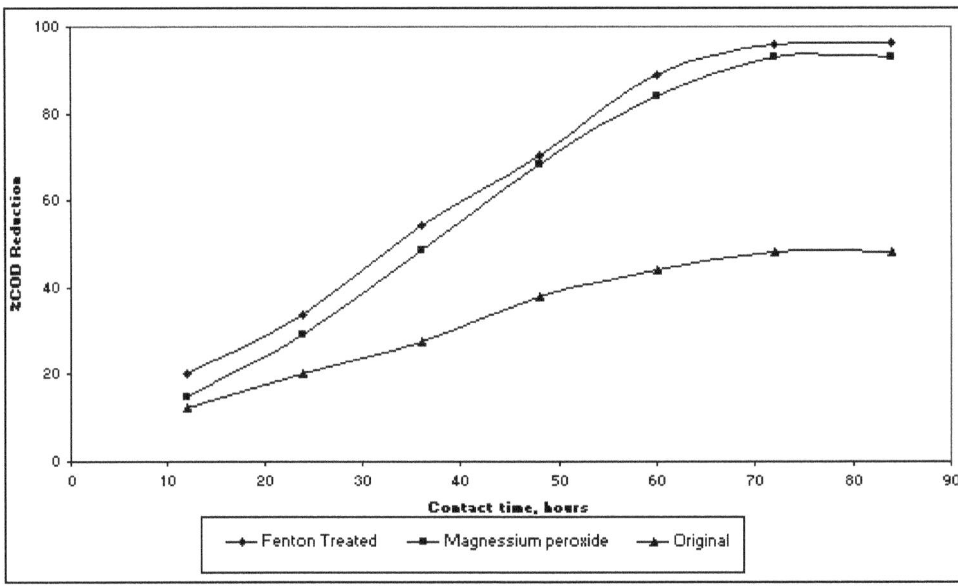

Figure 53.9: Comparative Evaluation of Biodegradability of Textile Wastewater with and without Preatreatment.

Discussion

The Fenton reaction employs ferrous sulphate and hydrogen peroxide as reactants. In the mechanism of Fenton reaction, the following reactions generally are reported to occur [4,5,6]

$$Fe^{2+} + H_2O_2 \rightarrow Fe^{3+} + HO^- + H^+$$

$$HO^- + Fe^{2+} \rightarrow HO^- + Fe^{3+}$$

$$HO + Rx \rightarrow Rx^- + HO^-$$

Organic Compound (Rx)

The results of present investigation indicate that a pH value of 3.0 is optimal and a pH range of 2.0 to 4.0 has been reported to be optimal for Fenton's reaction for treatment of different wastewaters. [4] Above pH 4.5, the ferrous ion gets converted into ferric ion, which combines with OH radical to form ferric hydroxide precipitate that settles down and reduces the ferrous ion availability in the reaction mixture. This minimizes the effectivity of oxidation reaction. A pH of 2.75 has been reported for 80 per cent TOC removal for creosote contaminated wastewater [7]. A pH value of 3.0 as optimal for treatment of colour and pigment containing wastewater has been reported [8]. A pH of 3.5 has been cited as optimal for dissociation of aqueous reactive dyes (Red and Blue). [8]

The results on optimization of contact time for Fenton's reaction indicate that a contact time of 180 minutes (3 hours) is optimal for treatment of textile wastewater. A contact time of 1 hour has been to be optimal for treatment of coloured and pigmented wastewater [8]. Further a contact time of 40 minutes for decolourization of reactive dyes using photo-assisted Fenton reaction has been cited in the literature [9]. The results on optimization of ferrous sulphate dose in the present investigation indicate that a ferrous dose of 0.70 g/l is optimal. A ferrous sulphate dose of 60.0 mg/l (0.06 g/l) resulted in more than 90 per cent transformation of creosote contaminated wastewater [7]. A ferrous sulphate dose of 4 g/l for treatment of reactive dyes viz. reactive red 2 and reactive blue 4 via photo-assisted Fenton reaction has been reported [9]. Further, a ferrous dose of 0.4 g/L for treatment of intensely coloured wastewater has been cited in the literature [8]. The results on hydrogen peroxide dose optimization indicate a dose of 3 ml/l to be optimal for the treatment of textile wastewater. A hydrogen peroxide dose of 0.4 ml/l has been reported to be optimal for treatment of colored wastewater [8]. Thus, the results obtained on optimization of different parameters for Fenton reaction in the present investigation fall well within the range reported for Fenton reaction previously.

The effectiveness of magnesium peroxide as an aqueous oxidant lies in the principle that, in aqueous medium, reactive radicals viz. O^{2-} and hydroxyl radical OH^- are generated which lead to a reaction between the organics in aqueous medium and the oxidants present in the slurry [10]. The results on pH optimization for magnesium peroxide treatment indicate a pH value of 3.0 is optimal for the present investigation. The results on contact time optimization for magnesium peroxide system indicate a one-hour contact time to be optimal for the treatment of textile wastewater. The results on optimization of percentage of slurry dose of magnesium peroxide for the treatment of textile wastewater indicate a 4 per cent slurry concentration is optimal.

The results on comparative evaluation of biodegradation of textile wastewater with and without Fenton treatment indicate that Fenton treated wastewater is more amenable to biodegradation as compared to the wastewater treated with magnesium peroxide reagent. Inspite of diluting the original textile wastewater to the same COD magnitude as that obtained after Fenton pretreatment, the available organic pollutants were not amenable to biodegradation to the extent as observed in the Fenton treated wastewater. Similar observations have also been reported for phenolic, creosote contaminated wastewater, wherein pretreatment by Fenton has reportedly altered the constituents of the wastewater in making it more amenable to biodegradation [11,6].

Wastewater from textile industry is very complex in nature with high COD and intense colour. This wastewater could be treated satisfactorily by applying Fenton oxidation under optimal conditions of pH, contact time and doses of ferrous sulphate and hydrogen peroxide to achieve a COD reduction of 69 per cent. The treated effluent can further be subjected to biodegradation with efficient removal of residual COD (96 per cent).

Further, the application of Fenton's reagent as a treatment option prior to biodegradation helps in the molecular modification of the indigenous pollutants occurring in textile wastewater which can easily be oxidized biologically, since the biodegradability of the wastewater without Fenton's treatment showed a COD removal of 48 per cent only inspite of having same concentration of input COD. The results obtained on biodegradation of textile wastewater pretreated with magnesium peroxide indicated that 93 per cent residual COD could be removed after treatment. However, comparative data could not be cited in the literature in context of textile wastewater. Thus, the results on the study indicate that Fenton's treatment is a better option for pretreatment of textile wastewater than magnesium peroxide treatment, since the former not only leads to a higher COD reduction but also further aids in much better COD removal during biodegradation.

Table 53.1: Physico-chemical and Biological Characteristics of the Textile Wastewater

Parameter	*Value/Range*	**Standard for Discharge into Inland Surface Water Body*
pH	10.95	5.5-9.0
Conductivity millisemens/cm	2.89-3.95	—
COD mg/l	1617.0	250.
BOD mg/l	688.0	30.0
Ammonia-nitrogen mg/l	20.32	50.0
Sulphide mg/l	#21.6	2.0
Heavy metal mg/l		
Cu	0.066	3.0
Cr	—	2.0
Fe	0.280	3.0
Pb	0.020	0.1
Cd	0.008	0.2
Mn	0.068	2.0
Zn	0.094	5.0

*: Inputs prescribed by MoEF (Ministry of Environment and Forest).

#: Values more than the prescribed standard.

Conclusion

The textile wastewater was characterized with high pH, and high concentration of BOD, COD, TDS, and intense colour (Table 53.1). Studies on pretreatment of the textile industry wastewater with aqueous oxidative treatment methods like Fenton's reaction and also magnesium peroxide treatment were conducted. Biodegradation of pretreated textile wastewater was carried out using a mixed biomass developed from bacterial cultures isolated on synthetic dyes. The results obtained on biodegradation of textile wastewater without any pretreatment indicate only 48 per cent removal of COD, while the results obtained on biodegradation of textile wastewater pretreated with Fenton's reagent indicated that 96 per cent residual COD removal whereas treatment with magnesium peroxide resulted in 93 per cent removal of residual COD. Aqueous oxidative pretreatment of textile industry wastewater with Fenton oxidation resulted in better biodegradation of the wastewater with minimum production of chemical sludge.

References

1. Upadhyay, R. and Pandey, G. N. Environmental pollution control in cotton textile industry, *J. Ind. Poll. Cont.*, 1991; 7: 67-75.
2. Ledakowicz, S. and Gonera, M. Optimization of oxidants dose for combined chemical and biological treatment of textile wastewater, *Water Res.*, 1999; 33: 2511-2516.
3. APHA. American Public Health Association. *Standard Methods for the Examination of Water and Wastewater*.18th Edition., Washington, DC: APHA, AWW and WEF, 1992.
4. Neodoloujko, A. and Kiwi, J. TiO_2 speciation producing mineralization of 4-Tert-Butylpyridine. Accelerated Mineralization on via Fenton photo-assisted reaction; *Water Res.*, 2000; 3277-3284.
5. Legrini, O., Oilveros, E. and Braun, A.M. Photochemical processes for water treatment; *Chem. Rev.*, 1993; 93: 671-698.
6. Arnold, S.M., Hickey, W.J., and Harris, R.F. Degradation of atrazine by Fentons reagent: condition optimization and product quantification, *Environ.Sci.Technol* 1995; 29: 2083-2089.
7. Engwall, A.M., Pegnatello, J.J. and Grasso, D. Degradation and detoxification of the wood preservatives cresote and pentachlorophenol in water by the photo-fenton reaction; *Water Res.* 1999; 33: 1151-1158.
8. Park, T. J., Lee, K. H., Jung, E. J. and Kin, C. W. Removal of refractory organic and color in pigment wastewater with Fenton oxidation, *Water. Sci. Technol* 1999; 39: 189-192.
9. Chang, J. S. Chou, C., Lin, Y. C., Lin, P. J. and Hu, T. L. Kinetic characteristics of bacterial azo-dyes decolorization by pseudomonas Luiteola, *Wat.Res* 2001; 35: 2841-2850.

10. Kabdasali, J., Gorel, H. and Tunay O. Characterization and treatment of textile printing wastewaters, *Env. Technol.*, 2000; 21: 1147-1155.

11. Preis S, Terentyeva Y, Rozkov A. Photocatalytic oxidation of phenolic compounds in wastewater from oil shale treatment. *Water Res.*, 1997, 35: 247-255.

Chapter 54

Coastal Zone Management

S.S. Daghe

CSIR-NEERI, Nagpur

Background

Water, as a resource, has multiple uses like recreational, hydroelectric power generation, transportation, agricultural, domestic, industrial, and commercial activities. Water supports all forms of life and constitutes one of the important and precious environmental components. More than three quarters of the earth's surface is made up of water, 97.2 per cent is salt water in the oceans. Out of 2.8 per cent of the Earth's freshwater only 0.6 per cent is available as water for human utilization and the remaining freshwater is frozen in polar ice caps, icebergs, and glaciers. There are several seas and oceans all over the world.

History of Marine Pollution

Ocean disposal of urban agglomeration's waste got its start indirectly long before the Agricultural Age when streams, lakes, and nearby estuaries were used as waste depositories. Historically, the disposal of wastes into any water body by humans was universally practiced. It was a cheap and convenient way to clear society's wastes. The advent of the Industrial Age brought with it the new problem of disposal of chemical wastes and by-products which were also commonly disposed of in the water resulting in potentially harmful effects.

As civilization developed at river deltas and estuaries adjacent to the ocean, disposal in the ocean became a popular alternative. It is presumed that ocean has large assimilative capacity. Hence, over the past 150 years, all types of wastes have been indiscriminately dumped in ocean. Most of the sources of marine pollution are land based. The pollutants also get access from non point sources like water runoff, windblown debris and dust particles. Ocean receives contaminants through the entry of chemicals, sewage (treated and untreated), industrial waste, military wastes (munitions and chemicals), entire ships, trash, garbage, dredged material, construction debris, residential waste and even radioactive wastes. Significant amount of wastes enter the ocean through river, atmospheric, and pipeline discharge; construction; offshore mining; oil and gas exploration; and shipboard waste disposal. Thus the ocean has become the ultimate dumping ground for all types of waste created due to civilization.

There are different ways to categorize the pollution into marine ecosystems. Three main types of inputs of pollution into the ocean are direct discharge of waste into the oceans, runoff into the waters due to rain, and pollutants that are released from the atmosphere. Point source pollution occurs when there is a single, identifiable, and localized source of the pollution. An example is directly discharging sewage and industrial waste into the ocean. Such pollution occurs particularly in developing nations. Non-point sources create pollution due to ill-defined and diffuse sources which are difficult to regulate.

Legislations for Protection of Coastal Zone: A Global Scenario

In past 50 years, Society understood the impacts on oceans from over fishing, mineral exploration, and construction activities in coastal region. The detrimental effects of ocean dumping are physically visible at trashed beaches, where dead fish and mammals entangled in plastic products may sometimes be observed. They are additionally reflected in the significant toxic chemical concentrations in fish and other sea life. In early 60s, few regulatory (or legal) actions were implemented to control or prevent the waste dumping activities. During 1960s numerous groups like global, regional, Governmental, and environmental organizations began to report on the detrimental impact of waste disposal in the ocean.

Late in the nineteenth century, the U.S. Congress enacted Section 10 of the River and Harbor Act of 1890, prohibiting any obstruction to the navigation of U.S. waters. In addition, the National Environmental Policy Act of 1969 (NEPA) required the review of policy issues pertinent to the public interest and an environmental impact statement on activities that might significantly affect the quality of the marine environment. Significant international laws to counter marine pollution with long history were enacted in the twentieth century. Marine pollution was a concern during several United Nations Conferences on the "Law of the Sea" beginning in the 1950s. Most scientists believed that the oceans were so vast that they had unlimited ability to dilute and get self purified, and thus render harmless, pollution. In the late 1950s and early 1960s, there were several controversies about dumping "radioactive waste" off the coasts of the United States by companies licensed by the Atomic Energy Commission, into the Irish Sea from the British reprocessing facility at Wind scale, and into the Mediterranean Sea by the French Commissariat à l'Energie Atomique. After the Mediterranean Sea controversy, for example, Jacques Cousteau became a worldwide figure in the campaign to stop marine contamination. "Marine pollution" made further international headlines after the 1967 crash of the "oil tanker Torrey Canyon", and after the 1969 "Santa Barbara oil spill" off the coast of California. Marine pollution was a major area of discussion during the 1972 United Nations Conference on the Human Environment, held in Stockholm. That year also saw the signing of the Convention on the Prevention of Marine Pollution by Dumping of Wastes and Other Matter, sometimes called the "London Convention".

In 1972 the U.S. Congress passed the Marine Protection, Research and Sanctuaries Act (Ocean Dumping Act or "ODA") and the Federal Water Pollution Control Act amendments (Clean Water Act or "CWA") that set a global standard for managing environmental restoration and protection, for maintaining the environment within acceptable standards, for prohibiting the disposal of waste materials into the ocean, and for regulating the discharge of wastes through pipelines into the ocean. International recognition of the need to regulate ocean disposal from land-based sources on a global basis was the result of the "UN Conference on the Human Environment" in June 1972 and the Inter-Governmental Conference on the "Convention of the Dumping of Wastes at Sea" in November 1972. These conferences resulted in a treaty entitled Convention on the Prevention of Marine Pollution by Dumping of Wastes and Other Matter–London Convention 1972 **(LC-72)**. The LC-72 came into effect in 1975. Another treaty addressing the issue of wastes disposed of from vessels, the "International Convention for the Prevention of Pollution" from Ships, 1973 (MARPOL), was adopted in 1973. The summary of regulations is presented in Table 54.1.

Countries signing MARPOL agree to enforce bans on dumping oil and noxious liquids into the ocean from ships, but the disposal of hazardous substances, sewage, and plastics remains optional. There are dozens of other international agreements dealing with ocean pollution, but the LC-72 and MARPOL are the most significant as far as dumping is concerned.

Global Habitation in Coastal Region

The 3rd United Nations "World Water Development Report (2009)" states that: "Every second, two persons are added to urban population." More than 50 per cent of world's population (almost 7 billion) lives in cities which was less than 15 per cent in 1900? The number is expected to grow to 60 per cent by 2030. Today, about 3.5 billion *i.e.* 50 per cent of urban population lives in coastal cities. Seventeen out of twenty five mega cities with 25 per cent world's population are located in coastal region and use ocean for waste disposal. By 2025, one third of World's population is projected to be in water stress or scarcity. These and other statistics found in the "Water and Cities Facts and Figures", UN-Water Decade Programme on Advocacy and Communication (UNW-DPAC) is shocking yet real. The information stated above warrants stopping the current practices of discharging domestic wastewater into the sea and recommends recycle and reuse after adequate treatment.

Indian Approach to Marine Ecosystem and Environmental laws

The Water (Prevention and Control of Pollution) Act was enacted in 1974 by Government of India for the prevention and control of water pollution, and for maintaining or restoring of wholesomeness of water in the country. The Act was amended in 1988. On February 19, 1991, the Ministry of Environment and Forests ("MOEF") issued a notification under Section 3 of the Environment Protection Act of 1986, seeking to regulate development activity on India's coastline. The guidelines

Table 54.1: Legislations for Prevention of Marine Pollution

MAJOR GLOBAL AGREEMENTS AND DOMESTIC LEGISLATIONS GOVERNING PROTECTION OF THE MARINE ENVIRONMENT

Key Global Agreements to Protect the Marine Environment from Dumping

Title	*Description*
Convention on the Prevention of Marine Pollution by Dumping of Wastes and other Matter (London Convention 1972) (entry into force: August 1975)	Provides measures to limit the use of the oceans as disposal area for wastes generated on land.
International Convention for the Prevention of Pollution from Ships, 1973, as modified by the protocol of 1978 relating thereto (MARPOL 73/78)	Provides measures for ships and national administrations to prevent pollution by oil (Annex I), noxious liquid substances in bulk (Annex II), and harmful substances in packaged form (Annex III), sewage (Annex IV), garbage (Annex V), and air pollution from ships (Annex VI).
1982 UN Convention on The Law of the Sea (UNCLOS 1982) (entry into force: November 1994);	Provides a framework for the determination of the rights and obligations of states relating to the oceans. Part XII contains provisions with regard to protection and preservation of the marine environment.

Key Domestic Legislation to Protect the Marine and Coastal Environment

Title	*Description*
National Environmental Policy Act of 1969 (NEPA)	To declare a national policy that will encourage productive and enjoyable harmony between people and the environment; to promote efforts that will prevent or eliminate damage to the environment and biosphere and stimulate human health and welfare.
Federal Water Pollution Control Act Amendments of 1972 (CWA)	To restore and maintain the chemical, physical, and biological integrity of the nation's waters.
Marine, Protection Research, and Santuaries Act of 1972 (ODA)	To regulate the dumping of all types of materials into ocean waters and to prevent or strictly limit the dumping into ocean waters of any material which would adversely affect human health, welfare or amenities, or the marine environment, ecological systems, or economic potentialities.

for classification of coastal Zone were developed during 1994 wherein the Indian coastal boundary was demarcated based on its "Best Designated Use" and environmental sensitivity. Four categories are created *viz.* CRZ I, CRZ II, CRZ III and CRZ IV. The approach adopted by the first notification was to define the 'High Tide Line' ("HTL") and 'Coastal Regulation Zone' ("CRZ") and thereafter specify the activities permitted and restricted in the vicinity of the CRZ. This regulated zone divided into four categories (CRZ I-IV) was decided on the basis of permitted land use.

There have been about 25 amendments to the notifications between 1991 and 2009, some of which have been based on the directions of the Supreme Court. In May 2008, the MOEF brought out a new draft CRZ notification that evoked much criticism from all sections of stakeholders. Eventually this notification was allowed to lapse and the Ministry brought out a fresh notification in September 2010, which after detailed discussions and deliberations, was finally passed as Coastal Regulation Zone Notification 2011 on January 6, 2011 (Annexure I).

Legislation Pertaining to Water Quality

The near coastal zone is being used for various activities. The Indian coast line is demarcated into five classes based on the best designated use. The classification of coastal zones is presented in Annexure II: Pollution due to organized discharges warrant legislation for both discharges themselves as well as the receiving water quality. A recreational water quality criterion is defined as a quantifiable exposure-effect relationship based on scientific evidence between the quality of water concerned and the associated potential human health risks. The concept of acceptability implies social, cultural, economic, political and medical factors (Annexure III). The water quality guidelines for the receiving sea water standards for different categories, applicable parameters and pertinent issues were presented in Annexure IV.

Description of the Indian Coast Line

India is bounded to the southwest by the Arabian Sea, to the southeast by the Bay of Bengal, and to the south by the Indian Ocean. Kanyakumari is the southern tip of the Indian peninsula. The southernmost point in India is Indira Point, in the Andaman and Nicobar Islands. The territorial waters of India extend into the sea to a distance of 12 nautical miles (13.8 mi; 22.2 km) measured from the appropriate baseline and a coastline of 7,517 km (4,671 miles) from the "baseline" (usually the mean low-water mark) of a coastal state. The territorial sea is regarded as the "sovereign territory" of the state, although foreign ships (both military and civilian) are allowed innocent passage through it; this sovereignty also extends to the airspace over and seabed below. The term "territorial waters" is also sometimes used informally to describe any area of water over which a state has jurisdiction, including internal waters, the neighboring zone, the exclusive economic zone and potentially the continental shelf.

(Map inDIA – sw status)

Current Status of Coastal Pollution

While availing the benefit of coastal location, the Municipal Corporations of the coastal cities discharge the treated as well as untreated wastewater into the near shore region. Though stringent pollutions control laws are implemented and consents for quantity and quality of waste to be disposed are issued by statutory bodies like Pollution control Boards of respective states, the norms are not met and the near shore marine region is receiving pollutants more than its assimilative capacity. The recent study by NIO on "Maharashtra Coastal water Quality" has brought out the extent of pollution in the coastal stretch of whole Maharashtra and places where the standards are violated (Annexure V).

The environmental impacts due to developmental activities are both beneficial and hazardous. With the urbanization, advancement, industrialization and modernization, the cities are growing at a faster rate leading to increased water demand which results in the release of gigantic contamination in the natural ecosystem which consequently alters original composition and natural behavior. The coastal cities release the liquid waste into marine water bodies through point and non-point discharges. Due to space limitations, conventional domestic wastewater treatment facilities like Aerated lagoons, Activated sludge process, Trickling filters, Oxidation ponds, Extended aeration, biological reactors etc. could not be preferred because of high cost of land and also non availability. Marine outfalls with multi port diffusers are one of the modern systems adopted for mega cities along the coast to dispose of large volume of domestic wastewaters. Several Marine outfalls are functional in the world *viz.* Boston Harbor area, North coast of Spain, Coast of Sao Paulo, South African Coasts, Australian coasts, Lisbon coastal area, NE Pacific (California), SW Pacific (Sedney), SE Pacific, Mediterrenian, Massachusetts Bay, Santander Bay (Biscay Gulf), Hong Kong, India and the conference held at Spain during January 2010 has brought out the merits and demerits of this modern system of waste disposal. (IWMOD, 2010; ISEST, 2009 and ISBT-2008).

Evaluation of Functional Behaviour of Marine Outfall

Once the marine outfall is operational, it is cumbersome to find its functional and operational behavior. Evaluation of dilution achieved is complicated due to many reasons like, variation in direction and buoyancy of upward flow, density, currents, tidal effects and also the submergence of the physical components of the diffuser. Hydrodynamic simulation through mathematical models and experimental application of tracers are the commonly adopted techniques for evaluation of mixing behavior, dilution and dispersion pattern of waste plume released through marine outfalls.

A tracer investigation is a valuable tool to obtain factual information for the dispersion of wastewater. Apart from natural tracers two externally added tracers are used in such studies *viz.* radiotracers with short half-life so that they decay rapidly and the impact on environmental components is insignificant and fluorescent dye tracer with low doses. Such experiments help to evaluate and monitor the operational factors for the diffuser system.

Simulation through mathematical models also predicts expected dilution and dispersion pattern of the waste plume. However these models have to be calibrated for hydrodynamic and water quality parameters. Model should have the relevant mathematical constants suitable to marine water quality. The Application of right mathematical models and use of different tracers, have been investigated by many scientists to find out functional behavior and the travel path of pollutant concentration through distribution of the tracers in longitudinal and transverse direction.

Kinetics of Biodegradation in Marine Waters

Ocean when used for waste disposal, experiences severe ecological stress. Though seawater has large assimilative capacity, the chemical reactions related to the biodegradation, decomposition and ratification of pollutants get altered and slow down due to salinity. The solubility of oxygen is limited due to high chlorides. In tropical region higher temperatures also have significant impact on DO and reaction rates. Biochemical oxygen demand (BOD), a bio-essay test indicating aerobic stabilization of the organic matter. and Dissolved Oxygen (DO) are the most important parameters considered in water pollution and self purification studies. Multiple factors like pH, osmotic conditions, temperature, time of incubation, seed acclimatization, salinity, chemical composition and availability of oxygen etc influence the rate of biodegradation and make the process complex. The decay of carbonaceous matter (CBOD) is considered as wet oxidation process with first order kinetics and expected to exert differently under varying salinities and temperatures. OECD test procedures for biodegradability recommend 28 days of the test period for the measurements of ultimate BOD in saline conditions as against 20 days for freshwater.

OSPAR convention for the "protection of marine environment of the North-East Atlantic" meeting of working group on priority substances (SPS [2] 01/4) and couple of other references have also cited the hypothesis that "degradation of organic matter in marine water has been observed to be slow with varied reaction rate constant". It is therefore essential to find out degradation rate of waste in marine conditions. The chloride concentrations are variable in creek, estuaries and ocean with 10-36 ppt of salinity.

The literature states that decomposition rate of organic contaminants, measure as BOD, is reduced in proportion to salinities. This phenomenon occurs due to restriction of osmoregulatory processes occurring within the cells of microorganisms responsible for breakdown of organic compounds. As a result, the kinetics and reaction rates of decomposition alter due to high chlorides. According to the first order equation of chemical kineticapplied for BOD, the rate of decomposition is directly proportional to the amount of undecayed material.

The formula adopted for BOD kinetics consideration states that the amount of pollutants remaining after any time t has elapsed, is fraction of L0 (Ultimate demand) corresponding to 10^{-kt} or the BOD that has not been exerted is percentage of L0 corresponding to $10^{-kt.}$ In statistics, a numerical measure of association between two variables is the correlation coefficient. The square of the correlation coefficient, r^2, is a useful value in linear regression.

Research is carried out to study the impact of salinity on biodegradation of carbonaceous waste. Domestic sewage, rich in biodegradable organic matter and organic chemicals like glucose-glutamic acid as standard chemical with known BOD, were subjected to the test [10]. The experiments were conducted at 20°C and 27°C to evaluate impact of hot climate in tropical regions for 10 days. The concentrations of chlorides were varied to resemble the conditions generally observed in the creeks and estuaries. The artificial seawater by B.Loganthan was used as dilution water. The existence and adequacy of heterotrophic microorganism under various chloride concentrations was evaluated, using Zobel/Nutrient agar media.

Experimental Observations

- ✰ BOD (Table 54.2) decreased with increase in chloride concentration thereby indicating the negative impact of chlorides on the oxidation of organic matter.

Table 54.2: Comparison of BOD of GG and Domestic Wastewater Exerted with Varying Vhloride Concentration in Dilution Media at 20ºC on 5th Day and 27ºC on 3rd Day

Sl.No.	*Chloride Concentrations (mg/L)/ Salinity ppt*	*Glucose Glutamic Acid - Ultimate BOD (L) as 480 mg/L*		*Domestic Wastewater Ultimate BOD (L) as 300 mg/L*	
		BOD (mg/L)/ Percentage Exerted at 20ºC on 5th Day	*BOD (mg/L)/ Percentage Exerted at 27ºC on 3rd Day*	*BOD (mg/L)/ Percentage Exerted at 20ºC on 5th Day*	*BOD (mg/L)/ Percentage Exerted at 27ºC on 3rd Day*
1.	0/0	336 (70%)	334 (69.6%)	167 (55.7%)	168 (56.0%)
2.	500/90	285 (59.3%)	281 (58.5%)	103 (34.3%)	127 (42.3%)
3.	1000/180	255 (53.1%)	282 (58.5%)	115 (38.3%)	126 (42.0%)
4.	15000/27	200 (41.7%)	249 (51.8%)	103 (34.0%)	104 (34.7%)
5.	20000/36	221 (45.8%)	240 (50%)	94 (31.3%)	110 (36.7%)

- ✰ The "k" values indicating the rate of reaction have reduced with increase in chloride concentrations. This finding confirms the slowing down of the rate of biodegradation of organic matter with elevation of chlorides (salinity). More time will be required to break down the pollutants. This aspect needs to be considered while planning waste discharges.
- ✰ Significant rise in oxidation of Glucose Glutamic acid and sewage at 27°C was observed when compared to 20°C indicating important role of temperature. This observation needs critical evaluation for tropical conditions.
- ✰ Sudden rise in BOD was not observed after 6th or 7th day when nitrogenous oxidation is expected. The finding supports the assumption *i.e.* "Degradation of only carbonaceous organic matter will occur under selected test conditions and consideration of calculations of second order kinetic mechanism will not be necessary". However, in sewage, about 20 mg/L of ammonical nitrogen is present which will require total 92 mg of oxygen per litre of sewage (1mg/L of NH_3–N requires 4.6 mg/L of oxygen for complete oxidation of nitrogen in two steps *i.e.* NH_3 to NO_2 and NO_2 to NO_3).
- ✰ The rate of reaction values ("k") obtained indicated reduction with increase in chloride concentrations confirming the slowing down of the rate of biodegradation of organic matter with elevation of chlorides (salinity). More

time will be required to break down the pollutants. This aspect needs to be considered while planning waste discharges (Table 54.3).

Table 54.3: Average "k" Values at 20ºC and 27ºC for GG and Domestic Wastewater

Sl.No.	Chlorides (mg/L)	Glucose Glutamic Acid (Ultimate BOD (L) as 480 mg/L)		Domestic Wastewater (Ultimate BOD (L) as 300 mg/L)	
		Av. K 20ºC	Av. K 27ºC	Av. K 20ºC	Av. K 27ºC
1.	0	0.24	0.50	0.20	0.30
2.	5000	0.25	0.45	0.12	0.22
3.	10000	0.18	0.37	0.13	0.21
4.	15000	0.11	0.21	0.13	0.20
5.	20000	0.11	0.20	0.13	0.20

- Appreciably higher reaction constant "k" were observed at 27°C as compared to 20°C indicating faster rate of oxidation at elevated temperature.
- The square of the correlation coefficient was negative and greater than 0.9 indicating good strength of the negative association between degradation rate and chloride concentration. This trend analysis indicates that as the chloride concentration increases, the rate of degradation of organic matter decreases.
- The findings of the study indicate the necessity for critical consideration of slow degradation rates while planning the disposal options for organic wastes in marine environment.

Thus the findings are supportive to the hypothesis of slow degradation in marine waters and reaction rates are inversely proportions to salinity values. BOD exertion at 20°C in five days and at 27°C in three days is comparable at selected chloride conditions thereby confirming validity of BOD test of shorter duration at elevated temperature for marine environment. The data set had excellent correlation between the selected variables.

A Case Study Supporting Delayed Biodegradation of Organic Matter in Marine Zones

Mumbai, a coastal metropolitan city and the commercial capital of India, accommodates 12.5 million people (2011 census). With ample water supply (about 150 lpcd), enormous volume of domestic wastewater (sewage) is generated. Due to space limitations, conventional domestic wastewater treatment facilities are not viable. The treated as well as untreated wastewater is released into coastal region through point and non-point discharges. The slow degradation rate of organic matter in marine waters was not well thought-off during planning and designing of the treatment and disposal schemes. As a result, the water quality at beaches, seafronts and near shore coastal region is showing high levels of BOD and microbial pollution with non-compliance of SW II standards stipulated by MoEF/CPCB, for designated class. The beaches, seafronts are not safe for recreation. Option of marine outfall was selected at three locations as preferred technology for effective disposal of the domestic wastewater.The wastewater released through diffuser section of the outfall at the bottom of a seabed generates waste field at the surface and pollutants get diluted and dispersed into the tidally influence coastal region.

The monitoring program was organised to establish the impact outfall discharges on water quality of coastal region and evaluate compliance of the stipulated standards. The mixing behavior, dilution and dispersion of released wastewater depend on interplay of ambient conditions of receiving water body. A number of factors like the quantity and quality of the waste discharge, currents, tides, approachability, relevant laws, stipulated receiving water standards were considered while planning the monitoring program. Concentration of non conservative natural tracers like BOD (250 mg L^{-1}), Fecal Coliform (FC), 10^7 (CFU)/100ml. Ammonical nitrogen ($NH_{3-}N$) 14.0mgL^{-1} were not providing comparable dilution factors either at the initial or far field location. Hence during field experiments, tools like tracer studies and mathematical simulation were used to test the performance and adequacy of such systems.

From the Tracer studies using fluorescent dye Rhodamine and radiotracer Bromine 82, it was observed that the wastewater released through the diffusers got diluted to the extent of 40 to 60 times in the near field areas up to the distance of 500 to 1000 m from the centre of diffuser. It was not adequate to meet the standards. Comparable performance with Dye and radiotracer was observed in near field region. Maximum dilution observed in far field region. The dilution and extent of degradation achieved was not adequate to meet the designated receiving water quality standard of 3 mg L^{-1} BOD near the diffuser location with the currently discharged wastewater quantity. The experiment also supported the hypothesis of slow degradation of organic matter. The counters of Dissolved Oxygen (DO) and BOD data for two tidal cycles (12 hrs) at near field diffuser region of one square km indicated noncompliance of BOD confirming slow biodegradation rates. Another elaborate experiment using Rhodamine fluorescent dye and Radiotracer Bromine 82 was conducted at outfall. The fluorescent dye and radiotracer was injected at in effluent channel from treatment works in "neap flooding tide." The results at different longitudinal distances and travel path also indicated noncompliance of water quality standards (SW II).

All these experiments have indicated delayed biodegradation of organic matter in the domestic wastewater. Therefore, it is essential to consider the appropriate BOD decay constants for marine environment in the design considerations for the current and proposed treatment systems for domestic wastewater discharges in the marine waters. Similar findings were recorded in the creeks like Thane and Malad, which are heavily polluted due to waste discharges. While planning for the rejuvenation of these badly affected water bodies, the slow rate of oxidation of organic matter must be emphasized.

Thus, the coast and near shore region of Mumbai at many places does not meet the requirement of a healthy water conditions, ecological diversity as well as aesthetics.

The ecosystem of Mumbai coast in the near shore region can be at risk due to increased load of pollutants through non-point discharges. If sewage collection and treatment do not match the growing needs, sewage toxicity can lead to aberration in micro anatomical structures of some fish species and the marine ecosystem will lead to serious ecological disturbances.

Annexure I

Coastal Regulation Zone Notification 2011: An Evaluation

Ashoo Gupta/Mumbai March 26, 2011, 16: 03 IST

On February 19, 1991, the Ministry of Environment and Forests ("MOEF") issued a notification under Section 3 of the Environment Protection Act of 1986, seeking to regulate development activity on India's coastline. The approach adopted by the first notification was to define the 'High Tide Line' ("HTL") and 'Coastal Regulation Zone' ("CRZ") and thereafter specify the activities permitted and restricted in the vicinity of the CRZ. This regulated zone was further divided into four categories (CRZ I-IV) as per permitted land use.

There have been about 25 amendments to this notification between 1991 and 2009, some of which have been based on the directions of the Supreme Court. In May 2008, the MOEF brought out a new draft CRZ notification that evoked much criticism from all sections of stakeholders. Eventually this notification was allowed to lapse and the Ministry brought out a fresh notification in September 2010, which after much discussions and deliberations, was finally passed as Coastal Regulation Zone Notification 2011 on January 6, 2011.

The current notification has several new positive features:

It widens the definition of CRZ to include the land area from HTL to 500 m on the landward side, as well as the land area between HTL to 100 m or width of the creek, whichever is less, on the landward side along tidal influenced water bodies connected to the sea. The CRZ also includes, for the first time, water area up to 12 nautical miles in the sea and the entire water area of a tidal water body such as creek, river, estuary without imposing any restrictions of fishing activities.Thus, the main change in the scope of regulation has been to expand the CRZ to include territorial waters as a protected zone. This may have been in response to the criticism that while the earlier CRZ notification regulated development on the coastal stretches, it did not per se deal with pollution of the sea in any direct terms.

The concept of a 'hazard line' has been introduced. While the notification merely states that the hazard line will be demarcated by the MOEF through the Survey of India, by taking into account tides, waves, sea level rise and shoreline changes, this concept owes its introduction to the realisation of natural disasters such as tsunami and floods that may take place in this zone. In May 2010, the MOEF signed a Memorandum of Understanding with the Survey of India for undertaking this exercise over a period of four-and-a-half years, at an estimated cost of Rs 125 crore.

The concept of classification of CRZ into four zones has continued in the 2011 notification with the following delineation:

CRZ I- Ecologically sensitive areas such as mangroves, coral reefs, salt marshes, turtle nesting ground and the inter-tidal zone.

CRZ II- areas close to the shoreline, and which have been developed.

CRZ III- Coastal areas that are not substantially built up, including rural coastal areas.

CRZ IV- Water area from LTL to the limit of territorial waters of India.

CRZ IV has been changed from the 1991 notification, which covered coastal stretches in the islands of Andaman and Nicobar and Lakshwadeep. The MOEF has issued a separate notification titled Island Protection Zone 2011 in relation to these areas.

A new category called areas requiring special consideration has been created which consists of (i) CRZ areas of Greater Mumbai, Kerala and Goa, and (ii) Critically vulnerable coastal areas such as Sunderbans.

Clearances for obtaining CRZ approval have been made time-bound. Further, for the first time, post-clearance monitoring of projects has been introduced in the form of the requirement to submit half-yearly compliance reports, which are to be displayed on the Ministry's website.

With respect to the list of prohibited activities, one of the most important changes has been that of expanding the list of exceptions to the rule prohibiting setting up of new industries and expansion of existing industries. While the earlier exception was limited to those activities which required access to the water front, four other exceptions have been now incorporated which include:

Projects of Department of Atomic Energy

- ✰ Facilities for generating non-conventional energy sources and desalination plans, except for CRZ-I zones on a case-by-case basis after doing an impact assessment study;
- ✰ Development of greenfield airport permitted only at Navi Mumbai; and
- ✰ Reconstruction, repair works of dwelling units of local communities including fishers in accordance with local town and country planning regulations.

Another important aspect is the introduction of the Coastal Zone Management Plans, which will regulate coastal development activity and which are to be formulated by the State Governments or the administration of Union Territories.

In Greater Mumbai, the redevelopment of approximately 146 existing slums in CRZ areas has been permitted, provided that the stake of the state government or its agencies in these projects is not less than 51 per cent. Redevelopment and reconstruction of old, dilapidated, and unsafe buildings in the CRZ-II area has also been permitted. Also, the floor space index (FSI) or floor area ratio (FAR) prevailing in the Town and Country Planning Regulations as on the date of the project being sanctioned, will apply. In order to ensure that the redevelopment of slums and dilapidated structures in Mumbai are done in the most transparent and accountable manner the Right to Information Act, 2005 will be applicable and auditing will be done by the office of the Comptroller and Auditor General (C and AG) of India.

The 2011 Notification also lists out certain measures that have to be taken to prevent pollution in the coastal areas/coastal waters.

While the CRZ Notification 2011 has introduced several positive concepts seeking to protect the interest of the local traditional communities, it does have a few drawbacks namely:

Although the no-development zone of 200 metres from the HTL is reduced to 100 metres, the provision has been made applicable to "traditional coastal communities, including fisher-folk", thereby giving the chance for increased construction on the coast and higher pressure on coastal resources.

Disallowing Special Economic Zone ("SEZ") projects in the CRZ.

There are no restrictions for expansion of housing for rural communities in CRZ III.

The CRZ Notification 2011 is a major step-up from the 1991 Notification and the MOEF has made special efforts to include specific provisions to benefit the fisher-folk community in all the coastal areas and address the shortcomings of the 1991 Notification such as time-bound clearances, enforcement measures, special provisions for specific coastal stretches etc. There is a significant change in the new notification but there is always need for further improvement.

The author is Partner at J. Sagar Associates, a Mumbai law firm. Views expressed are personal.

Annexure II

Categories of Coastal Regulation Zone In India

Category	*Description*
CRZ I	Ecologically sensitive and important areas, such as national parks/marine parks, sanctuaries, reserve forests, wildlife habitats, mangroves, corals coral reefs, areas close to breeding and spawning grounds of fish and other marine life, outstanding natural beauty historical heritage, areas rich in genetic diversity; areas likely to be inundated due to rise in sea level consequent upon global warming and others as may be declared by the concerned authorities
CRZ II	The areas that have already been developed up to or close to the shoreline. Developed area is referred to as that area within the municipal limits or in other legally designated urban areas, already substantially build up and provided with drainage, approach roads and infrastructure facilities, such as water supply and sewerage mains
CRZ III	Areas that are relatively undisturbed, which do not belong to either Category-I or II *viz.* coastal Zone in the rural areas (developed and undeveloped) within Municipal limits or in other legally designated urban areas, which are not substantially building
CRZ IV	Coastal stretches in the Andaman and Nicobar, Lakshadweep and small island except those designated as CRZ-I, CRZ-II or CRZ-III–CRZ-IV

Annexure III

Classification of Indian Marine Zones as per "Best Designated Use

Class	*Details*
SW I	Salt pans, shell fish, ecologically sensitive zones
SW II	Commercial fishing, contact recreation, bathing water
SW III	Industrial Cooling
SW IV	Harbour
SW V	Navigation, controlled Waste Disposal

Annexure IV

Table 54.4: Receiving Sea Water Standards for different Categories, Applicable Parameters and Pertinent Issues

Class	*Categories*	*Major Parameters*	*Rationale/Remarks*	*Issues*
SW I	Salt pans, shell fish, ecologically sensitive zones	pH-6.5-8.5, no colour and floating matter	General broad range, conductive for propagation of aquatic lives, is given. Value largely dependent upon soil-water interaction.	This category could also be applicable to the bathing water quality
		DO- 5.0 mg/l	Never less than 3.5 mg/l at any time of the year for protection of aquatic lives.	No value assigned to bacterial density
		Oil and Grease-0.1 mg/l	Concentration should not exceed 0.1 mg/l as because it has effect on fish eggs and larvae.	
		Heavy metals-0.01mg/l (Hg, Pb and Cd)	Value depends on: i. Concentration in salt, fish and shell fish. ii. Average per capita consumption per day. iii. Minimum ingestion rate that induces symptoms of resulting diseases.	
SW II	Commercial fishing, contact recreation, bathing water	pH- 6.5-8.5	Range does not cause skin or eye irritation and is also conducive for propagation of aquatic lives.	Commercial fishing generally not undertaken near shore where bathing, recreational activities are common. Length (seaward) upto which bathing is undertaken should be also taken into consideration.
		DO- 4.0 mg/l	Not less than 3.5 mg/l at anytime for protection of aquatic lives.	
		No noticeable colour, odour and floating matters	Specially caused by chemical compounds like creosols, phenols, naphtha, benzene, pyridine, toluene etc. causing visible colouration of water and tainting of and odour in fish flesh.	
		Turbidity	30 NTU (Nephelo Turbidity Unit) measured at 0.9 depth.	Sampling schedule suggested (average 200/100ml not to exceed in 20 per cent of the sample and 3 consecutive samples in monsoon) may give varied results depending upon tide timings and location of sampling
		BOD_5–3 mg/l	Restricted for bathing (aesthetic quality of water). Also prescribed by IS: 2296-1974.	
		Fecal Coliform–100/100 ml	The average value not exceeding 200/100 ml in 20 per cent of samples in the year and in 3 consecutive samples in monsoon months.	

Contd...

Table 54.4–*Contd...*

Class	*Categories*	*Major Parameters*	*Rationale/Remarks*	*Issues*
SWIII	Industrial Cooling	pH–6.5-8.5	The range is conducive for propagation of aquatic species and restoring natural system.	The presence of fecal coliform for the industrial cooling is not very relevant.
		DO–3.0	To protect aquatic lives.	
		No noticeable colour and floating matter, odour	None in such concentration that would impair usages specifically assigned to this class.	
		Turbidity	30 NTU	The presence of fecal coliform for the industrial cooling is not very relevant.
		Fecal Coliform-500/100 ml	Reasonably clear water for Recreation, Aesthetic appreciation and Industrial cooling purposes	
			Not exceeding 1000/100 ml in 20 per cent of samples in	
			the year and in 3 consecutive samples in monsoon months.	
SW IV				
SW V				

Annexure V

"Monitoring of Coastal Marine and Estuarine Ecology of Maharashtra" NIO Report

Summary

NIO had conducted the coastal survey at 120 locations covering the area from Dahanu in north and Redi in south boundary of Maharashtra. The sampling locations were distributed in three zones *viz.* coastal, near shore and inner, middle and outer portion of creek/estuarine regions. The offshore locations were about 5 to 7 km away from the shoreline; near shore locations at 2 to 3 kms and remaining locations were within creek/estuarine area within the distance of 0 to 0.5 km. Tidal variations were also monitored

The information about the sampling locations, their distribution in the marine zones and type of waste discharges is presented in Table 54.5.

Observations

The results of the monitoring during pre and post monsoon are discussed transect wise as "Prevailing Environment under various sub heads. The parameter wise ecological assessment for the study area was done for coastal and creek/estuarine waters separately for two major regions *viz.* North Maharashtra (Dahanu to Murud) and South Maharashtra (Savitri to Redi). Parameter wise status of water quality was also evaluated.

The findings are summarized below:

- The pH of coastal and creek segments was comparable.
- The north coast revealed higher suspended solids than the south coast.
- DO of more than 3 mg/L was observed in coastal waters meeting SW II standards at most of the places. However, the creek regions had DO < 3 mg/L.
- Av. BOD of 5.8 mg/L was observed in the northern region. The southern coastal region had BOD mostly < 3 mg/L.
- Severe water quality deterioration was observed at Transects like Bassin, Manori, Versova, Mahim, Thane, Patalganga, Kundalika, Savitri, Vashishti and Ratnagiri bay.
- The critical areas of high environmental stress identified by this study are creek and estuarine regions. Domestic Wastewater releases in the inshore region are the major contributors for degradation. It is vital to free these inshore areas from such unplanned discharges and such waste discharges should be released in Arabian sea through marine outfalls for selected critical areas of Mumbai.

Table 54.5: Details of the Sampling Stations in Coastal Region of Maharashtra

Sl.No.	Transect	Wastewater Sources	Stations	Coastal Stations	Near shore Stations	Creek/ Estuarine Stations
1.	Dahanu	Industrial and domestic wastes	DH1, DH2, DH3, DH4, DH5, DH6	2	1	3
2.	Tarapur	Industrial and domestic wastes	TP1, TP2, TP3, TP4, TP5, TP6	2	1	3
3.	Bassein	Industrial and domestic wastes	BS1, BS2, BS3, BS4, BS5, BS6, BS7	3	–	4
4.	Manori	Industrial and domestic wastes	BYMa4, BYMa5, BYMa6	–	–	3
5.	Versova	Industrial and domestic wastes	BYV4, BYV5	–	–	2
6.	Mahim	Industrial and domestic wastes	BYM4,BYM5, BYM6	–	–	3
7.	Bandra	Domestic waste through marine outfall	BYB1, BYB2	2	–	–
8.	Worli	Domestic waste through marine outfall	BYW1, BYW2	2	–	–
9.	ThaneCreek	Industrial, domestic and port based wastes	BY1, BY2, BY3, BY4, BY5, BY6	3	–	3
10.	Patalganga	Industrial waste	PT4, PT5, PT6, PT7, PT8, PT9	–	–	6
11.	Amba Estuary	Industrial and port based wastes	AB4, AB5, AB6, AB7	–	–	4
12.	Thal DP	Industrial waste (RCF, DP)	DP	1	–	–
13.	Alibaug	Domestic (minor) waste	A1, A2, A3	2	–	1
14.	Kundalika Estuary	Industrial and domestic (minor) wastes	K1, K2, K3, K4, K5, K6, K7	2	1	4
15.	Murud	Domestic (minor) waste	MR1, MR2, MR3, MR4, MR5, MR6, MR7	2	1	4
16.	Savitri Estuary	Industrial and domestic (minor) wastes	S1, S2, S3, S4, S5, S6, S7	2	1	4
17.	Vashishti Estuary	Industrial and domestic (minor) wastes	VS1, VS2, VS3, VS4, VS5, VS6, VS7	2	1	4
18.	Enron DP	Industrial waste	ENDP	1		
19.	Jaigad Creek	Domestic (minor) and port based wastes	J1, J2, J3, J4, J5, J6	2	1	3
20.	Ratnagiri	Industrial (minor), port based and domestic wastes	R1, R2, R3, R4, R5	3	2	–
21.	Bhatye Creek	Domestic (minor) waste	B4, B5, B6	1	–	2

Contd...

Table 54.5–*Contd...*

Sl.No.	*Transect*	*Wastewater Sources*	*Stations*	*Coastal Stations*	*Near shore Stations*	*Creek/ Estuarine Stations*
22.	Pawas Creek	Domestic (minor) and port based wastes	P4, P5, P6	–	2	1
23.	Vijaydurg	Domestic (minor) and port based wastes	VJ1, VJ2, VJ3, VJ4, VJ5, VJ6	2	1	3
24.	Deogad	Domestic (minor) and fishery based wastes	D1, D2, D3, D4, D5, D6	2	2	2
25.	Malvan	Domestic (minor) and fishery based wastes	M1, M2, M3, M4	2	2	–
26.	Vengurla	Domestic (minor) waste	V1, V2, V3	2	1	–
27.	Redi	Port based wastes	RD1, RD2, RD3	2	1	–
		Total	**120**	**42**	**19**	**59**

☆ The coastal water has defined capacity to dilute and disperse contaminants released directly or indirectly via innumerous creeks, estuaries and bays. The present sets of results indicate that coastal see off Maharashtra is by and large unpolluted except for the pockets off Mumbai wherein some degradation limited to a few km off the shoreline is evident.

Site specific corrective measures had been suggested for improvement in water environment (Table 54.6).

Table 54.6: Site Specific Suggestions for Improvement in Environmental Status

Sl.No.	*Transect*	*Site Specific Suggestions*
1.	Dahanu	Periodic monitoring of the creek should be considered.
2.	Tarapur	Disposal system should be shifted from the creek and Periodic monitoringshould be considered.
3.	Bassein	Corrective measures and intensive monitoring are needed.
4.	Manori	Corrective measures and intensive monitoring are needed.
5.	Versova	Corrective measures and intensive monitoring are needed.
6.	Mahim	Intensive monitoring is needed.
7.	Bandra	Periodic monitoring
8.	Worli	Periodic monitoring
9.	Thane Creek	Upper and mid segment needed corrective measures and intensive monitoring.
10.	Patalganga	Corrective measures and intensive monitoring are needed.
11.	Amba Estuary	Periodic monitoring
12.	Thal DP	Periodic monitoring
13.	Alibaug	Existing marine environment should be maintained.
14.	Kundalika Estuary	Corrective measures and intensive monitoring are needed.
15.	Murud	Periodic monitoring
16.	Savitri Estuary	Corrective measures and intensive monitoring are needed.
17.	Vashishti Estuary	Intensive monitoring is needed.
18.	Enron DP	Periodic monitoring
19.	Jaigad Creek	Periodic monitoring
20.	Ratnagiri	Corrective measures and intensive monitoring are needed.
21.	Bhatye Creek	Existing marine environment should be maintained.
22.	Pawas Creek	Existing marine environment should be maintained.
23.	Vijaydurg	Periodic monitoring
24.	Deogad	Periodic monitoring
25.	Malvan	Periodic monitoring
26.	Vengurla	Existing marine environment should be maintained.
27.	Redi	Existing marine environment should be maintained.

Chapter 55

Reuse and Recycling of Wastewater through Soil Aquifer Treatment: A Water Resource Management Strategy

P. Nema

CSIR-NEERI, Nagpur

ABSTRACT

Water supply and sanitation are important basic needs for improvement of the quality of life and enhancement for productivity efficiency of the people. In many parts of the world, specially the arid ones, the availability of surface water is constrained due to increasing urbanization and industrialization. To meet the growing demand, exploitation of groundwater has become necessary. However, this has resulted in substantial lowering of groundwater table. To sustain the pace of development, it has become necessary to search for alternative sources of water. In this context, wastewater renovation and reuse have emerged as a practical supplemental strategy for water resource development and management. The advantages associated with the wastewater, specially municipal wastewater, are its assured year round availability, stable composition, and easy amenability to renovation/treatment. The effluent quality requirement, depending upon the intended use, can also be met through application of one or a combination of treatment technologies. This treated water can meet a variety of non-potable water demands such as irrigation, industrial and consequently can reduce the demand pressure on freshwater resources to a considerable degree. This presentation highlights various possibilities of reuse of wastewater in different areas along with likely adverse health impacts and associated risks. Various wastewater treatment approaches generally applicable prior to any reuse have been discussed. A field study carried out on the dry bed of river Sabarmati at Ahmedabad for renovation of primary treated municipal wastewater has been presented. Finally, reuse research areas encompassing technical, socio-economic, health and cultural aspects as also public concerns and attitudes towards acceptance have been indicated.

Introduction

Urbanization in the developing countries poses special threats with regard to availability and management of water for drinking, domestic use, sanitation and hygiene. According to Population Council, 95 per cent of population growth will occur in urban areas over the next 25 years in developing countries. The urban population will roughly double in size and it will reach the figure of 4 billion. There will be more than 500 cities with more than one million populations. By 2015 the urban population in these countries will be 20 per cent of the total population. The growth outnumbers the capacities of conventional service provision.

The cumulative effect of rapid population growth and industrialization coupled with a non-uniform distribution of water resources throughout the country will result in increasing water shortage in developmental planning. There are two logical approaches to this dilemma:

- Reduce consumption through conservation, and/or
- Increase the availability of supply through wastewater reuse

Municipal wastewater is the most widespread and generally available source of reusable wastewater. Municipal wastewater is the main source of water pollution in India especially in urban centres which is often let out untreated or only partly treated. This wastewater either sinks into the ground as a potential pollutant of groundwater or discharged into the natural water bodies causing pollution in downstream areas. Urban wastewater, when properly treated and recycled is an alternate water source, which can reduce the demands for freshwater. Recycled wastewater can also reduce stress on the environment.

Factors affecting the quality of municipal wastewater include:

- Type of treatment system,
- Quality of water supply (primarily effects of TDS),
- Water usage,
- Impact of any industrial waste discharges, and
- Infiltration into collection systems

Water reuse is inherently a water conservation measure. Consequently, agricultural water reuse applications may, at certain locations, play a significant role in water conservation. The economics of water reuse determines whether a program will be feasible. The value of reclaimed water depends on its constant availability and its cost of production and delivery; *viz.* capital and O and M expenses, additional treatment expenses, conveyance and distribution cost, storage cost and water quality monitoring expenses. Cost of using non-potable water may include on-site connections, facilities to adjust and monitor quality, piping for dual system, steps to assure human safety, and changes in conventional operations.

The reclaimed wastewater can be used either directly or indirectly. Direct reuse indicates planned deliberate, or intentional use of treated wastewater for some beneficial purpose. However, there are only a few examples for direct reuse of wastewater for drinking mainly because of public acceptance and thus is limited in extent. Indirect reuse of reclaimed wastewater, on the other hand is practiced on a wider range from potable reuse to agriculture, industry etc. Major categories of wastewater reuse applications are discussed hereunder:

Wastewater Reuse

Groundwater Recharge

Recharging underground aquifer with treated wastewater and specifically municipal wastewater is one of the most generally accepted forms of wastewater reuse. The wastewater disappears from sight and thus looses the connotation as wastewater. A long travel time through the aquifer provides dilution through dispersion of contaminants.

Groundwater recharge with reclaimed municipal wastewater is a promising area that results in the planned augmentation of groundwater supplies. The main purpose of artificial recharge of groundwater has been to supplement groundwater supplies and to protect underground freshwater in coastal aquifers against saltwater intrusion from the ocean; groundwater recharge is also incidentally achieved in land treatment and disposal of municipal and industrial wastewater through infiltration and percolation. There are several advantages of this technique:

- ☆ The cost of artificial recharge may be less than the cost of surface reservoirs of equivalent capacity.
- ☆ The aquifer serves as an eventual distribution system and may reduce the need for surface pipelines or canals.
- ☆ Water stored in surface reservoirs is subject to evaporation, potential taste and odor problems, and to pollution, which may be avoided by underground storage.
- ☆ Suitable sites for surface reservoirs may not be available or environmentally acceptable.
- ☆ The inclusion of groundwater recharge in a water reuse project may also provide psychological and aesthetic secondary benefits as a result of the transition between reclaimed water and groundwater. The infiltration and percolation of wastewater through the soil and the groundwater zone will typically provide additional treatment and create a loss of identity between reclaimed water and groundwater.

Irrigation

Agriculture is the largest user of the wastewater. Wastewater applied to the soil is used in parts by the plant growing on the surface. A portion will penetrate the nearest aquifer and may be recovered as treated water.

Industry

Treated municipal wastewater can be an attractive source of water for many industries, particularly in water-short areas. The quantity of water used in most manufacturing processes and power generation is very large and the availability of adequate quantities of water is considered to be a pre-requisite.

The greatest single demand for industries is for cooling. Boiler feed water is another consumptive area but with more stringent quality requirement than the cooling. The quantity of cooling water used by a modern large power station producing 100 MW of electricity is about 60 million liters per day which would be comparable to the water uses of a city with 3,00,000 people (supply rate@ 200 lpcd). Water shortages and discharge regulations have forced many industries to reduce the net quantity of water that they consume by as much as an order of magnitude.

One of the greatest potential applications for water reuse is to supplement or replace the potable and/or freshwater demand of industries. In many instances, the industries' water quality requirements are less than drinking water quality, and the quantity is often, the critical concern.

Storage and Distribution System

A common misconception in planning for wastewater reuse is that reclaimed water represents a low cost new water supply. The assumption is generally true only when the wastewater reclamation facilities are conveniently located near large industrial or agricultural users and the need for storage and distribution is obviated. Storage and distribution of reclaimed water represents the principal cost of most proposed reuse projects.

Health Effects

The effluent quality for reclaimed water depends on intended use and the regulatory requirements of the concerned state. The degree of treatment is dictated by the safety of reuse from both microbiological as well as chemical quality considerations. The health risks of water reuse associated with the effect of exposure to the microbial or chemical contents of the reclaimed water may not be known, and even if known, its health effects may not have been studied. Both potable and non-potable water reuse applications may lead, either directly or indirectly, to some type of human exposure by ingestion of contaminated water or food supplies exposed to or mixed with recycled water, inhalation of aerosols, or through bodily contacts.

Risk Assessment and Risk Management

In recent years the need to quantify the health risks associated with exposure to environmental contaminants has generated a new interdisciplinary methodology referred to as risk assessment and risk management. Risk assessment determines the qualitative nature of the risk posed by exposures to contaminants and quantifies the dimensions of that risk. The assessment relies on information from epidemiological, clinical, toxicological, and environmental research. Risk management refers to the selection and implementation of the most appropriate regulatory action based on the results of the risk assessment, available control technology, and economic, social, and political factors.

Wastewater Treatment

Prior to any reuse, the wastewater will have to be treated. Conventional treatment of wastewater has been the accepted method of treatment world over. The treatment sequence comprising of preliminary, primary and secondary processes has been specifically developed to remove various pollutants in the wastewater namely, suspended solids, biodegradable organics, and microorganisms. However, the objective of treatment has so far remained mostly confined to disposal in accordance with the stipulated quality requirements. This scenario has now changed in view of wastewater being considered as a resource due to growing demand of water as a consequence of population growth, expansion of industry, and much greater demand for recreational and aesthetic purposes.

Table 55.1 illustrates the performance which can be expected of a typical well operated conventional activated sludge treatment plant. As can be seen, this type of plant is effective in removal of suspended solids and organic but cannot provide significant removal of phosphorus, nitrogen, or microorganisms which is a pre-requisite in planning a reuse programme.

Table 55.1: Performance of a Conventional Secondary Treatment System (Typical)

Pollutant	*Influent (mg/L)*	*Per cent Removal*
COD	250-550	70-80
BOD	80-300	85-90
Suspended Solids	100-400	80-90
Nitrogen (TKN)	40-70	30-40
Phosphorus-P	5-10	30-45
Coliforms	$\sim 10^7$	60-90

Wastewater Renovation through Soil Aquifer Treatment–Indian Experience

A pilot study was carried out in the Sabarmati River Bed at Ahmedabad to examine the feasibility of Soil Aquifer Treatment (SAT) system for renovation of municipal wastewater. A unique feature of the study was that the primary treated wastewater was directly used as influent for renovation by SAT process in contrast to the other studies wherein secondary treated effluent was used. Through the direct application of primary treated wastewater cost savings upto 30 per cent can be realized.

The study was conducted on two scales. The first one was a column lysimeter study and the second one was a pilot SAT system (2 MLD capacity) constructed in the river bed. The systems were operated for a period of about six months during 1995-96 and data collected. The study indicated that SAT treated wastewater was much superior in quality (COD, BOD, SS, N, P as well as Bacteria and Viruses) as compared to conventional wastewater treatment methods and conformed to effluent quality norms for unrestricted irrigation (Table 55.2). A detailed cost analysis in respect of a 55 MLD SAT system revealed the cost competitiveness of SAT system with the conventional aerobic and anaerobic wastewater treatment systems. It was found that the cost of SAT system is 20-70 per cent less than the cost of conventional aerobic and anaerobic treatment systems (Table 55.3). It has been reported elsewhere that the cost of SAT system is less than 40 per cent of the cost of equivalent above ground treatment.

Table 55.2: Average Quality of Influent and Renovated Wastewater (Pilot SAT System)

Parameter	*Influent*	*Effluent*
BOD	100	8-10
COD	230	14-24
SS	110	12-18
TKN	40	22
Total Phosphate	6	0.2
Bacteria (FC)	3×10^{-6}	72
Viruses	44	~ Nil

System operated at an average HLR of 0.08m/day (30m/year)

Table 55.3: Cost Analysis of SAT System with Other Conventional Wastewater Treatment

Systems (Capacity–55 MLD)

Treatment System	*Capital Cost**	*Annualized Cost**	*O and M Cost**	*Total Annualized Cost**	*Treatment Cost (Rs./m³)*	*Cost Ratio (TAC Basis)*
Conventional ASP	1450	204	290	494	2.45	1.55
Extended Aeration	1290	180	340	520	2.60	1.65
Trickling Filter	1397	193	350	543	2.70	1.70
Anaerobic Filter	1300	169	260	429	2.13	1.35
UASB	1100	175	200	375	1.86	1.17
SAT	900	126	192	318	1.58	1.00

* Rs. in Lakh.

Issues of Concern

The various modes of treatment involved in treating wastewater can be divided into three broad categories: physical, chemical, and biological processes. Though most of the information in regard to mode of operation and efficacies of these processes is available, there are still many other areas for investigation and research in the wastewater reclamation and reuse field. Some of these areas can be listed as below:

- ☆ Groundwater contamination
- ☆ Toxic contaminants
- ☆ Monitoring methods and instrumentation
- ☆ Land and soil systems
- ☆ Aquaculture
- ☆ Epidemiological studies
- ☆ Standards and regulations

One of the major quality consideration in wastewater reclamation and reuse include exposure to pathogenic bacteria, viruses, protozoa, and helminths. These pathogenic organisms can be transmitted by direct contact, ingestion – such as by consumption of food irrigated with the reclaimed wastewater, drinking contaminated groundwater, or by inhaling spray aerosols.

Societal Issues

In general, the technical research aspect of wastewater reclamation and reuse can be quantified, explained, evaluated and solutions to the problems developed. However, the entire spectrum of the societal issues cannot be quantified specifically those pertaining to public opinion, users concerns and public acceptance as also emotional and culture attitudes towards the use of reclaimed wastewater. A World Health Organization report (1989) entitled "Health Guidelines for the Use of Wastewater in Agriculture and Aquaculture" outlined the following research needs in wastewater reclamation and reuse areas.

Water Quality Assessment

- ☆ To develop and improve a low-cost, reliable monitoring method for helminth egg detection.

Wastewater Treatment Technologies that should Receive Special Attention

- ☆ Waste stabilization pond technology and its efficacy in bacterial removal.
- ☆ Helminth egg removal systems such as biological filters, activated sludge plants.
- ☆ Effectiveness of disinfection with particular reference to helminth egg inactivation.
- ☆ Treatment to inactivate pathogens in sludges, especially for use on a small scale.

Epidemiology

An urgent need for wastewater reuse projects is an evaluation of the improvement in the health of the agricultural workers and their children. Studies are also needed to fill in the gaps in existing knowledge about excess morbidity and excess infections, especially the more violent infections such as typhoid fever, as they relate to reclaimed wastewater reuse.

Sociological and Economical Considerations

Research into the public's and users' attitudes to wastewater reuse is needed, both in areas where it is practiced and where it may be introduced. The social and cultural factors affecting the acceptability and compliance with the health protection measures suggested (*e.g.* crop restriction) need to be evaluated in a variety of different cultural settings.

Future of Water Reuse

More stringent effluent quality requirements with entailing treatment/costs will raise an obvious question in peoples mind as to why discharge reusable quality effluent when it is a cost-effective potential source of water supply. Moreover, urban centres in arid and semi-arid regions will want to retain and reuse wastewater locally rather than throwing it in drainage system. The choice however may be between a central facility and small size localized system. Therefore, the future will be for improved technology for small-scale on-site water reclamation and reuse systems.

Extensive studies have been conducted to address scientific and technical concerns about water reuse. These studies have shown the acceptability of reclaimed water. However, concerns about indirect – potable reuse continue and more issues have emerged with the improvement in monitoring and analysis methods. According to US Environmental Protection Agency, the preferential area in future will be small scale on site water reclamation and reuse, ranging from gray water recycling systems to membrane reactors.